Coastal and Estuarine Studies

57

Richard A. Valigura, Richard B. Alexander,
Mark S. Castro, Tilden P. Meyers, Hans W. Paerl,
Paul E. Stacey, and R. Eugene Turner (Eds.)

Nitrogen Loading in Coastal Water Bodies
An Atmospheric Perspective

American Geophysical Union
Washington, D.C.

Published under the aegis of the AGU Books Board

John E. Costa, Chair; Andrew Dessler, Jeffrey M. Forbes, W. Rockwell Geyer, Rebecca Lange, Douglas S. Luther, Darrell Strobel, and R. Eugene Turner, members.

Library of Congress Cataloging-in-Publication Data
Nitrogen loading in coastal water bodies : an atmospheric perspective
/ Richard A. Valigura ... [et al.], editors.
     p. cm -- (Coastal and estuarine studies ; 57)
  Includes bibliographical references.
  ISBN 0-87590-271-5
    1. Chemical oceanography. 2. Atmospheric nitrogen compounds. 3. Ocean-atmosphere interaction I. Valigura, Richard A. II. Series.
GC117.N5 .N47   2000
551.46'01--dc21   00-045109

ISSN 0733-9569
ISBN 0-87590-271-5

Cover image: USGS stream monitoring locations and river reaches in the conterminous United States, from Alexander et al., this volume.

Copyright 2001 by the American Geophysical Union, 2000 Florida Ave., NW, Washington, DC 20009, USA.

      Figures, tables, and short excerpts may be reprinted in scientific books and journals if the source is properly cited.
      Authorization to photocopy items for internal or personal use, or the internal or personal use of specific clients, is granted by the American Geophysical Union for libraries and other users registered with the Copyright Clearance Center (CCC) Transactional Reporting Service, provided that the base fee of $1.50 per copy plus $0.35 per page is paid directly to CCC, 222 Rosewood Dr., Danvers, MA 01923. 0733-9569/01/$01.50+0.35.
      This consent does not extend to other kinds of copying, such as copying for creating new collective works or for resale. The reproduction of multiple copies and the use of full articles or the use of extracts, including figures and tables, for commercial purposes requires permission from AGU.

Printed in the United States of America

CONTENTS

Preface
Richard A. Valigura vii

An Introduction to the First Assessment of Nitrogen Loads to US Estuaries with an Atmospheric Perspective
Richard A. Valigura 1

Atmospheric Deposition of Nitrogen in Coastal Waters: Biogeochemical and Ecological Implications
Hans W. Paerl, Walter R. Boynton, Robin L. Dennis, Charles T. Driscoll, Holly S. Greening, James N. Kremer, Nancy N. Rabalais, and Sybil P. Seitzinger 11

Atmospheric Nitrogen Deposition to Coastal Estuaries and Their Watersheds
Tilden Meyers, Joseph Sickles, Robin Dennis, Kristina Russell, James Galloway, and Thomas Church 53

Contribution of Atmospheric Deposition to the Total Nitrogen Loads to Thirty-four Estuaries on the Atlantic and Gulf Coasts of the United States
Mark S. Castro, Charles T. Driscoll, Thomas E. Jordan, William G. Reay, Walter R. Boynton, Sybil P. Seitzinger, Renée V. Styles, and Jaye E. Cable 77

A Comparison of Independent N-loading Estimates for U.S. Estuaries
R. E. Turner, D. Stanley, D. Brock, J. Pennock, and N. N. Rabalais 107

Atmospheric Nitrogen Flux From the Watersheds of Major Estuaries of the United States: An Application of the SPARROW Watershed Model
Richard B. Alexander, Richard A. Smith, Gregory E. Schwartz, Stephen D. Preston, John W. Brakebill, Raghavan Srinivasan, and Percy A. Pacheco 119

Uncertainties in Individual Estuary N-loading Assessments
D. A. Brock 171

Contributions of Atmospheric Nitrogen Deposition to U.S. Estuaries: Summary and Conclusions
Paul E. Stacey, Holly S. Greening, James N. Kremer, David Peterson, and David A. Tomasko 187

An Annotated Summary of Nitrogen Loading to US Estuaries
D. Stanley 227

List of Contributors 253

PREFACE

Every time it rains, biologically active nitrogen compounds are transferred from the air to whatever surface lies underneath; what a farmer once called "the poor man's fertilizer." In fact, nitrogen containing compounds are transferred between surface and atmosphere even when it is not raining. That this atmospheric nitrogen deposition impacts on the environment is a scientifically accepted fact, established during the acid rain debates led by the National Acid Precipitation Assessment Program. The extent to which this deposition contributes to the decline of coastal waters around the United States due to over fertilization, however, is still under debate. In response to and as a continuation of this debate is the current work before you: the first attempt to consistently and comprehensively estimate the relative contribution of atmospherically delivered nitrogen to the total amount of nitrogen entering coastal estuaries around the United States.

The study of chemical transfers between the earth and the atmosphere is not a new science; historically, however, it has been a rather specialized one. Currently, the use of this knowledge by a variety of terrestrial, estuarine, and marine ecologists and biogeochemists to investigate the degradation of coastal waters does constitute a new application. And as we now understand, the first nine years of this application have yielded widely variable results with appreciably different interpretations. Be that as it may, most conclusions drawn at this point in time suggest that the amount of reactive, nitrogen-containing compounds entering our coastal waters through the atmosphere may be significant in some areas.

Concurrent with the increase in scientific interest here, we have witnessed an increase in the number of management programs that wish to investigate the importance of atmospheric deposition to their water body, with an increasing number of funding sources willing to provide resources for such studies. This funding environment, especially from the federal stand point, is problematic; it can promote the misuse of funds to support well-meaning efforts that effectively "re-invent the wheel." In 1997, both the United States Environmental Protection Agency (EPA) and the National Oceanic and Atmospheric Administration (NOAA) recognized this situation and sought a solution. Subsequently, the EPA's Office of Water and Office of Air jointly funded the subject of this book. The effort was led by the NOAA Air Resources Laboratory (ARL).

In September of 1996, ARL set up a steering committee to develop a framework for the assessment. The steering committee compiled a list of 40 potential participants: scientists from a variety of federal, state, and academic institutions. Of the 40 scientists that the steering committee contacted to enlist their involvement, all 40 agreed to participate. This unanimity is itself testimony to the interest surrounding the issue.

In the end, as careful readers may note, not all the scientists who agreed to participate in the assessment felt sufficiently involved in the actual writing of the chapters to list themselves as authors. Nonetheless, as the material was largely a result of debate and consensus, I owe a debt of gratitude to all who participated: certainly the 32 named authors, and to those others who contributed along the way. Their names follow:

Bruce Hicks, Richard Artz, Margaret Kerchner, Sharon Brown, Maureen McMahon, and Suzanne Bricker at NOAA; Bill Matesuski, Rona Birnbaum, Doris Price, Darryl Brown, Jonathan Garber at EPA; Lisa Biegel at the Sun Prairie Environmental Commune; Scott Nixon at the University of Rhode Island; Don Bosch and Tom Simpson at the University of Maryland; Bob Twilley at the University of Southwest Louisiana; Joe Schubauer-Berigan formerly with the University of South Carolina; Cliff Randall at Virginia Polytechnical Institute; Jonathan Pennock at the University of South Alabama; Jan Newton at the University of Washington.

Richard A. Valigura
Editor

1

An Introduction to the First Assessment of Nitrogen Loads to US Estuaries with an Atmospheric Perspective

Richard A. Valigura

Abstract

A scientific assessment of nitrogen inputs to 42 coastal waterbodies around the United States was executed in 1997-1998, the results of which are published in this volume. This introduction provides a brief background into why there is concern over the amount of nitrogen entering our estuaries, presents a few of the unique concerns about atmospheric nitrogen, and provides an overview of how the assessment was conducted.

Introduction

The issue of how much atmospheric deposition contributes to the declining health of U.S. coastal ecosystems was brought to the public's attention by a single analysis of the Chesapeake Bay by scientists of the Environmental Defense Fund /EDF; Fisher et al., 1988]. The news was met with cynicism, because the research community had long operated on the assumption that the causes for over-enrichment were entirely related to discharges of nitrogenous chemicals from the land and from point sources into rivers and streams. However, subsequent analyses confirmed the main tenet of the EDF analysis — that a significant fraction (20-30%) of the nitrogen entering the Chesapeake Bay is derived as a result of atmospheric transport from distant as well as local sources. There have been approximately 50 studies around the world, the majority of which were published since 1990, which directly addressed at least one aspect of atmospheric loadings. However, the measurement and modeling techniques used vary considerably among individual studies, making intercomparisons difficult. Many of these analyses have been or are being refined; however, two-thirds of these analyses suggest air contributions of 20% or higher. There are several caveats related to existing studies. The first of these is that agreement in numbers should not be overly encouraging since it may reflect the fact that understanding of how atmospheric deposition enters the aquatic biosphere has not expanded greatly in the past decade. Of greater concern are the large uncertainties in overall understanding of the amount of nitrogen that enters our estuaries. Biogeochemical cycling of nitrogen on local, regional, and global scales is a complex system of emissions, transformations, dispersion, and deposition. This system is most complex in coastal environments where landscape processes affect coastal processes and vice versa.

From the perspective of coastal ecosystem eutrophication, nitrogen compounds from the air, along with nitrogen from sewage, industrial effluent, fertilizers, and other "natural" sources becomes a source of nutrients to the receiving ecosystem. Normally in short supply, nitrogen plays an important role in controlling the productivity, dynamics, biodiversity, and nutrient cycling of estuarine and marine ecosystems. Though the availability of nitrogen normally limits biological productivity in estuaries, overabundance of nitrogen is of concern in areas that have developed nutrient enrichment problems (i.e., eutrophication). In addition to increasing productivity, nutrient enrichment generally alters the normal ratios of nitrogen to phosphorus and to other elements such as silicon. This alteration may induce changes in phytoplankton community structure. Species that normally occur in low abundances may be favored, and, in some cases, toxic and/or noxious algal blooms may result. Furthermore, in coastal areas with weak tidal exchange or stratified water masses (e.g., Chesapeake Bay, Long Island Sound, the Louisiana Continental Shelf) the "overproduction" of algae tends to sink to the bottom and decay, using all (anoxia) or most (hypoxia) of the available oxygen in the process, causing loss of habitat. In extreme cases, the increase in suspended matter due to overproduction decreases light infiltration, in turn causing a loss of submerged aquatic vegetation. [ESA/SBI, 1997].

As a direct result of scientific efforts to date, atmospheric nitrogen input to coastal areas has become a specific topic of speculation amongst coastal management organizations. In fact, between 1994 and 1997 three major workshops were held to address this issue [AQCG, 1995, 1997; ESA/SBI, 1997]. One of the major conclusions these workshops had in common was the need for a consistent, holistic assessment of atmospheric inputs along both the East and Gulf Coasts of the United States. In 1997 funding was secured to perform the first such assessment. The primary goal of the assessment was to evaluate waterbodies around the Atlantic and Gulf coasts to identify those with enough information to estimate total nitrogen entering each water body, and, subsequently, the relative atmospheric inputs to those waterbodies which improves upon past estimates. The results are presented in this volume as a series of related chapters. This intent of this introduction is to i) provide a brief background into why there is concern over the amount of nitrogen entering our estuaries, ii) present a few of the unique concerns about atmospheric nitrogen, and iii) provide an overview of the how the assessment was conducted.

Sources of Nitrogen to our Estuaries

During the past century, human activities have doubled the amount of nitrogen available annually to living organisms. In many estuaries, human sources of nitrogen now rival or exceed natural inputs of nitrogen. The growth of the population and the accompanying increase in per capita resource use has significantly changed the N cycle on local, regional, and global scales. In terms of absolute amounts of additional N mobilized due to human activity, food production is more important than energy production. In terms of impacts on critical earth systems, both are important [Vitousek et al., 1997].

N fertilizer is produced by industrial fixation (INF), utilizing the Haber Bosch process to convert unreactive N_2 to NH_3. With this process, fertilizer production doubled approximately every eight years from 1950 to 1973. The next doubling took 17 years (Figure 1). From

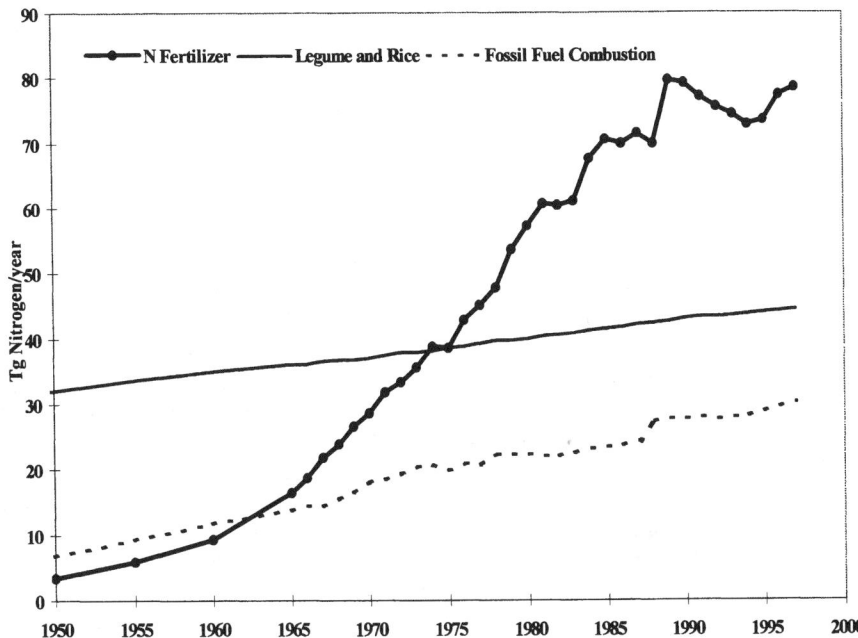

Figure 1. Temporal trends (1950-1997) in anthropogenic production of reactive N (Tg N yr^{-1}) from fertilizer, fossil fuel combustion, and legume and rice cultivation (reprinted from Galloway 1998).

1990 to 1995, N fertilizer production decreased, in large part due to the removal of subsidies from countries of the former Soviet Union. Increases in fertilizer production occurred in 1995-1997 due to large demands in Asia.

In addition to INF, human activities increased food production by promoting biological nitrogen fixation (BNF) with the cultivation of legumes and the production of rice. The amount of N created by BNF is dependant on the extent of area planted to a particular crop and the fixation rate per hectare. There are several estimates of current global biofixation, ranging from about 30 to 40 Tg N yr^{-1} Using 40 Tg N yr^{-1} for the current estimate, Galloway (1998) estimated that over the last 47 years BNF has increased by about 25%. Thus up until the 1960's or 1970s human-induced BNF was the most important mechanism of introducing new N into agricultural systems. As mentioned above, with the Haber-Bosch process, commercial production of N fertilizer increased rapidly and by 1970 was the dominant process.

Food production introduced new reactive nitrogen into the environment on purpose; energy production through the combustion of fossil fuels introduces it by accident, as a waste byproduct - nitrogen oxide and nitrogen dioxide (in combination commonly referred to as NO_x). Over the period 1950 to 1997, the amount of NO_x emitted into the atmosphere from fossil fuel production increased from 0.8 to 30 Tg N (Figure 1). Up until 1960, N mobilization by fossil fuel combustion was more important than the production of commercial fertilizer. Fossil fuel combustion was never as important an N source as human induced BNF. The total amount of N fixed by human activities in 1997 was about 150 Tg

N, of which 80% was due to food production (with fertilizer production being about twice as important as cultivation related BNF) and 20% was due to energy production. While there are uncertainties about these estimates (especially for cultivation related BNF), the uncertainty is small relative to the magnitude of change in the introduction of additional N to the global cycle.

Airsheds and Watersheds

Atmospheric nitrogen compounds are ubiquitous. Diatomic nitrogen (N_2) is the dominant component of the atmosphere, and is comparatively unreactive. However, nitrogen is also present in the atmosphere in other chemical forms, some of which are relatively reactive. Most atmospheric nitrogen compounds (other than N_2 and nitrous oxide [N_2O]) fall into two categories: reactive nitrogen (composed primarily of nitrogen oxides), and reduced nitrogen (typically dominated by NH_3). There are also organic nitrogen species that are typically referred to as a subset of reactive nitrogen. These chemical species arise in the atmosphere from the interaction between nitrogen oxides and biogenic or anthropogenic hydrocarbons. Though available evidence suggests that organic nitrogen originates from both anthropogenic and natural sources, speciation of organic nitrogen in the atmosphere is very poorly understood. The relative portions of the different forms nitrogen can take--i.e., nitrate, nitrite, ammonia, dissolved organic nitrogen--vary widely based on proximity of sources to receptors, receiving waters, and atmospheric transformations as well as on the land surface and transfer through the watershed via surface runoff [Paerl, 1997]. Note that there was a standardized distinction of the terms "deposition" and "loadings" used in this assessment. *Deposition* is the flux of nitrogen from the atmosphere to whatever surface is beneath. *Indirect loading* is the portion of nitrogen deposited onto the terrestrial watershed which is transmitted to the water body itself, defined in terms of flow to the tidal waters. Deposition to the water surface itself constitutes a *direct loading*.

Sources of nitrogen to the atmosphere, and subsequently into the estuaries are particularly difficult to quantify. They span broad geographical areas, cross environmental media and management authorities, and involve diverse scientific disciplines. Several studies argue that long range atmospheric transport of nitrogen from sources located in the mid-continental regions is a significant source of nitrogen deposition to surface waters on the east coast. It is important to note that there is a high degree of uncertainty in the estimates of how much is actually transported (the uncertainties reported with each estimate varying from a factor of two to an order of magnitude). This is largely attributable to the lack of a standardized or even widely accepted method for measuring or monitoring the average rates of emission of compounds from the surface to the atmosphere (and vice versa) over long periods of time. Short term intensive field measurements are possible for many compounds, but these are at the mercy of the conditions during the measurement period. There has also been progress made towards continuous emissions monitoring at large stationary sources through Title IV of the Clean Air Act. The only alternative to direct measurement is to use a limited amount of information on localized exchange, and subsequently extrapolate averages of the measured values over the area of consideration. This limitation applies especially to the case of natural emissions, but also extends to the cases of automotive and industrial sources.

Computer modeling is considered the most viable method for extrapolating site-specific, local measurements of emissions to large areas, and for then assessing the impact of these emissions on distant areas downwind [Pielke, 1984; Fisher, 1992; Koch, 1993]. Discussions of atmospheric transport have evolved to the level where it is has proven useful to describe/define airsheds for a given target area, such as a watershed. An airshed defines the predominant emissions area that influences the air concentrations over or deposition to a specified area, such as an estuary or watershed. Unlike a watershed, an airshed does not have clearly defined boundaries; the longer the range of influence (pollutant lifetime) the fuzzier the boundary. Dennis (1997) developed an operational definition of an airshed, based on calculated ranges of "significant" influence from emissions source regions. This results in defining an area of emissions that include sources that are expected to make an important contribution to deposition loading and will *en toto* account for approximately 70% of the deposition loading to the specified region of interest.

Following the procedure outlined in Dennis (1997), airsheds for oxidized N were developed for many of the estuaries in this assessment [Chapter 2 this volume]. These airsheds provide a sense of the area from which emanate the important air emissions of NO_x emanate, that has a principal influence on the deposition of oxidized N to these estuaries. The airsheds for oxidized N are the order of several hundred thousand km^2 or tens of millions of hectares; most are roughly 10-100 times larger than their associated watersheds (Figure 2).

However, the inputs are generally diffuse inputs at low concentrations leading one to question the biological significance of this nitrogen source. In addition, changing demographic and settlement patterns in the United States will affect the relative contribution of atmospheric and other nitrogen sources to estuaries over time.

This Assessment

The goal of this assessment was to develop an approach that would allow the calculation of i) the total amount of nitrogen entering each of the identified waterbodies from all sources, ii) the amount of this total load which came through the atmospheric pathway and iii) the amount of uncertainty associated with each of these estimates. The only limitations were that the assessment must be reported after one year and be completed with minimal funds. Therefore, this work should be considered the foundation for future, more intense assessments. A team of scientists was recruited and brought together for two week-long workshops held for coordination and conclusion. In the end, 32 scientists contributed enough to this effort to be considered authors.

The first objective of the assessment was to evaluate waterbodies around the U.S. coast and identify those with enough information to allow an assessment of the relative atmospheric inputs to those waterbodies which improves upon past estimates. Primary criteria used include the existence of nitrogen flux data, a previous nitrogen balance study, the existence of a major coastal management or scientific program, and geography/morphologic coverage and consistency. Forty waterbodies around the east and gulf coasts were chosen for primary study and two waterbodies on the Pacific Coast were chosen for partial study (Figure 3). Although the hypoxic zone in the northern Gulf of Mexico is a vast area and poses a serious problem to living resources [Rabalais et al., 1998], it was not considered with the other U.S. estuaries in this analysis. The open coastal region adjacent to the effluents of the Mississippi and Atchafalaya Rivers, while having characteristics of estuaries (a stratified water column and gradients of suspended sediments, salinity and nutrients), is not an enclosed body of water and is not an estuary.

Once the waterbodies were chosen, the objective was to develop a consistent method to quantify the nitrogen loads to each water body. The current level of data collection and

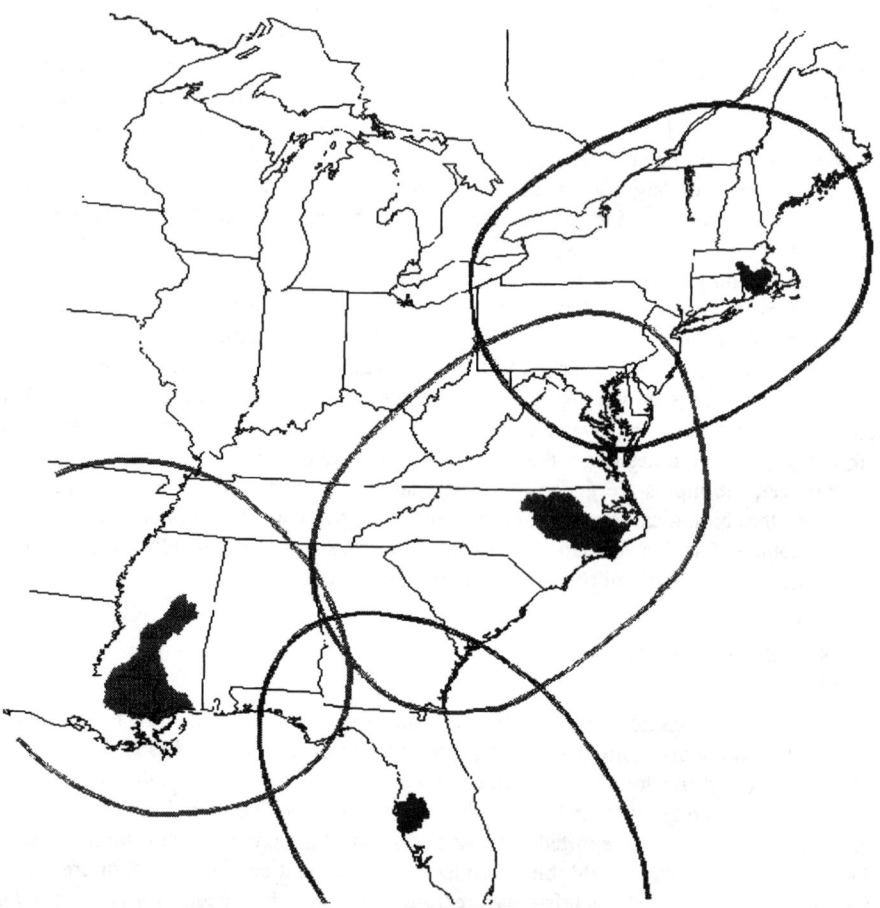

Figure 2. Principal nitrogen oxide airsheds for Narragansett Bay, Pamlico Sound, Tampa Bay, and West Mississippi Sound. Developed by R. Dennis, Atmospheric Sciences Modeling Division, NOAA/ARL and NERL USEPA.

scientific understanding is such that no consensus was reached pertaining to the one best method to use. The result is that three different approaches were used the estimate nitrogen input to the chosen estuaries. Three important points should be kept in mind when reviewing the results presented in this manuscript:

  1. while each method employed in this assessment is consistent by itself; no attempt was made in advance to constrain outcomes between the different methods, which sometimes lead to disparate results for a given water body;
  2. the range in results for a given water body is closely related to the amount of information available for that particular water body;
  3. the estimates developed in this assessment represent long-term averages and therefore do not address issues of seasonality and episodicity.

The three approaches used to estimate nitrogen loading form the backbone of this document

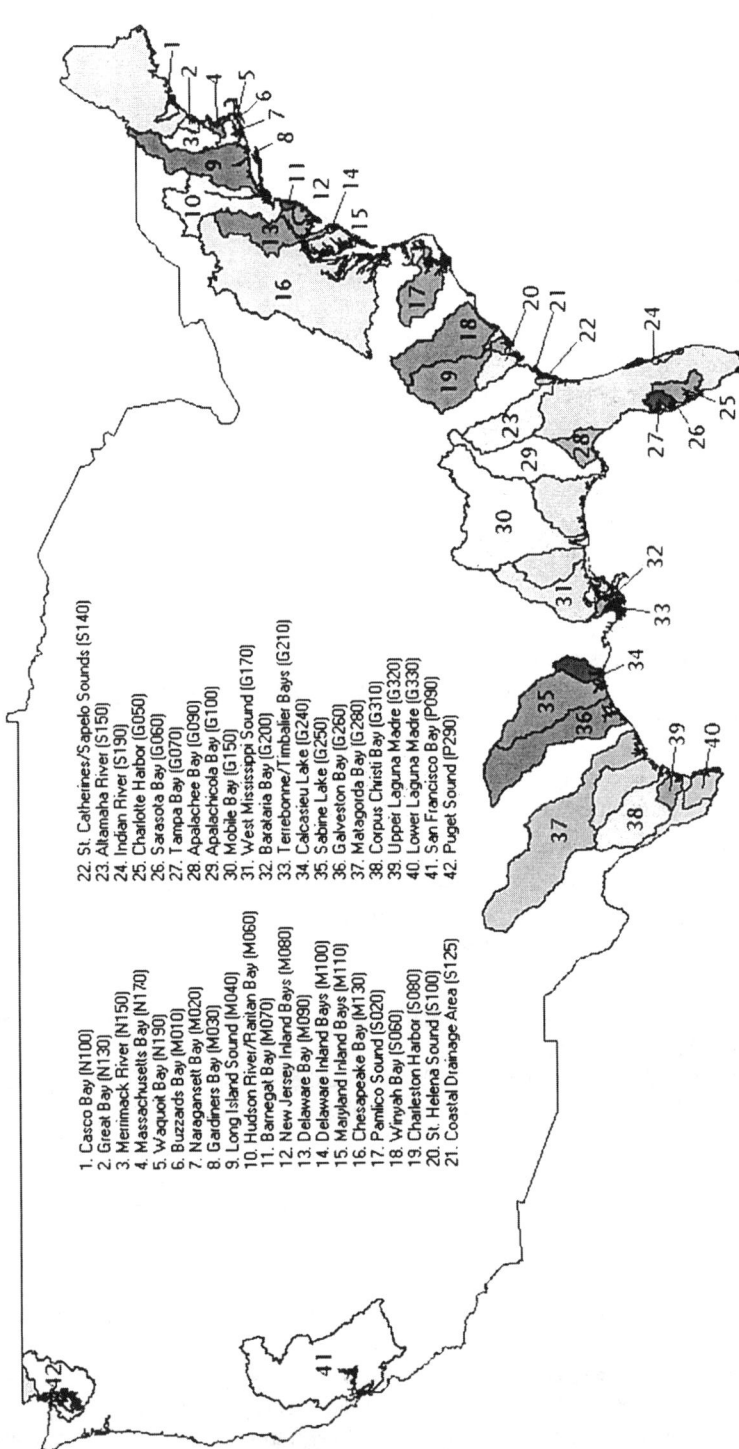

Figure 3. The 42 watersheds addressed by this Assessment

which is organized in much the same manner as the assessment was executed. A brief description of the chapters in this volume follows.

Chapter 2. Atmospheric Deposition of Nitrogen in Coastal Waters: Biogeochemical and Ecological Implications

Because atmospherically deposited nitrogen is quantitatively important and capable of stimulating primary production in N-sensitive estuaries, there is concern that increasing rates of new atmospheric N deposition may be linked to symptoms of accelerating eutrophication, including expansion of harmful (toxic, hypoxia/anoxia generating) algal blooms impacting estuaries and beyond. This chapter presents a synthesis of current knowledge concerning N inputs from atmospheric sources, and an assessment of the sensitivity of the biogeochemical and trophic responses of coastal ecosystems to this new N source.

Chapter 3. Atmospheric Nitrogen Deposition to Coastal Estuaries and Their Watersheds

As part of the overall process of evaluating the impacts of total atmospheric deposition of nitrogen on coastal ecosystems and estuaries, estimates of wet and dry deposition of nitrogen to both watershed and water bodies are required. Currently, most bays, estuaries, and watersheds lack onsite deposition monitoring data. In this first order attempt to assess the impact and fate of nitrogen delivered from atmospheric processes, estimates of wet and dry nitrogen deposition are derived using data from either regional monitoring activities or calibrated atmospheric deposition models.

Chapter 4. Assessment of the Contribution Made by Atmospheric Nitrogen Deposition to the Total Nitrogen Load to Thirty-four Estuaries on the Atlantic and Gulf Coasts of the United States

In this chapter, we calculate the net anthropogenic N inputs (N fixation, N fertilization, atmospheric nitrate deposition, net food and net feed import) to 34 watersheds on the Atlantic and Gulf coasts of the United States (U.S.), estimate the retention and export of N from the watershed to each estuary, and determine the contribution made by atmospheric N deposition to the total N loads to each estuary.

Chapter 5. A Comparison of Independent N-Loading Estimates for US Estuaries

The result of many recent local efforts to quantify nitrogen loadings in individual estuaries is the subject of this analysis. As many independent estimates of nitrogen loading to estuaries as could be found were gathered, and those data were converted into comparable units and categories. An annotated bibliography of these data sources is provided in Appendix 1. Twenty-seven recent and independently-derived total nitrogen (TN) loading estimates for watersheds of east and Gulf of Mexico estuaries were gathered. These results were compared to those from two other recent studies that used national water quality databases and synthetic or statistical analyses to estimate loadings.

Chapter 6. Atmospheric Nitrogen Flux from the Fluvial Drainages of Major Estuaries of the United States: An Application of the SPARROW Watershed Model

To assess the atmospheric contributions of total nitrogen (TN) in riverine export to the coastal and estuarine ecosystems of the United States, an empirical watershed model SPARROW (Spatially Referenced Regression on Watershed Attributes) was applied to a selected set of 40 major estuaries. Intercomparisons of the SPARROW model with other large-scale watershed models indicate general agreement in the predictions of TN export over a wide range of watershed sizes, but illustrate the intrinsic difficulties of comparing the flux rates of models having different temporal and spatial scales of measurement and prediction and specifications of nitrogen supply, transformation and transport processes.

Chapter 7. Uncertainties in Individual Estuary N-Loading Assessments

Both modeling and measurement techniques have been used in an attempt to understand the nitrogen inputs to the complex and open systems that are estuaries. The breadth of confidence bounds assignable to typical loading estimates should help us to see the limitations in our knowledge from both a scientific and a management point of view. The uncertainties involved in compilation of nitrogen loading estimates are reviewed here, with examples from Texas estuaries.

Chapter 8. Contributions of Atmospheric Nitrogen Deposition to U.S. Estuaries: Summary and Conclusions

This chapter addresses the consistencies/inconsistencies of the multiple methods reported in this assessment. Loading estimates derived by each of the three methods are compiled and presented in tabular form. The goal of this intercomparison is to highlight the differences among the estimates and clarify any unresolved inconsistencies. A summary discussion of the major findings of this assessment is presented along with an evaluation of the overall process used to execute the assessment.

References

AQCG, Airsheds and watersheds -- the role of atmospheric nitrogen deposition, Proceedings, Airlie Conference Center, Warrenton, VA, 11 - 12 October. Available from the Chesapeake Bay Program Air Subcommittee, 410 Several Avenue, Suite 109, Annapolis, MD 21403. 32 pp. 1995

AQCG, Airsheds and watersheds II -- a shared resources workshop, Proceedings, Brownstone Hotel, Raleigh, NC, 5-7 March, Available from the Chesapeake Bay Program Air

Subcommittee, 410 Several Avenue, Suite 109, Annapolis, MD 21403. 34 pp. 1997

Dennis, R.L., Using the Regional Atmospheric Deposition Model to Determine the Nitrogen Deposition Airshed of the Chesapeake Bay Watershed. In: J.E. Baker (ed). Atmospheric Deposition of Contaminants to the Great Lakes and Coastal Waters. Denver, Co.: Society of Environmental Toxicology and Chemistry. 1997

ESA/SBI, Atmospheric Nitrogen Deposition to Coastal Watersheds: Workshop Report. Sustainable Biosphere Initiative Project Office of the Ecological Society of America. University of Rhode Island Coastal Institute. June 2-4, 1997. 1997

Fisher, B.E.A., The long range transport of atmospheric acidity. In: Radojevic, M. and R.M. Harrison, eds. Atmospheric acidity: Sources, consequences and abatement. New York: Elsevier Applied Science, pp. 139-166. 1992

Fisher, D.C., J. Ceraso, T. Matthew, and M. Oppenheimer, Polluted coastal waters: the role of acid rain. New York: Environmental Defense Fund. 1988

Galloway, J.N., The global nitrogen cycle: changes and consequences. Environmental Pollution. Vol 102:S1, 15-24pp. 1998

Koch, S.E., Proceedings of a colloquium and workshop on mesoscale coupled modeling. February 22-25, Calverton, MD. NASA Conference Pub. 3217, 112 pp. 1993

Paerl, H.W., Atmospheric nitrogen deposition in coastal waters. In: J.E. Baker (ed). Atmospheric Deposition of Contaminants to the Great Lakes and Coastal Waters. Denver, Co.: Society of Environmental Toxicology and Chemistry. 1997

Pielke, R.A., Mesoscale meteorological modeling. New York: Academic Press Inc., 612 pp. 1984

Rabalais, N.N, R.E. Turner, W.J. Wiseman Jr. and Q. Dortch, Consequences of the 1993 Mississippi River flood in the Gulf of Mexico. Regulated Rivers: Research and Management. 14:161-177. 1998

Vitousek, P.M., J.D. Aber, R.W. Howarth, G.E. Likens, P.A. Matson, D.W. Schindler, W.H. Schlesinger and D.G. Tilman. 1997. Human alterations of the global nitrogen cycle: Sources and consequences. Ecol App. 7:737-750.

2

Atmospheric Deposition of Nitrogen in Coastal Waters: Biogeochemical and Ecological Implications

Hans W. Paerl, Walter R. Boynton, Robin L. Dennis, Charles T. Driscoll, Holly S. Greening, James N. Kremer, Nancy N. Rabalais, Sybil P. Seitzinger

Abstract

Atmospheric deposition of nitrogen (AD-N, as wet deposition and dry deposition), is a significant and growing source of biologically-available nitrogen (NO_x, NH_3/NH^+_4, and dissolved organic N (DON)) entering nitrogen-limited estuarine and coastal waters (jointly termed coastal). AD-N ranges from 400 to >1000 mg N m^{-2} yr^{-1}, and represents from <10 to >40% of new N inputs in North American and European coastal waters downwind of emission sources. The relative contribution of AD-N to total external N loading depends on land use, watershed and airshed size, and hydrological and morphological characteristics (i.e., water retention time) of receiving waters. In heavily-impacted, N-sensitive waters, the ecological impacts of AD-N include accelerating primary production (eutrophication), which may yield a variety of negative impacts including increased algal bloom activity, toxicity, oxygen depletion (hypoxia) events, and food web alterations. Depending on their sources (i.e., agricultural, urban, industrial) certain forms of AD-N are increasing relative to others, leading to qualitative changes in deposition and biogeochemical response in receiving waters. Because phytoplankton and bacteria differentially utilize different forms of N, changes in the ratios of NH_4^+ to NO_x and DON in AD-N may usher in community compositional changes. One example is intensive animal operations in Western Europe and the U.S. Mid-Atlantic States, which are linked to regionally elevated NH_4^+ deposition rates. Experimental evidence indicates that increasing levels of AD-NH_4^+ enhance primary production, while favoring growth of specific phytoplankton functional groups. In addition, AD (as well as other sources of new N) enrichment alters the stoichiometric nutrient ratio (N:P:Si) which may impact phytoplankton community composition and growth potentials. Both quantitative and qualitative changes in AD-N inputs may be linked to eutrophication and algal bloom dynamics.

Introduction

Non-recycled, externally-supplied nitrogen (N), termed "new" nitrogen (much of it originating from man-made sources), is a key factor controlling primary production in N-sensitive estuarine and coastal waters (jointly termed coastal waters) [Ryther and Dunstan, 1971; Nixon 1986). Enrichment of the coastal zone with N and other nutrients (e.g., phosphorus, P) is a well-established phenomenon [Rosenberg, 1980; D'Elia, 1987; de Jonge, 1990; Smetacek et al., 1991; Forsberg, 1994; Nixon, 1995; Howarth et al., 1996]. There is a general consensus that N-driven eutrophication of shallow estuaries and enclosed coastal seas has increased over the last several decades. Nutrient flux to coastal systems, while essential for maintaining overall productivity, has increased to the extent that degradation of coastal systems is now a widespread occurrence. The deleterious effects of eutrophication include accelerating rates of primary production, increasing phytoplankton biomass, and changes in phytoplankton community composition, including increasing frequencies and magnitudes of algal blooms, with a shift to harmful forms. These changes can be accompanied by a reduced photic zone, loss of aquatic habitat, and decreased concentrations of dissolved oxygen in bottom waters (hypoxia or anoxia), all of which yield negative impacts on biological resources.

Evidence from nutrient-enriched coastal waters having good historical water quality records, including the Baltic Sea, Kattegat, Skagerrak, the Dutch Wadden Sea and northern Adriatic, suggests a long-term increase in the frequency of phytoplankton blooms, including noxious forms [Smayda, 1990]. Parallel increases in the areal extent and/or severity of hypoxia have been observed in some areas of the Baltic Sea [e.g., Andersson and Rydberg, 1987], the northern Adriatic Sea [Justic et al., 1987], mid-Atlantic estuaries, including the Chesapeake Bay [Officer et al., 1984], the Neuse River and Pamlico Estuaries, NC [Copeland and Gray, 1991; Paerl et al., 1998], and the continental shelf adjacent to the Mississippi and Atchafalaya River discharges [Turner and Rabalais, 1994a; Eadie et al., 1994].

Inputs of new N, associated with expanding human use and development of coastal watersheds and airsheds are increasing [Vitousek et al., 1997]. One of the most rapidly-growing (both in terms of amount and geographic scale) sources of anthropogenic N loading is atmospheric deposition (AD). It has been estimated that ~1 to >40% of new N inputs into coastal waters are now of atmospheric origin (Table 1), much of it attributable to growing agricultural, urban and industrial emissions [Martin et al. 1989, Loye-Pilot et al. 1990, Prado Fiedler 1990, Duce et al. 1991, Paerl 1993, 1995, Valigura et al., 1996]. This represents at least a ten-fold increase in atmospheric N emissions over pre-industrial and Agricultural Revolution (early 1800s) [Howarth et al., 1996; Holland et al., 1999]. In North America, coastal AD-N ranges from 400 to over 1000 mg N m^{-2} y^{-1}, while in the highly urbanized, industrialized and intensively-farmed regions of Western Europe deposition commonly exceeds 1 g N m^{-2} y^{-1} [Prospero et al.; 1996; Holland et al., 1999]. The relative contribution of AD-N to coastal N budgets is expected to increase substantially in the early part of the next century [Galloway et al., 1994], when nearly 70% of the US and European populations will reside within 50 km of the coast. On regional and global scales, AD-N constitutes a significant input of new N to the oceanic environment; accounting for ~40 Tg N y^{-1}, compared to ~30 Tg N y^{-1} from runoff and riverine discharge, ~10 Tg N y^{-1} for groundwater, and ~15 Tg N y^{-1} from biological N fixation [Prospero et al., 1996; Codispoti et al., in press; Paerl and Whitall, in press;

TABLE 1. Estimated contributions of atmospherically-derived N (ADN) to "new" N inputs in diverse estuarine, coastal and open ocean waters. When identified, the sources [wet (W) and dry deposition (D)] and chemical forms [inorganic (I) and organic (O)] of ADN are indicated. Informational sources and references are provided. [Adapted from Paerl, 1997].

Receiving Waters	Percent of "new" N as ADN	Sources and Forms Included in Study	References
Baltic Sea (Proper)	>25	W+D, I	Rodhe et al. 1980, Elmgren 1989, Ambio 1990
Western Baltic Sea (Kiel Bight)	60	W+D, I	Prado-Fiedler 1990
North Sea (Coastal)	20-40	W+D, I	GESAMP 1989
Western Mediterranean Sea	10-60	W, I	Martin et al. 1989, Loye-Pilot et al. 1990
North Pacific Ocean stratified surface waters	40-70	W+D, I (NO_x only)	Prospero and Savoie 1989
water column	<5		Prospero and Savoie 1989
Sargasso Sea surface waters	25	W, I	Duce 1986
water column	10	W, I	Michaels et al. 1993
Waquiot Bay, MA, USA	29	W, I+O	Valiela et al. 1996
Narragansett Bay, USA	12	W, I+O	Nixon 1995
Long Island Sound, USA	20	W, I+O	Long Island Sound Study 1996
New York Bight, USA	38	W, I+O	Hinga et al. 1991
Barnegat Bay, USA	40	W, I+O	Moser et al. Unpublished
Chesapeake Bay, USA	27	W, I+O	Chesapeake Bay Program 1996
Rhode River, MD, USA	40	W, I+O	Correll and Ford 1982
Neuse River-Pamlico Sound NC, USA	20 to >38	W+D, I	Paerl and Fogel 1994
Sarasota Bay, FL, USA	26	W+D, I	Sarasota Bay NEP 1996
Tampa Bay, FL, USA	28	W+D, I	Tampa Bay NEP 1996
Mississippi River Plume, USA	<5	W+D, I+O	Goolsby et al. (in review)

Capone, unpublished]. Note that there is a standardized distinction of the terms "deposition" and "loadings" used in this manuscript. *Deposition* is the flux of nitrogen from the atmosphere to whatever surface is beneath through precipitation (wet deposition) or through gaseous/particulate exchange (dry deposition). *Indirect loading* is the portion of nitrogen deposited onto the terrestrial watershed than is subsequently transmitted to the water body itself, defined in terms of flow to the tidal waters. Deposition to the water surface itself constitutes a *direct loading*.

In developed regions, AD-N may be the single most important source of new N now entering the coastal zone [Valigura et al., 1996; Paerl, 1997; Jaworski et al., 1997]. As such, it is imperative to assess the biogeochemical and trophic ramifications of this new N source. Because AD-N is quantitatively important and capable of stimulating primary production in N-sensitive coastal waters [Paerl, 1985, 1995; Willey and Paerl, 1993], there is concern that increasing rates of new atmospheric N deposition may be linked to symptoms of accelerating eutrophication [Paerl 1995, 1997]; including expansion of harmful algal blooms impacting estuarine and coastal waters [Paerl, 1988; Hallegraeff, 1993; Richardson, 1997; Anderson and Garrison, 1997]. Here, we present a synthesis of current knowledge concerning N inputs from atmospheric sources, and an assessment of the sensitivity of the biogeochemical and trophic responses of coastal ecosystems to this new N source.

AD-N Production and Deposition

Atmospheric deposition of N introduces a variety of biologically-available inorganic (NO_3^-, NH_4^+; DON) and organic compounds (amino acids, organoniles, urea; DON) into estuarine and coastal waters [Timperley et al., 1985; Duce et al., 1991; Mopper and Zika, 1987; Paerl, 1995], largely from human activities [Likens et al., 1974; Galloway et al., 1994]. Emissions of nitrogen containing compounds into the atmosphere from fossil fuel combustion, agricultural activities, and industrial land-uses is a significant and growing percentage of total nitrogen input into aquatic ecosystems [Duce, 1986; Luke and Dickerson, 1987; Asman, 1994; Paerl, 1995; Holland et al., 1999]. Depending on the ratio of watershed to estuary surface area, the degree of watershed N-retention, and the proximity in relation to atmospheric sources, an important fraction of AD-N is directly deposited to estuarine and coastal waters. Current estimates of the contribution of atmospheric direct loadings to the total (i.e., natural + anthropogenic) N loading to a given water body range from less than 5% to over 50% [Duce, 1991; Fisher and Oppenheimer, 1991; Paerl, 1995, 1997; Valigura et al., 1996; Dennis, 1997; Holland et al., 1999].

A. *AD-N Composition and Sources*

Atmospheric nitrogen compounds are ubiquitous. Diatomic nitrogen (N_2) is the dominant component of the atmosphere, and is comparatively unreactive. However, nitrogen is also present in the atmosphere in other chemical forms, some of which are relatively reactive. Most atmospheric nitrogen compounds (other than N_2 and nitrous oxide [N_2O]) fall into two categories: reactive nitrogen (composed primarily of nitrogen oxides), and reduced nitrogen (typically dominated by NH_3). There are also organic nitrogen species that are typically referred to as a subset of reactive nitrogen. These chemical species arise in the atmosphere from the interaction between nitrogen oxides and biogenic or anthropogenic hydrocarbons. Though available evidence suggests that organic nitrogen originates from both anthropogenic and natural sources, speciation of organic nitrogen in the atmosphere is very poorly understood. The primary "reactive" nitrogen compounds of concern in the atmosphere are nitric acid (HNO_3) and aerosol nitrate (p-NO_3). Nitric acid is a by-product of the oxidation of nitrogen oxide (NO_X = NO + NO_2) and volatile organic compounds (VOC's) which occurs through atmospheric photochemistry [Seinfeld and Pandis, 1998; U.S. E.P.A., 1996]. Aerosol NO_3^- stems from a partitioning of a fraction of HNO_3 into particulate form due to heterogeneous chemistry in the presence of ammonia (NH_3), producing ammonium nitrate [Seinfeld and Pandis, 1998]. Nitric acid and aerosol NO_3^- are secondary products; they are not emitted directly. Their principal source in North America is the combustion of fossil fuels (e.g., by coal and oil-fired power plants, automobiles and other forms of transportation) that produces precursor emissions of NO_X, primarily NO. Globally, biomass burning is also an important source of NO_X emissions. The primary "reduced" nitrogen compounds of concern in the atmosphere are in the form of gaseous ammonia (NH_3) and aerosol ammonium (NH_4^+). Ammonium stems from heterogeneous chemistry involving sulfuric acid droplets, HNO_3, and NH_3 that partitions a fraction of NH_3 into fine particulates: ammonium sulfate, ammonium bisulfate, ammonium nitrate and other complex aerosol mixtures [Seinfeld and Pandis, 1998]. Ammonia is directly emitted into the atmosphere, largely from agricultural sources of fertilizer application and animal operations, particularly confined animal operations [Buijsman et al., 1987].

These soluble oxidized and reduced N species can be utilized directly by a variety of microorganisms and higher aquatic plants, including suspended microalgae (phytoplankton), macroalgae and rooted macrophytes [Morris 1974, Glibert et al., 1991]. Dissolved organic nitrogen is another important fraction of AD-N, accounting for ~15 to 30% of wet deposition in coastal and open ocean regions [Correll and Ford, 1982; Paerl, 1995; Cornell et al., 1995; Mopper and Zika, 1987; Peierls and Paerl, 1997; Russell et al., 1998; Timperley et al., 1985]. Bioassays indicate that the DON fraction in rainfall can also stimulate phytoplankton growth and therefore should be included as a biologically-reactive from of *new* N input [Seitzinger and Sanders, 1997, 1999; Peierls and Paerl, 1997].

The processes by which these AD-N sources are generated and enter coastal waters are changing, both temporally and spatially. For example, recent urbanization in southwest Florida (e.g., the Tampa-St. Petersburg-Sarasota region) has been accompanied by rapidly increasing NO_X emissions from power plants and automobiles. In the Tampa Bay-Sarasota Bay region, oxidized N deposition associated with these activities accounts for approximately 30% of the new N input into adjacent near-shore waters [Greening et al., 1997]. Another area of concern is nitrogen generated from expanding agricultural activities, specifically animal operations. During the past decade, proliferating commercial swine and poultry operations have made the Southeast U.S. the largest poultry producers in the U.S. and a leader of pork production, with North Carolina alone ranking only second to Iowa. Unlike human waste, swine waste is stored in open lagoons and periodically sprayed onto fields as a fertilizer. Substantial amounts (30 to >80%) of lagoon and field-applied N are lost via NH_3 volatilization [O'Halloran, 1993]. Regionally,

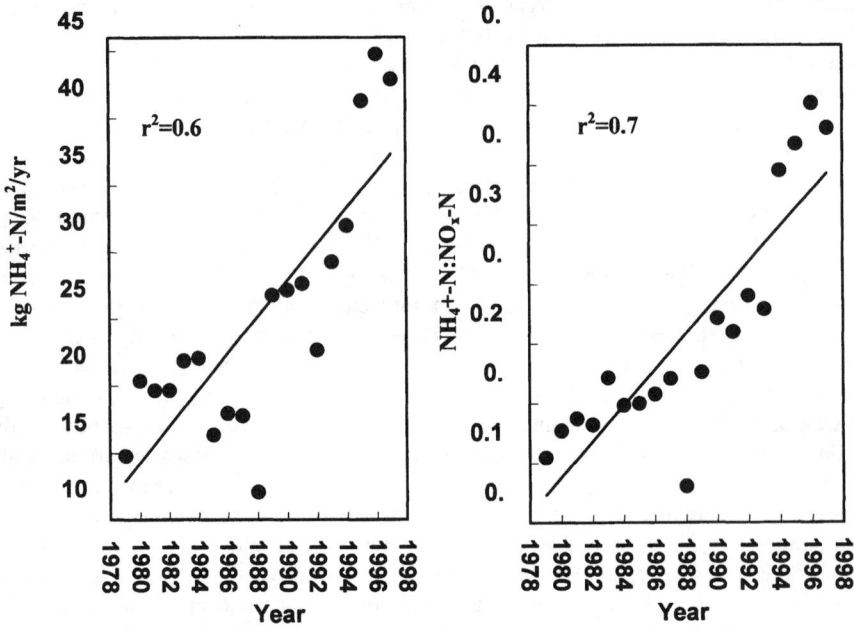

Figure 1. Long-term (1977-1998) record of NH_4^+ and NO_3^- wet deposition in Sampson County, eastern North Carolina. Data are adapted from the National Atmospheric Deposition Program site NC 35.

volatilization of NH_3 from animal wastes is a major and rapidly growing source of biologically-available N [Asman, 1994]. There is evidence in the National Acid Deposition Program (NADP) wet deposition data that, relative to oxidized N, reduced N concentrations and total N deposition have increased markedly ($\approx$ 50% since 1980) in eastern North Carolina bordering N-sensitive estuarine and coastal waters (Figure 1) [Paerl, 1997].

Spatial and temporal scales of N emissions, transport, and fate are highly variable, yet functionally linked (Figure 2). Because N-containing air pollutants can originate from urban, industrial, and agricultural regions that transcend watershed boundaries, management of estuarine and coastal water quality cannot be accomplished solely on the watershed level. Sources and sinks of N pollutants operate on spatially explicit scales; all of which are essential for understanding how changing human activities and land-use affect water quality. For example, fossil fuel NO_X emissions from large urban areas can yield substantial impacts on N deposition dynamics in remote (>500 km) N sensitive coastal waters (RADM Model output, US EPA). In North Carolina, volatilized NH_3 emitted from animal waste generated outside of a watershed but within its airshed (e.g., Neuse River) exemplifies the growing complexity and concern over anthropogenic new N sources originating over a range of spatially- and temporally- explicit scales. In North Carolina coastal waters, spring and summer N enrichment often stimulates the potential for algal blooms, hypoxia, and fish kills more than equivalent mid-winter enrichment [Paerl et al., 1995, 1998]. Inputs to slow-moving downstream waters are more likely to

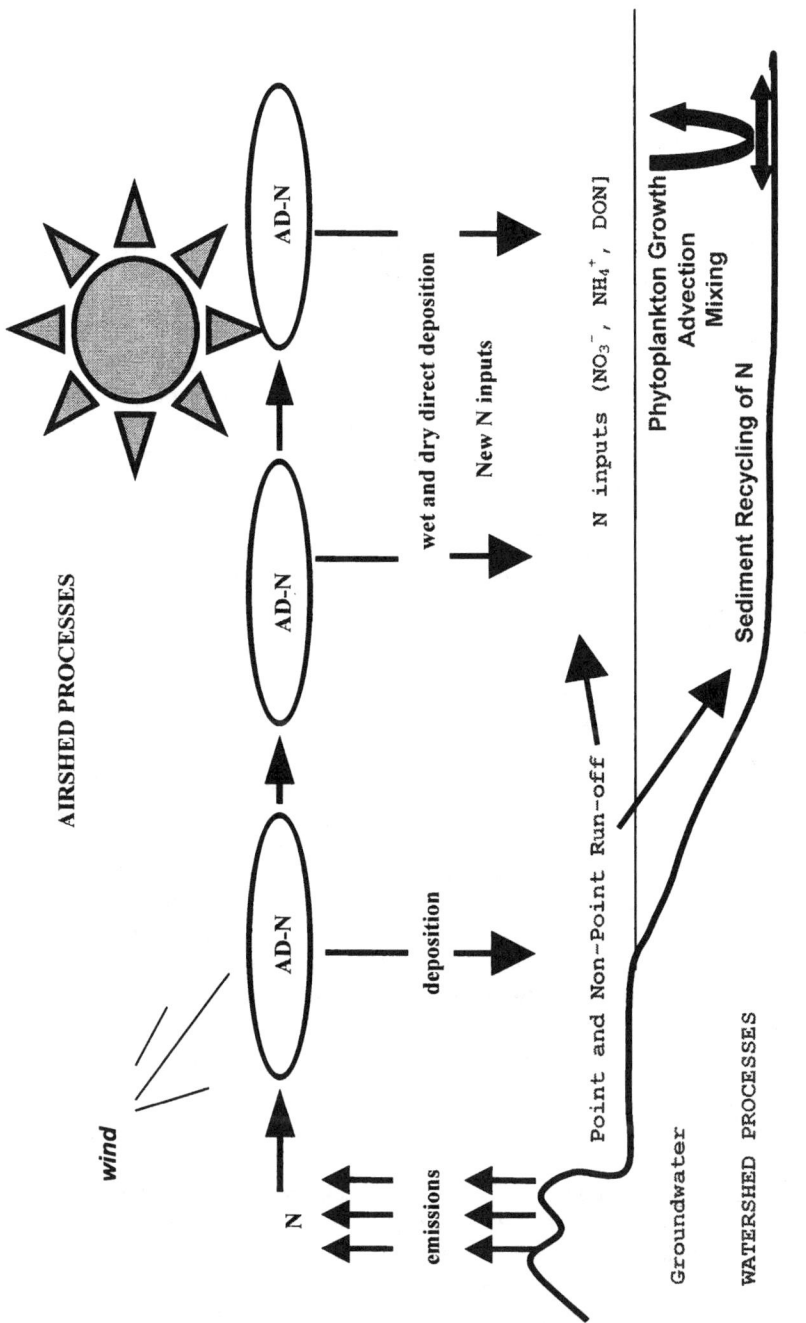

Figure 2. Conceptual diagram linking watershed and airshed-based external or "new" N loading to estuarine production and nutrient cycling dynamics.

negatively impact these indicators than to fast-moving upstream waters. Therefore, the timing and location of anthropogenic N inputs are critical determinants in determining ecosystem sensitivity, water quality and habitat responses to upstream (watershed) and upwind (airshed) N enrichment.

B. AD-N Deposition Processes

Once incorporated into precipitation, all forms of inorganic N deposit equally well. In contrast, the form of the inorganic N and whether the deposition is to land surfaces or to open water significantly influence dry deposition. Dry deposition rates for gaseous HNO_3 and NH_3 are the same order of magnitude at moderate to high concentrations of NH_3 but with a faster rate for HNO_3 [Meyers et al., 1998; Duyzer, 1994; Fowler et al., 1998]. At low concentrations of NH_3 there is a compensation point where plants can become a source of NH_3 rather than a sink [Langford and Fehsenfeld, 1992; Duyzer, 1994; Fowler et al., 1998; Wyers and Erisman, 1998]. Though fine particles of a similar size deposit at the same rate, gaseous and aerosol deposition rates of inorganic N are very different; the gaseous rate is approximately 10 times higher than the aerosol rate. Also, for both gases and aerosols, the rate of deposition to open estuarine and coastal water surfaces is roughly 3 to 5 times lower than to terrestrial surfaces.

Because of the large difference in gaseous and aerosol deposition rates, the dry deposition rates of oxidized and reduced N are different, affecting ratios of wet to dry deposition for the two types of inorganic N. Modeling analysis, using an extension of the Regional Acid Deposition Model [Chang et al., 1987], termed the Extended RADM (ExtRADM). The ExtRADM incorporates the particle physics (for SO_4^{-2}, NO_3^-, NH_3) of the Regional Particulate Model [Binkowski and Shankar, 1995], and indicates that the fractions of $NO_3^-/(NO_3^- + HNO_3)$ and $NH_4^+/(NH_4^+ + NH_3)$ for the coastal areas of the eastern U.S. are basically reversed, with medians of 0.23 and 0.74, respectively, for atmospheric concentrations over the East Coast and Gulf Coast estuaries. This same pattern was observed in Northern Europe [Sorteberg et al., 1998] and supported by CASTNet data for oxidized N in the eastern U.S. CASTNet monitoring data are described in Clarke et al. [1997]. Hence, the dry deposition of oxidized N is significantly larger than that of reduced N, whereas wet deposition is hardly or not at all influenced by the differences in partitioning. Model analyses with the ExtRADM and CASTNet data suggest wet and dry deposition of oxidized forms of N is roughly comparable on an annual basis. Model analyses with ExtRADM (no monitoring data available) suggest wet and dry deposition of reduced forms of N are unequal, with wet deposition, in general, being roughly a factor of 2 to 3 larger on an annual basis.

C. AD-N Residence Times and Airsheds

Model analyses of atmospheric residence time [Wojcik and Chang, 1997] and the northern and southern range of influence of emission source regions [Dennis, 1997] indicate that residence times of nitrogen compounds in the atmosphere over the eastern U.S. appear to be controlled by wet scavenging, not by dry deposition rates. Wet scavenging is an efficient cleansing mechanism for the atmosphere. The transport distances at which two-thirds of the continental total (wet + dry) deposition of oxidized N from a region of NO_x emissions deposits (its range of influence) is roughly 400-600 km [Dennis, 1997]. The transport distances for reduced N emissions are still under study for the eastern U.S., although it is expected to be shorter than those of oxidized N. Modeling studies have been conducted for Northern Europe [Hov et al., 1994; Hov and Hjøllo, 1994] that indicate reduced N does have a shorter residence time than oxidized N.

An *airshed* defines the predominant emissions area that influences the air concentrations over or deposition to a specified area, such as an estuary or watershed. Unlike a watershed, an airshed does not have clearly defined boundaries; the longer the range of influence (pollutant lifetime) the fuzzier the boundary. Dennis (1997) developed an operational definition of an airshed, based on calculated ranges of "significant" influence from emissions source regions. This results in defining an area of emissions that include sources that are expected to make an important contribution to deposition loading and will *en toto* account for approximately 70% of the deposition loading to the specified region of interest.

Following the procedure outlined in Dennis (1997), airsheds for oxidized N were developed for many of the estuaries in this assessment. These airsheds provide a sense of the area from which the important air emissions of NO_x emanate, that has a principal influence on the deposition of oxidized N to these estuaries. The airsheds for oxidized N are on the order of several hundred thousand km^2 or tens of millions of hectares; most are roughly 10-100 times larger than their associated watersheds. Examples are shown in Figure 3 and tabulated in Table 2. Model evaluation results [e.g., Cohn and Dennis, 1994] suggest that the estimates of airshed areas using RADM should be biased small. Areas of the airsheds over the ocean involve some "artistic" license.

Most watershed areas used in this study are in the thousands to tens of thousands of km^2; however, the Chesapeake Bay and Mobile Bay watershed areas are both more than one hundred thousand km^2. Airshed areas are in the hundreds of thousands of km^2 and their size ranges by a factor of two. For the airsheds in Table 2 the continental areas for all but the two largest are all within ~30% of 600,000 km^2 and the two largest are one million km^2. There is significant overlap of the airsheds along each coast. Figure 3 also shows that emissions from one watershed can often affect both itself as well as other watersheds. The airsheds for even very small watersheds are substantial (Fig. 3). Extreme differences in airshed and watershed areas are represented at one end by Narragansett and Barnagat Bays, with ratios of airshed to watershed in the 100's, and at the other end by Chesapeake Bay and Mobile Bay, with ratios less than 10. Figure 3, in all its parts, shows that the main sources of NO_x emissions affecting East Coast and Gulf Coast estuaries with oxidized N deposition emanate from a substantial portion of the eastern U.S., and southern Ontario, Canada. This same analysis has yet to be done for reduced N, although airsheds for reduced N are expected to be smaller. This is an area of debate but it is expected that NH_3 airsheds will be still quite a bit larger than watersheds. No attempt has yet been made to estimate airsheds for DON emissions. More information on sources and factors controlling transport and removal are required before these analyses can be initiated.

Biogeochemical and Ecological Impacts of AD-N

It is critical to conceptually and functionally link current knowledge of sources, transport and fate of AD-N with biogeochemical (i.e. fluxes and budgets) and trophic

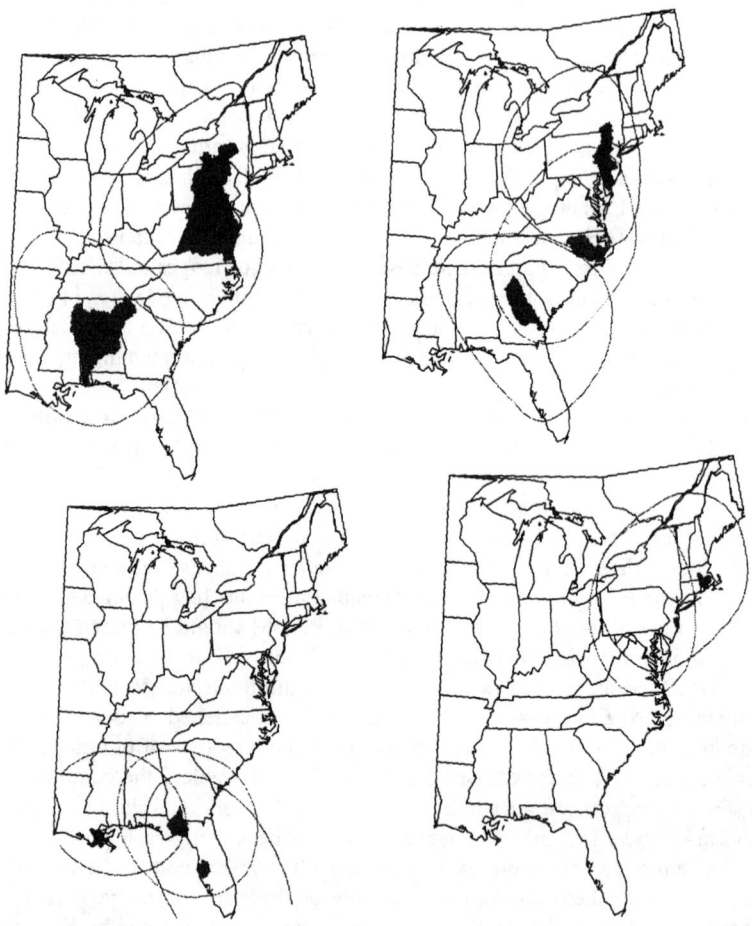

Figure 3. Aerial relationship between watershed and airshed sizes for selected estuaries. Watersheds are in solid colors, whereas airsheds are delineated with colored lines. Chesapeake Bay and Mobile Bay are examples of estuaries with relatively large watersheds and hence small airshed to watershed ratios (shown upper left). Delaware Bay, Pamlico Sound and Altamaha Sound are examples of estuaries with moderately large watersheds and large airsheds (shown upper right). Tampa Bay, Apalachee Bay and Barataria-Terrebone are examples of proximate estuaries having non-overlapping, moderate to small watersheds, but overlapping airsheds (shown bottom left). Narragansett Bay and and Barnegat Bay are examples of relatively small watershed, but large, overlapping airshed estuaries (shown bottom right). Developed by R. Dennis, Atmospheric Sciences Modeling Division, NOAA/ARL and NERL USEPA.

(productivity, food web structure and function) responses in N-sensitive watersheds and receiving waters (Fig. 2). Accordingly, the biogeochemical and ecological ramifications for community structure, trophodynamics and water quality in N-limited coastal ecosystems are discussed in this section.

TABLE 2. Comparison of Watershed and Airshed Areas (km^2).

Estuary	Watershed Drainage Area (W)	Total Airshed Area	Continental Airshed Area (CA)	Ratio* CA/W	Percent NOx Load Explained	Airshed NOx Emissions as a % of Eastern US NOx Emissions
Barnegat Bay	1,404	563,200	505,600	360	67%	16%
Narragansett Bay	4,321	803,200	595,200	140	73%	18%
Tampa Bay	5,725	972,800	256,000	45	76%	5%
Barataria-Terrebonne	7,446	601,600	409,600	55	63%	8%
Apalachee Bay	14,342	793,600	441,600	31	50%	9%
Pamlico Sound	26,829	883,200	665,600	26	63%	18%
Delaware Bay	33,381	832,000	729,600	22	75%	26%
Altamaha Sound	37,105	1,004,800	678,400	18	68%	13%
Lake Pontchartrain	39,598	851,200	659,200	17	63%	11%
Mobile Bay	114,858	1,062,400	992,000	8.6	68%	17%
Chesapeake Bay	167,443	1,177,600	1,081,600	6.5	76%	34%

A. Watershed Impacts and Processing: Effects of Elevated N Inputs to Forests and Soils

AD-N enters coastal waters either by direct input to the water surface, or indirectly where deposition is intercepted and processed by the watershed prior to discharge to coastal waters. Deposition to the watershed has been shown to impact nitrogen cycling within receiving watersheds. This impact has best been characterized for forests and their soils.

Forests respond to increased N availability with a gradual increase in forest productivity [Aber et al., 1989]. Forests that are repeatedly harvested or burned, and so experience a periodic removal of N, would probably become healthier and more productive in response to chronic N inputs. Unmanaged forests continue to show slow increases in the total N content of vegetation and soils.

Under prolonged conditions of elevated atmospheric N deposition, forests may become N saturated. These conditions reflect the shift of forests from being N limited to being carbon, phosphorus or water limited. As N deposition to the saturated system continues, NH_4^+ concentrations in soils increase, leading to nitrification even at low soil pH [Aber et al., 1989, 1998]. The N cycle is then dominated by nitrification and NO_3^- leaching rather than by uptake. This change is critical since nitrification is a strongly acidifying process. Increased emissions of nitrous oxide occur from forest soils as a result of increased nitrification.

Under these conditions of watershed N loss, the seasonal (spring-summer) onset of N limitation is further delayed. Further, NO_3^- leaching occurs even in the growing season. If the biological demand is sufficiently depressed during the growing season, dissolved inorganic N (DIN) begins to percolate below the rooting zone, resulting in elevated groundwater NO_3^- concentrations. This process in turn leads to elevated baseflow concentrations of NO_3^-. Episodic NO_3^- concentrations are as high as observed in earlier stages of N saturation, but under these conditions the seasonal pattern is damped by an increase in baseflow concentrations to levels as high as those found in deposition [Stoddard, 1994].

In the final stages of N saturation, Aber et al. [1989, 1998] suggest the forest begins to decline. Excessive N availability leads to nutrient imbalances, yellowing of foliage, reduced fine-root biomass which in turn result in reduction of net photosynthesis and forest productivity. Nitrification rates are substantial. The combined inputs of N from AD, mineralization, and nitrification can produce concentrations of NO_3^- in surface waters that exceed inputs from deposition alone. The key characteristics of this condition are extremely high NO_3^- concentrations and the lack of a seasonal pattern in NO_3^- concentrations [Stoddard, 1994].

A further consequence of an increase in N inputs is an increase in nitrous oxide (N_2O) emissions from forest soils. Experimental additions of N to a mixed hardwood forest in the northeastern U.S. by Bowden et al. [1991] resulted in small increases in annual N_2O fluxes. Nitrogen fertilization in forest soils has also been implicated in decreasing the amount of CH_4 taken up by microbes in aerobic soils [Steudler et al., 1989]. These effects have not been well established and warrant further study because of the importance of N_2O and CH_4 as greenhouse gases.

B. Biogeochemical and trophic responses of surface waters

1. Freshwater impacts

Increased N inputs through leaching to surface-waters may result in surface-water acidification. Surface waters are considered acidic if values of acid neutralizing capacity (ANC) are less than zero. The ANC is a measure of the ability of a water body to neutralize inputs of strong acids, and largely results from the presence of bicarbonate and organic anions in water. The acidification processes in surface waters are conventionally separated into chronic (long-term) and episodic (event-based) effects. The N cycle does not result in the production of excessive H^+ as long as NH_4^+ inputs are low and NO_3^- is not readily leached from soils (as much H^+ in the plant-soil system are consumed as produced). Under conditions of NO_3^- leaching, net H^+ production equals NO_3^- loss. Ammonium will only contribute to the acidification if it is nitrified, resulting in the leaching of NO_3^- and H^+.

Chronic acidification resulting from N deposition is much more common in Europe than in North America. Several upland regions in Great Britain show elevated NO_3^- concentrations in surface waters due to high N inputs [Allott et al., 1995]. Henriksen and Brakke [1988] reported regional chronic increases in surface-water NO_3^- in Norway. These increases in NO_3^- concentrations were associated with increasing concentrations of Al, which is toxic to many fish species.

Except in extreme cases, N loss from watersheds is more likely to occur as an episodic or seasonal process rather than by chronic leaching. It is estimated that 1.4-1.7 times as many streams in the eastern United States undergo episodic acidification than are chronically acidic [Eshleman, 1988]. A number of processes seem to contribute to the timing and severity of acidic episodes [Driscoll and Schaefer, 1989]. These processes include dilution concentrations of base cations [Galloway et al., 1980], increases in SO_4^{2-} concentrations [Johannes et al., 1985], and increases in NO_3^- concentrations [Driscoll et al., 1984]. The likelihood of an acidic episode is also influenced by the chemical conditions prevailing before the episode begins. Strong NO_3^- pulses in both lakes [Driscoll and Schafran, 1984] and streams [Driscoll et al., 1987] are apparently the primary factor contributing to depressed ANC and pH during snowmelt in the Adirondack mountains of New York. Schaefer et al. [1990] examined intensive monitoring data from 11 Adirondack lakes and concluded that the magnitude of the episodes experienced by lakes was strongly dependent on their base-cation concentration. They concluded that lakes with high base-cation concentrations (and therefore high ANC values) undergo episodes that are largely the result of dilution by snowmelt. Low-ANC lakes, on the other hand, undergo episodes that result largely from increases in NO_3^- concentrations. At intermediate ANC levels, both base-cation dilution and NO_3^- increases affect lakes. Therefore these lakes may undergo the greatest increases in acidity and loss of ANC during snowmelt episodes.

As is the case with soils, an additional consequence of increased N inputs to rivers is an increase in the production of nitrous oxide emissions. A recent model of world rivers suggests that rivers are an important source of nitrous oxide, with approximately 1 Tg N y^{-1} of nitrous oxide emitted from rivers globally [Seitzinger and Kroeze, 1998]. The model of Seitzinger and Kroeze [1998] also suggests that approximately 20% of the DIN exported by world rivers globally (21 Tg N y^{-1}) is attributed to atmospheric deposition, indicating the potential importance of AD-N to river nitrous oxide emissions.

Eutrophication is the accelerated rate of production of organic matter in aquatic ecosystems impacted by enhanced nutrient inputs [Nixon, 1995]. The potential for AD-N (among other N inputs) to contribute to the eutrophication of freshwater lakes is restricted to those surface waters that are chronically N-limited. This condition occurs in some surface waters that receive substantial inputs of anthropogenic P and in some surface waters where both P and N are found in low concentrations, such as ultra-oligotrophic Lake Tahoe, CA-NV [Jassby et al., 1994]. In the former case, the primary dysfunction of the lakes is an excess supply of P, and controlling N deposition would be an ineffective method of water-quality improvement. In the latter case the potential for eutrophication by N addition is limited by low P concentrations; additions of N to these systems would soon lead to N-sufficient and P-deficient, conditions [Goldman, in press]. In the Adirondacks an increase in only 0.3 mol L^{-1} of N would quickly lead to P-limitation in the few lakes that actually show N-limitation [Kanciruk et al., 1986].

2. Coastal N utilization dynamics

"Nitrogen is nitrogen" is a common assumption when relating inputs to trophic and biogeochemical responses in N-depleted waters. However, linking N inputs to *specific* trophic responses in natural systems has proven difficult and unpredictable. It is known that timing, modality, and composition of N loading play key roles in determining growth responses of the dominant primary producers, suspended algae or phytoplankton [Dugdale and Goering, 1967; Harrison et al., 1987; Bronk et al., 1994]. Anthropogenic and natural N inputs display chemical individuality and different bioreactivities. Contrasting N compound reactivities may lead to physiologically and taxonomically distinct responses of phytoplankton growth. These differential responses (especially as blooms) could translate into large inputs of newly-fixed carbon to coastal waters, yielding cascading impacts on community productivity and structure, O_2 dynamics, nutrient cycling, and water and habitat quality.

Nutrient acquisition is a major factor determining the outcome of competitive interactions and phytoplankton community structure [Titman, 1976; Tilman, 1977; Huisman and Weissing, 1995]. In N-limited systems, ambient N concentrations govern phytoplankton growth rates and primary productivity [Goldman et al., 1979]. Nutrient uptake and assimilation by phytoplankton varies among species and N sources [Eppley et al., 1969a; Dortch and Conway, 1984]. In some species, NH_4^+ inhibits NO_3^- and/or urea uptake [Eppley et al., 1969b; Molloy and Syrett, 1988; Collos, 1989; Harrison and Turpin, 1982; Dortch, 1990]. Uptake rates are influenced by the nutritional state, and N-starved individuals uptake N at higher rates [Morris, 1974; McCarthy, 1981; Syrett, 1981]. Nitrogen formulations may play an important role in structuring the size distribution of phytoplankton communities. Stolte et al. [1994] showed that large diatoms dominate when NO_3^- is supplied as the only N source at 3-day intervals. In NH_4^+-pulsed cultures, both large and small species coexisted. The larger species may have the capacity for more internal storage of nutrients and become dominant in fluctuating nutrient regimes [Margalef, 1978; Turpin and Harrison, 1979]. In general, smaller species have a higher preference for NH_4^+ uptake over NO_3^- than larger phytoplankton species [Stolte et al., 1994].

Based on culture [Neilsen and Lewin, 1974; Antia et al., 1991] and natural community studies [Peierls and Paerl, 1997; Seitzinger and Sanders, 1997, 1999], dissolved organic N (DON) warrants consideration as a biologically-reactive constituent of AD-N. To understand the trophic ramifications of this fraction, it is necessary to first determine which DON components are biologically available and what members of the microbial and higher plant communities most effectively utilize these components. While the chemical composition of atmospherically derived DON (ADON) is not well known, it is established that a number of different types of organic nitrogen compounds are present in atmospheric deposition (e.g., amino acids, organoniles, urea, etc.). Bacteria are known to utilize each of these compounds. Phytoplankton also can use some small molecular weight organic compounds (amino acids, urea, etc.) following decomposition by extracellular enzymes [Antia et al., 1991]. Some of the DON compounds identified in atmospheric deposition are toxic, although the threshold for toxicity may not be reached at typical atmospheric flux rates. The bioavailability or toxicity of the identified ADON compounds is inferred mostly from laboratory studies using isolated cultures in controlled bioassays. The chemical composition of ADON is a likely modulator of microbial activity and composition. For instance, certain species may have a competitive advantage in assimilating organic N forms leading to changes in dominant taxa, including those responsible for harmful algal blooms [Berg et al., 1997]. Bioassays of natural planktonic communities have shown that ADON affects community production and structure [Peierls and Paerl, 1997; Seitzinger and Sanders, 1999]. However, little is known about which of the specific compounds in ADON are available to organisms in the environment, and what their effects (e.g., as a nutrient or as a toxicant) are at the species, community or ecosystem level.

There are a number of pathways by which DON can be incorporated into biogeochemical cycles. Typically, bacteria utilize organic N for growth, mineralizing ammonia in the process. Protozoa and other microorganisms can feed on the bacteria, transferring bacterial N to higher trophic levels that also regenerate readily available DIN (typically as ammonia). Phytoplankton and higher plants can assimilate DIN to produce new biomass, but bacteria also compete for DIN. Phytoplankton also are able to use some DON compounds after mineralization by extracellular enzymes or by direct assimilation of the compounds for use as N or C in cellular metabolic pathways [Antia et al., 1991]. An additional pathway of incorporation of organic N into terrestrial and aquatic ecosystems is through utilization by fungi. Finally, non-biological pathways such as photochemical reactions may degrade DON to DIN.

Overall, the implication is that long-term changes in the sources and amounts of AD-N compounds in N-limited systems may alter both the species composition and relative size distribution of the phytoplankton community. This disturbance will have cascading effects on the trophic structure and biogeochemical cycling of impacted ecosystems.

3. Estuarine and coastal eutrophication: The role of nitrogen

Eutrophication of estuaries and coastal waters is a serious human threat to the integrity of coastal ecosystems. In contrast to the phosphorus limitations of a majority of temperate zone lakes, estuaries and coastal waters tend to be nitrogen limited [Nixon, 1986; D'Elia et al., 1986]. This is largely because the natural supply of N by drainage inflow to coastal waters and the rate of N_2 fixation by planktonic organisms are relatively low compared to high rates of denitrification by sediment microbes that release N back to the atmosphere [Vitousek et al., 1997]

The effect of N inputs on estuarine and coastal surface waters through precipitation are significant in coastal areas such as the coastal mid-Atlantic, Baltic Sea, W. Mediterranean (Figure 2). AD-N inputs in these areas can stimulate phytoplankton growth and influence community composition [Martin et al., 1989; Prado-Fiedler, 1990; Paerl, 1985, 1995; Paerl and Willey, 1995]. Although primary consumers recycle a portion of phytoplankton carbon, bloom events (including noxious/toxic algal species) triggered by *new* N inputs (including AD-N) result in the deposition of phytoplankton carbon to the benthos. This input of N can produce hypoxia (< 2-3 mg O_2 L^{-1}) and anoxia (no detectable dissolved O_2) of bottom waters over a large spatial scale. Microbial degradation of deposited phytoplankton results in a high oxygen demand that depletes the overlying water of dissolved oxygen.

Bottom-water hypoxia and anoxia result from the interaction of excessive C loading with several non-biological variables such as freshwater discharge, vertical stratification, salinity intrusions resulting in establishment of salt-wedges, and meteorological conditions (primarily wind). Anoxia and hypoxia strongly influence biogeochemical cycling in affected habitats. Benthic release of NH_4^+ and PO_4^{-3}, following regeneration is often enhanced under low dissolved oxygen concentrations [Nixon, 1986; Boynton et al., 1995; Rabalais, 1998], further fueling phytoplankton blooms, perpetuating the oxic/anoxic cycle in impacted ecosystems. As the frequency, duration, and areal coverage of these perturbations increase, estuarine trophic structure and function may be radically altered [Justic et al., 1996]. In waters having relatively long hydraulic residence times, low flushing rates and periodic vertical stratification (e.g., Chesapeake Bay, Long Island Sound, Neuse and Pamlico Rivers), hypoxic/anoxic conditions can persist for weeks and cover large areas [Cooper and Brush, 1991; Parker and O'Reilly, 1991; Van Dolah and Anderson, 1991; Paerl et al., 1998; Rabalais et al,. 1998].

Hypoxia occurs naturally in many aquatic environments, yet there is growing concern for anthropogenically-enhanced hypoxia in U.S. estuarine and coastal waters. Seasonally depleted oxygen levels occur in over 50% of the U.S. estuaries [Rabalais, 1998]. The occurrence of hypoxia in shallow coastal and estuarine areas worldwide appears to be increasing [Diaz and Rosenberg, 1995], most likely accelerated by anthropogenic nutrient enrichment. Prolonged oxygen depletion can not only disrupt benthic and demersal communities but can also cause mass mortalities of aquatic life [Diaz and Rosenberg, 1995]. When oxygen levels fall below critical values, those organisms capable of swimming [e.g., demersal fish, portunid crabs and shrimp) evacuate the area. The effects on less motile fauna vary, but they also experience stress or die as oxygen concentrations fall to zero. Important fisheries resources are variably affected by direct mortality, forced migration, reduction in suitable habitat, increased susceptibility to predation, changes in food resources and disruption of life cycles.

Changes in land-use have been associated with oxygen depletion due to an increase in nutrient enrichment of receiving waters. Concurrently, increases in nutrient inputs have also been linked to increases in population density within coastal watersheds [Peierls et al., 1991; Howarth et al., 1996]. These increases in loading come from higher rates of fertilizer application [Vitousek et al., 1997], sewage inputs [Nixon et al., 1986], conversion of forests and prairies to agriculture land [Turner and Rabalais, 1991], and atmospheric deposition [Prado-Fiedler, 1990; Paerl , 1995].

Globally, coastal watersheds receive 103 Tg y^{-1} of N from synthetic fertilizer (73.6 Tg), atmospheric deposition of oxidized nitrogen (22.5 Tg), and human sewage (9.1 Tg) [Seitzinger and Kroeze, 1998]. Approximately 20% of the DIN exported by world rivers to the coastal zone globally (21 Tg N y^{-1}) is attributed to atmospheric deposition [Seitzinger and Kroeze, 1998], underscoring the importance of AD-N to coastal N loading and eutrophication globally.

4. The impacts of N enrichment on nutrient concentrations and ratios

The three nutrients most often cited as important for controlling phytoplankton growth and production in marine ecosystems are N, P, and silicon (Si). From studies in marine and freshwater ecosystems, the spatial and temporal interactions of N and P limitation are known to strongly influence phytoplankton species composition, possibly in ways that will affect the function of the ecosystem [e.g. Riegman, 1992]. In addition, the presence of certain phytoplankton taxa may be determined by N and P interactions with Si. Diatoms, a quantitatively important estuarine and marine food group for invertebrates to fish, require Si for their cell walls (frustules). If Si supplies are limited, other non-siliceous forms, such as dinoflagellates or cyanobacteria, may become proportionally more important; some of these are harmful (anoxia-inducing, toxic, food web altering) and form blooms. Recent studies point to the concentrations and relative proportion (ratios) of N, P and Si to each other as important determinants of phytoplankton production, composition, food web structure, energy flow and trophodynamics [Dortch and Whitledge, 1992; Justic et al., 1995a,b; Turner et al., 1998]. In waters experiencing accelerating N and P loading, the potential for Si limitation and associated structural and functional changes in food webs appears to be increasing.

Seasonal variations in N loading can also affect nutrient availability and shifts in phytoplankton communities. There is nearly a two-fold difference in NO_3^- supply over the course of the year in the Mississippi River due to elevated inputs during the spring [Turner and Rabalais, 1991], but only small annual variations in the Si and total P supply. Consequently, the ratios of nutrient supply vary on a seasonal basis, with Si and P in the shortest supply during the spring and N more likely to be limiting during the rest of the year. Fluctuations in the Si:N ratio within the major distributataries of the Mississippi River system are thought to be major determinants in estuarine and coastal food web structure [Turner et al., 1998]. Here, the availability of dissolved Si and its ratio to total inorganic N are important in controlling diatom community production and composition, with implications for carbon flux and control of oxygen depletion [Dortch and Whitledge, 1992; Nelson and Dortch, 1996].

The water quality changes in the Mississippi River are not unique among world rivers [Justic et al., 1995a; Howarth et al., 1996]. The Mississippi is among a suite of large rivers in which the concentrations and proportions of N, P and Si have changed over many decades in response to anthropogenic activities in the watershed [Peierls et al., 1991]. Justic et al. [1995a] examined nutrient data from ten large world rivers. Pristine rivers [Yukon, Mackenzie, Amazon and Zaire, for example) deliver Si to the coastal ocean in great stoichiometric excess over N and P, relative to the Redfield ratio (Si:N:P = 16:16:1), and nutrient requirements of diatoms. Consequently, pristine rivers tend to have N- or P-deficient plumes and support N or P limitation in their coastal receiving waters. In anthropogenically affected rivers, such as the Po, Mississippi, Rhine and Seine, proportions of the nutrients have changed in such a way (mostly increases in N and P and declines in Si) that they now approximate the Redfield ratio. Thus, in the coastal waters affected by these rivers, the Si excess has diminished and more nutrient-replete conditions prevail, with expected consequences to include increases in planktonic primary productivity and increased frequencies of hypoxia. These effects have been demonstrated in the northern Adriatic Sea adjacent to the Po River [Justic, 1991], the North Sea, near the entrance of the Rhine River, and the northern Gulf of Mexico waters influenced by the Mississippi River. Those rivers with increased N and P in relation to Si (e.g., Seine, Rhine) may affect a limitation to the productivity of diatoms while increasing total productivity. One result would be a shift in dominance from diatoms to non-siliceous taxa, a scenario unfolding in N-enriched coastal systems supporting growing riverine and atmospheric N inputs (North and Baltic Seas, W. Mediterranean). The increasing frequency of non-diatom blooms in the coastal waters of northwestern Europe [Smayda, 1990] is consistent with the progression towards Si deficiency. Decreased Si inputs in relation to increasing inputs of other nutrients could stimulate harmful algal blooms (HABs) in geographically-diverse coastal waters, including the Baltic and North Seas, the Western Mediterranean and the U.S. Atlantic coast and coastal Gulf of Mexico [Smayda, 1990].

In terms of impacts at higher trophic level effects, Turner et al. [1998] demonstrated that, in addition to primary production, zooplankton, carbon flux and respiration in the Mississippi River bight were sensitive to the Si:DIN ratio in waters delivered. The percentage of copepods in the total mesozooplankton assemblage changed dramatically as the average Si:DIN ratio approached 1:1. Copepods made up about 30% of the mesozooplankton at a Si:DIN ratio of 0.5 and 75% at 1.0. Both the percentage of fecal pellet carbon in an upper water column particle trap and the estimated primary production captured as fecal pellets were high when the average Si:N atomic ratio in the river was greater than 1:1. The patterns identified by Turner et al. [1998] are consistent with the hypothesis of Officer and Ryther [1980] that a shift in the Si:N atomic ratio from above 1:1 to below 1:1 would alter the marine food web by reducing the diatom-to-zooplankton-to-higher trophic level food web, and increasing the proportion of flagellated algae, including those that are potentially harmful in that they may contribute to hypoxic water formation if sufficient ungrazed biomass sinks from the surface to the bottom layer.

Case Studies of Coastal Ecosystem Responses to Nitrogen Loading

A. Chesapeake Bay

The Chesapeake Bay is the largest estuary in the U.S., having an area of 6,500 km^2, a length > 300 km, a width of 5-30 km, a mean depth of 8.4 m, and an extensive shoreline (drainage basin surface area: bay surface area = 28:1). The surface area of the bay system is equally divided between the mainstem bay and the numerous (~30) tributary rivers and bays; however, about 66% of the volume is contained in the deeper mainstem.

The mainstem bay is stratified from late winter through early fall; stratification in tributaries is generally weaker and less persistent [Boicourt, 1992]. Water column stratification is in part responsible for chronic hypoxic and anoxic conditions in deeper regions of the system. It appears that the volume of hypoxic water has increased since colonial times [Cooper and Brush, 1991], especially in the last three to four decades [Boicourt, 1992]. This disturbance is likely in response to increased nutrient, specifically N, loads which have accelerated primary production, increased phytoplankton blooms, organic matter deposition to deep waters and oxygen depletion during decomposition of deposited organic matter [Boynton et al., 1995]. Relationships between N, as well as P, loading rates and oxygen conditions are highly relevant to management issues.

The Bay and its watershed lie in the coastal corridor of high population density between New York and Virginia. Current population in the watershed is 13.6 million and is projected to soon be 16.2 million [Magnien et al., 1995]. Current N and P loading rates averaged for the entire system are about 13 g N m^{-2} y^{-1} and 1 g P m^{-2} y^{-1} respectively. However, loading rates to distinct portions of the bay system range both a factor of 5 higher and lower than these values. Estimates indicate that since European settlement, loading rates of N and P to the Chesapeake system have increased about 6 and 17-fold, respectively [Boynton et al., 1995]. It has been estimated that as a whole, the bay receives approximately 27% of its *new* N from atmospheric deposition [Fisher and Oppenheimer, 1991; Chesapeake Bay Program, 1996]. In some of its tributaries, AD-N constitutes the largest single source of *new* N. One example is the Rhode River, where AD-N accounts for 40% of *new* N loading.

Temporal variations in flow and associated N loading exert a strong influence on algal productivity and growth responses. Nitrogen enrichment during the spring freshet has been examined as the key regulator of annual phytoplankton production. The flow parameter capable of explaining the most variability is an average of annual flow from the current and the previous years. This suggests some nutrient retention or memory over time scales of a year rather than seasonal periods as indicated by bay water residence times. Given the shallow depths of the bay, water retention time is short, suggesting sediments play a key role in nutrient storage and memory. In years of particularly high flow, relatively high amounts of algal biomass are generated during the spring bloom. Recycling of this biomass supports production through summer which serves to conserve nutrients within the bay, facilitating a large fall bloom. The deposition of the fall bloom to sediments, coupled with decreasing water temperatures, slows nutrient regeneration through winter. Regenerated nutrients become available the next spring to support algal production at higher than expected levels.

In the mid-1970s limnologists developed a series of statistical models relating nutrient loading rates and algal biomass for a large sampling of lakes [e.g. Vollenweider, 1976] and ultimately used these relationships to estimate the degree to which changes in nutrient loading rates would shift a particular lake to a higher trophic state. In Chesapeake Bay, an attempt was made to directly duplicate the Vollenweider [1976] model. Both average annual surface water and depth-integrated chlorophyll *a* showed annual average P loading rate as the independent variable. However, when N was substituted for P, results improved substantially. These results suggest that the Chesapeake Bay responds in an understandable fashion to N loading rates. Further, there is some indication that different systems respond in a similar fashion when loading rates are scaled for local conditions of depth and flushing rates [Boynton et al., 1995].

In general, sediment-water fluxes (e.g. sediment oxygen consumption, Si) examined within the study were consistently better correlated with TN loading than with TP loading. Even sediment fluxes of PO_4^{-3} exhibited a stronger relationship with TN loads than with TP loads. In part, this may result from the fact that there is a considerably broader range in TN loads than TP loads [Boynton et al., 1995]. It may not be possible to resolve TP influences on sediment flux over this relatively narrow loading range. Alternatively, the poor correspondence with TP loads may indicate that most of the phytoplankton detritus that reaches sediments, and eventually supports sediment-water fluxes, was produced more in response to N rather than P availability in the water column.

B. *Albemarle-Pamlico Sound-Neuse River Estuary*

The Albemarle-Pamlico Sound System (APSS) of North Carolina and its tributaries (Neuse, Pamlico and Chowan/Roanoke River Estuaries) are the second largest estuarine complex on the mainland U.S. The APSS is considered to be N-limited and is undergoing eutrophication in response to expanding use and development of coastal watersheds.

Much of the blame for the adverse impacts of eutrophication including hypoxia, anoxia, toxicity, and food web alterations, has been placed on rapidly-growing and diversifying non-point sources of N entering this system [Paerl, 1983, 1987; Stanley, 1983; Christian et al., 1991; Rudek et al., 1991].

The Neuse River estuary is a major tributary of the APSS and typifies progressive eutrophication [Paerl, 1983; Christian et al., 1991; Boyer et al., 1994; Paerl et al., 1995]. The APSS is enclosed by the Outer Banks, and as a result water exchange with the Atlantic Ocean is severely constrained by narrow inlets. The minimal oceanic exchange results in long water residence times; approximately 1 year for the Pamlico Sound and from 1 to over 4 months for the Neuse River Estuary. Thus, nutrients entering the system remain for relatively long periods. Nitrogen has repeatedly been shown to be the nutrient limiting primary production [Hobbie and Smith, 1975; Paerl et al., 1990; Rudek et al., 1991], and as such N inputs modulate phytoplankton growth and community composition [Paerl, 1987; Paerl et al., 1995; Boyer et al., 1994; Pinckney et al., 1998]. Although primary consumers recycle a portion of phytoplankton C and N, bloom events triggered by N loading, together with the long residence time in the system, result in the delivery of substantial quantities of organic C associated with phytoplankton biomass to the bottom waters. Microbial degradation combined with strong vertical stratification promotes bottom water anoxia, which enhances sediment nutrient release (NH_4^+ and PO_4^{-3}). Hypoxic and anoxic conditions persist for weeks, cover large areas and promote finfish and shellfish kills [Paerl et al., 1998; Lenihan and Peterson, 1998]. The shallow water depths of the Neuse (average depth ~ 2.5 m) allow for efficient sediment- water column nutrient cycling. Episodic nutrient fluxes support phytoplankton production during and following anoxic events [Christian et al., 1991; Rizzo et al., 1992]. Elevated nutrient levels may further fuel phytoplankton blooms, perpetuating the oxic/anoxic cycling in this system [Paerl et al., 1998].

In the Neuse River estuary and other tributaries of APSS, non-point sources contribute a large share (>70%) of the external or "new" N loading. Agricultural discharge, atmospheric deposition and groundwater dominate non-point N loading. Agricultural expansion, including creation of new farm and forest land, widespread use of N-based fertilizers, proliferating livestock (swine, cattle) and poultry (chicken, turkey) operations, and coastal urbanization have led to a doubling in surficial and atmospheric N loading over the past two decades [Dodd et al., 1993; Paerl, 1997]. Livestock operations account for approximately 30% of air N emissions in the entire State (NC Dept. Environ. Nat. Resources), while in the Coastal Plain region which includes the Neuse River air- and watershed, these operations account for nearly 50% of emissions (Fig 4).

The atmosphere is a significant contributor of externally-supplied N to APSS. Paerl and Fogel [1994] identified atmospheric emissions of fossil fuel combustion (NO_x) and volatilization of NH_3 from stored and applied animal waste as a major source (~38%) of new N loading to the APSS. The contribution of atmospheric N to N inputs to the Neuse River watershed is approximately 30% [Whitall, unpublished]. Paerl et al. [1995] estimated that, depending on seasonal rainfall, discharge and flow patterns, direct loadings from the atmospheric contributed from 10 to over 35% of new N loading to the Neuse River Estuary (Table 1).

In the Neuse River, phytoplankton groups (e.g., diatoms, dinoflagellates, cryptomonads, cyanobacteria) exposed to a range of N compounds in varying ratios and supply rates reveal community shifts that respond to different N loading scenarios [Pinckney et al.,

1996 NC N Emissions (Total = 325,322

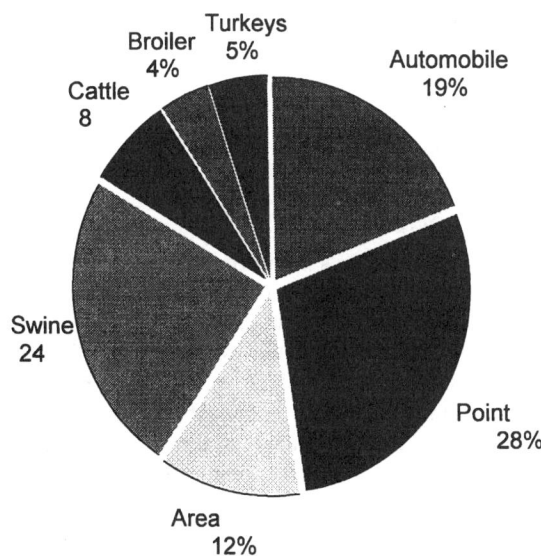

1996 Coastal N Emissions (Total = 140,111

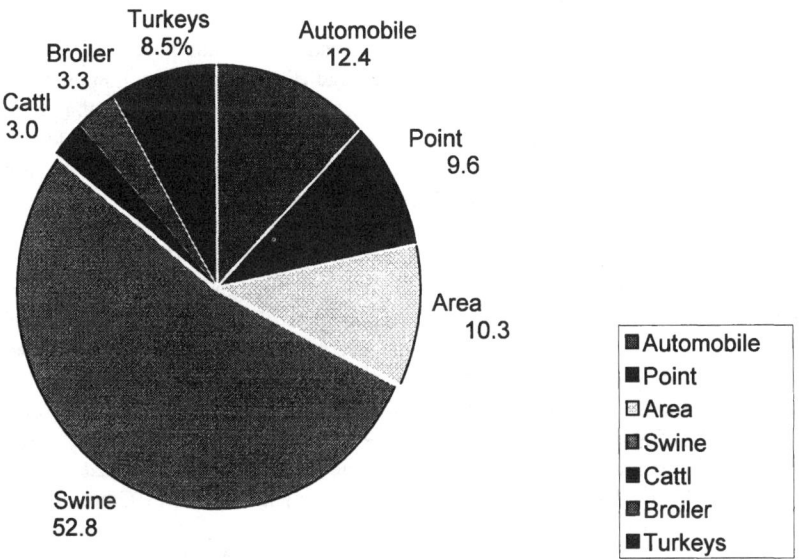

Source: NC DENR, Div. Air Quality

Figure 4. Estimates of North Carolina statewide (upper frame) and coastal N emissions for 1996. Coastal is defined as emissions east of the Piedmont, including the coastal plains and coastal regions. Emissions are partitioned among dominant sources and given as percent of total emissions. Data were provided by the North Carolina Department of Environment and Natural Resources, Division of Air Quality.

1998]. NO_3^- and NH_4^+ uptake rates vary spatially and seasonally in the Neuse River, suggesting differential community responses to these N sources [Boyer et al., 1994]. *In situ* bioassays and field surveys showed that enrichment with the major AD-N constituents NH_4^+ and NO_3^- at natural dilutions results in enhanced phytoplankton primary production and increased biomass [Paerl, 1985; Willey and Paerl, 1993; Paerl and Fogel, 1994]. The addition of DON as urea, amino acids and possibly other organic N compounds contained in AD-N, may selectively stimulate growth of facultative heterotrophic and photoheterotrophic phytoplankton taxa (especially dinoflagellates and cyanobacteria) [Paerl, 1991]. Taxa-selective phytoplankton responses to specific AD-N inputs, and changes in stoichiometric C:N ratios resulting from these inputs may induce changes at the zooplankton, herbivorous fish, benthic invertebrate and higher trophic levels (i.e., zooplanktivores, predacious fish) [Mallin and Paerl, 1994; Crowder, unpublished data]. The proportion of primary production that settles to the sediment depends on phytoplankton speciation and the rate at which various taxa are grazed and consumed (i.e., palatability, size). Sedimentation is an important mechanism for storing externally-loaded N in the estuary and perpetuating blooms in the Neuse, similar to that observed in Chesapeake Bay [Boynton et al., 1995]. Decomposition of bloom-derived sedimentary organic matter serves as a major source of recycled N and P to the water column during the summer and is the dominant oxygen sink fueling bottom water hypoxia/anoxia during the summer and fall [Paerl et al., 1998].

C. Shallow Coastal Embayments

Large portions of the North American coastline are composed of small, shallow adjacent embayments. Some of these discharge directly into continental shelf waters, while others constitute the shoreline of larger estuaries. In both cases, embayments are the proximate site of anthropogenic pollution associated with changing usage of adjacent lands. Increasing N loads, including a significant contribution from the atmosphere ($\approx 30\%$), have been implicated in biogeochemical and ecological changes in these ecosystems.

The shallow depths increase the potential contribution of other aquatic primary producers, such as eelgrass and macroalgae to the energy flow of these coastal embayments. Shifting patterns of relative abundance of these producer groups are among the symptoms of incipient eutrophication. Declining abundance of submerged aquatic vegetation has been noted in many shallow coastal systems. Eelgrass beds, which covered most of the bottom of Waquoit Bay [Cape Cod, U.S.), have nearly disappeared from this system within forty years [Valiela et al., 1992; Short and Burdick, 1996]. Declines in eelgrass habitat have also been seen in a range of sites along the coast of Long Island Sound [Koch and Beer, 1996] as well as embayments of North Carolina's Albemarle Pamlico Sound, Tampa Bay, and Florida Bay.

1. Waquoit Bay, Massachusetts

Waquoit Bay is a shallow embayment fronted by barrier beaches on Vinyard Sound, and backed by upland forests and marshes in a system of seven sub-watersheds undergoing various degrees of suburban development [Geist, 1996]. The area is a

Massachusetts state Area of Critical Environmental Concern, and the bay was in 1988 designated a National Estuarine Research Reserve (WBNERR). Waquoit was the site of a NSF Land Margin Ecosystem Research (LMER) project from 1990-95 (LMER, 1992) studying terrestrial inputs and retention of N, and the estuarine responses to groundwater N enrichment.

Atmospheric deposition is the second largest source of N to Waquoit Bay [Valiela et al., 1997]. Atmospheric deposition accounts for a third of the N load to the estuary, while wastewater contributes 48% and fertilizer contributes 15%. Most of the atmospheric inputs to Waquoit Bay are indirect, with less than 10% of the atmospheric inputs directly deposited to the water surface.

In large part due to nutrient (specifically N) enrichment, the bay has seen a dramatic decline of eelgrass and water clarity, increases of macroalgae, and frequent anoxic events especially in sub-watersheds with the highest number of houses. It may be an especially dramatic example of the changes underway in many of our shallow coastal embayments and estuaries. With the loss of vascular plants, abundant macroalgal mats have become common. In Waquoit Bay, macroalgal biomass averaged 335 g m^{-2} [Valiela et al., 1992], with mats accumulating to more than 50 cm in systems of only 1-m depth. Standing stocks and various physiological responses of dominant macroalgae paralleled the N load [Hersh, 1995; Peckol et al., 1994]. Changes have also been documented at higher trophic levels. Scallops and quahogs have declined in the Waquoit system [Valiela et al., 1992]. Fish biomass and species diversity are depressed when preferred eelgrass habitats are replaced by dense algal cover [Deegan et al., 1997].

Finally, the large accumulations of macroalgal biomass have pushed oxygen dynamics to a critical point, with very large oscillations between daytime oxygen supersaturation and nighttime oxygen depletion. When consumption exceeds atmospheric input, the oxygen stock in the shallow water column may be consumed rapidly relative to deeper waters with many cubic meters of water for each square meter of bottom. Hypoxia during pre-dawn hours is common, and unfavorable meteorological conditions may drive nearly the entire water column anoxic [D'Avanzo and Kremer, 1994].

2. Barnegat Bay, New Jersey

Barnegat Bay, New Jersey, is one in a series of coastal lagoons lining the east coast of the state. The bay was designated as one of the National Estuary Program (US EPA) sites in 1995. Bay water exchanges with offshore continental shelf water through the permanent breach at Barnegat Inlet. Barnegat Bay (exclusive of Little Egg Harbor) is about 48 km long, 2 to 6 km wide, with an average depth of 1.4 m (range <1 to 6 m) [Chizmadia et al., 1984] and a water surface area of approximately 190 km^2. Salinity throughout the main bay (exclusive of the river and stream mouths) is fairly uniform spatially and ranges temporally from 8-32 psu, depending on freshwater inflow. The two major surface flows into Barnegat Bay are the Toms River (5.7 m^3 s^{-1}) and Cedar Creek (3.1 m^3 s^{-1}) which comprise about half of the total freshwater surface flow into the bay (18 m^3 s^{-1}). Groundwater input is estimated to add an additional 3.3 m^3 s^{-1} [USGS, unpublished data] of freshwater. There is a considerable amount of residential development, particularly in the northern portion of the bay watershed. However, most of the population within the watershed is serviced by wastewater treatment plants that discharge directly to the ocean, resulting in approximately 1.7 x 10^6 kg N y^{-1} [Ocean Country Utilities Authority, unpublished data] which bypasses the estuary.

Seagrasses (primarily *Zostera marina*) are found throughout much of Barnegat Bay, although models of seagrass distribution suggest that light levels in the bay are becoming marginal for the continued occurrence of seagrasses [Seitzinger et al., unpublished]. In addition, benthic microalgal production in summer is markedly light limited [Seitzinger et al., 1991]. Annual phytoplankton production in Barnegat Bay (~480 g C m^{-2} y^{-1} [calculated from Moser 1997]) is high compared with the range of 20-360 g C m^{-2} y^{-1} for many lagoonal systems [Nixon et al., 1986]. Mesocosm experiments indicate that overall N is more important than P in limiting phytoplankton production in the bay [Seitzinger et al., 1993].

A recent budget of N inputs to Barnegat Bay indicates that atmospheric deposition is a major source of N to the bay [Moser, 1997; Seitzinger and Sanders, 1999]. Atmospheric deposition of N directly to the bay surface is estimated to account for over a third of N inputs to the Bay, compared to approximately 60% from rivers and 5% from groundwater. This is an underestimate of the total contribution of atmospheric deposition to the Barnegat Bay N budget as it does not include deposition to the watershed, a portion of which is transported to the bay through rivers. Direct plus indirect atmospheric deposition is estimated to account for over 70% of the N inputs to the bay [Seitzinger and Sanders, 1999]. The above estimates include DIN plus DON in atmospheric deposition. Long-term (20+ days) bioassay experiments with the DON component of AD-N indicate that a substantial portion (>40%) of the AD-DON is bioavailable to the planktonic microbial communities of Barnegat Bay. AD-DON also appeared to affect the phytoplankton species composition in those experiments [Seitzinger and Sanders, 1999].

Clearly, controlling AD-N inputs to Barnegat Bay will be an important management issue in the near future, given that the bay is primarily N limited, the current N loading to the bay is dominated by atmospheric deposition (direct and indirect), there is already high phytoplankton production, and water quality conditions are nearing limitations for the survival of seagrasses and benthic microalgal production.

3. Tampa Bay, Florida: Reversing N-Driven Eutrophication

The depth and extent of seagrass beds in Tampa Bay appear to be closely associated with external N loadings. Historical (pre 1930s) seagrass meadows in Tampa Bay are believed to have covered 31,000 ha of the shallow bay bottom [Lewis et al., 1985]. However, impacts to the bay from increasing population and industrial development of the Tampa Bay area resulted in increased N loading and chlorophyll concentrations, and subsequent large seagrass reductions due to reductions in light penetration. By 1982, approximately 8,800 ha of seagrass remained, suggesting a 72% loss from the early estimate [Johansson and Greening, in press].

Starting in 1985, however, water quality monitoring programs have indicated that chlorophyll concentrations have returned to levels observed in 1950 [Boler, 1998]. Furthermore, since 1988, seagrass monitoring programs have shown that the trend of seagrass loss has been reversed. The bay-wide seagrass coverage in 1996 was estimated at 10,930 ha, a 25% increase since 1982 [Johansson and Ries, 1997]. The Tampa Bay chlorophyll concentration reductions and seagrass expansion apparently were in response to water quality improvements that occurred in the late 1970s and mid 1980s, which included reductions in phytoplankton biomass and light attenuation. These improvements followed a nearly 50% reduction in external N loading from domestic wastewater treatment plants that occurred in the early 1980s [Johansson and Greening, in press].

To maintain the improvements in water quality and seagrass coverage obtained from the point source reductions in the 1980s, participants in the Tampa Bay National Estuary Program have adopted N loading targets for Tampa Bay based on the water quality and related light requirements of restoring turtle grass (*Thalassia testudinum*) to depths observed in 1950 [TBNEP, 1996]. Based on monitoring data and modeling results, it appears that light levels can be maintained at necessary levels by "holding the line" at existing nitrogen loadings. However, the "hold the line" goal may be difficult to achieve given the 20% increase in the watershed's human population and associated 7% increase in nitrogen loading that are projected to occur over the next 15 years. In addition, the sources of nitrogen are varied, with no obvious single "fix" to control nitrogen. External nitrogen sources and their relative contribution to existing loads include industrial wastewater (4%), municipal wastewater (10%), fertilizer spillage (7%), residential runoff (10%), commercial/industrial runoff (5%), intensive agriculture (6%), pasture/rangeland (13%), undeveloped land (7%), mining (4%), groundwater and springs (5%), and atmospheric deposition (29% direct deposition) [Janicki and Wade, 1996].

A Nitrogen Management Consortium has been established to develop, and implement, a Consortium Action Plan to address the target atmospheric load reduction needed to "hold the line" at 1992-1994 levels (a total reduction of 1.54×10^4 kg N y^{-1} each year through the year 2010). This consortium is comprised of a local electric utility, industries and agricultural interests, as well as local governments and regulatory agency representatives. At the time of this publication, the Consortium has both planned and implemented projects that are expected to reduce existing nitrogen loads by 1.27×10^5 kg y^{-1} by the Year 2000, potentially exceeding the agreed-upon reduction goal [Greening et al., 1997; Greening, 1998].

Reductions from local power plant emissions have contributed significantly to the total reduction estimates. In the Tampa Bay area, one of the power plants has chosen to meet Title IV of the Clean Air Act Amendments of 1990 requirements by reducing emissions early, and will have reduced 1995 NO_x emissions by $1.05\text{-}1.81 \times 10^7$ kg per year by the end of 1999. Using a conservative 400:1 ratio (NO_x emissions units to nitrogen units entering the bay), these expected emissions reductions are expected to result in a $2.63\text{-}4.54 \times 10^4$ kg y^{-1} nitrogen load reduction to Tampa Bay [Greening, 1998].

D. *Gulf of Mexico: Mississippi River plume*

The largest zone of oxygen-depleted coastal waters in the U.S. is in the northern Gulf of Mexico on the Louisiana-Texas continental shelf. From 1993 to 1997, the size of the hypoxic zone was consistently greater than 16,000 km^2 in mid-summer [Rabalais et al., 1998], but covered 12,480 km^2 in mid-summer of 1998 [Rabalais et al., unpublished data]. Hypoxic waters (< 2 mg L^{-1}) extend from the bottom to as much as 20 m from the bottom over extensive areas. Oxygen depletion begins in the spring, reaches a maximum in mid-summer and dissipates in the fall and winter. The timing, duration and spatial extent of hypoxia are related to the freshwater discharge and nutrient flux from the Mississippi River system. The northern Gulf of Mexico adjacent to the discharge of the Mississippi River is an example of a coastal water body that has undergone eutrophication as a result of increasing N loading. The linkages between riverine N loads, production and subsequent oxygen depletion on short-time scales are strong [Lohrenz et al., 1997; Justic et al., 1993, 1997], as are the long-term (decadal and century) changes in N load and indicators of environmental stress [Rabalais et al., 1996]. As N loads increased (tripling since the 1960s), phytoplankton production increased, carbon accumulation increased, oxygen stress increased, phytoplankton community shifts occurred, some harmful algal species increased, and food web trophodynamics likely changed [Turner and Rabalais, 1991, 1994 a, b; Eadie et al., 1994; Nelsen et al., 1994; Rabalais et al., 1996; Sen Gupta et al., 1996; Dortch et al., 1997; Turner et al., 1998].

The best current knowledge is that the outflows of the Mississippi and Atchafalaya rivers dominate the nutrient loads to the continental shelf where hypoxia is likely to develop [Rabalais et al., in review]. The Mississippi River ranks among the world's top ten rivers in length, freshwater discharge and sediment delivery [Milliman and Meade, 1983] and drains 41% of the lower forty-eight U.S. One-third of the combined flow from the Mississippi and Red Rivers is diverted through a control structure and joins the Atchafalaya River for discharge to the Gulf. Of all streams discharging to the northern Gulf of Mexico that most likely influence the zone of hypoxia, the combined flows of the Mississippi and Atchafalaya Rivers account for 96% of the annual freshwater discharge, 98.5% of the total N load, and 98% of the total P load [Rabalais et al. 1998, unpublished, calculations from Dunn 1996].

The relative contribution of direct atmospheric deposition of N (wet and dry, NO_3^- and organic N) to the N load affecting the area of hypoxia was estimated by R. Artz [Goolsby et al., in review] to be only a few percent of the N input from the Mississippi-Atchafalaya River system. There are no studies of groundwater discharge to coastal waters for the coastlines of Texas, Louisiana, Mississippi or Alabama. On the scale of contributions from the Mississippi and Atchafalaya Rivers, estimates of groundwater to Apalachee Bay in the northeastern Gulf (11 kg ha^{-1} y^{-1}; J. Cable, personal communication, [calculated from data in Fu and Winchester 1994]) would be a minimal contribution (if similar values were to be expected from the central and northwestern Gulf of Mexico) compared to the Mississippi and Atchafalaya River flux of 1,500 Gg ha^{-1} y^{-1}. Groundwater sources to the area affected by hypoxia are unlikely to be important because of the lack of shallow aquifers along the Louisiana coast. Any contributions of groundwater to the coastal waters of the northern Gulf would probably become entrained in the Louisiana Coastal Current where strong, shore-parallel frontal boundaries develop and where net flow is generally to the west along shore except for reversals in mid-summer. Thus, the contribution of groundwater is unknown, but expected to be small, and the potential for transfer in a cross-shelf direction is unknown. The relative contribution of offshore sources of nutrients from upwelled waters of the continental slope is also unknown. Onwelling of NO_3^- from deeper waters may be important in shelf edge (100 m depth range) cycling of C and N [Walsh, 1988, 1991], but these N sources are physically and geographically removed from the area subject to hypoxic water formation.

The Mississippi and Atchafalaya Rivers contribute (based on available information) the major sources of nutrients to the northern Gulf of Mexico, and atmospheric contributions to the surface waters where hypoxia is likely to occur are minimal. The AD-N that ultimately influences the productivity of the northern Gulf and subsequent oxygen depletion comes from the Mississippi River watershed for which AD-N (wet and dry NO_3^- and ON, plus wet NH_3) is estimated to be 9% of the total N load to the system

[Goolsby et al., in review]. The water quality degradation and associated ecological problems in this system are river driven, and the AD-N component of new N inputs this system is at the low end when compared to other US estuarine and coastal ecosystems (Table 1). Attempts to manage excess N loads to the northern Gulf of Mexico, therefore, would likely differ in relative effort or focus from other areas where AD-N is a much greater component of new N inputs.

E. Atlantic Continental Shelf

A recently constructed N budget for the North Atlantic basin provides insight into the magnitude of atmospheric deposition as a source of N in continental shelf regions. As part of that effort, the North Atlantic watershed was divided into ten regions and estimates were developed for:
1. total N export from land to the coastal zone by rivers [Howarth et al., 1996],
2. total N export from estuaries to shelf regions plus total N transport by large rivers directly to shelf regions (hereafter referred to as river input) [Nixon et al., 1996],
3. direct deposition of atmospheric NOy and NHx [Prospero et al., 1996],
4. particulate N burial in shelf sediments [Nixon et al., 1996], and
5. denitrification [Seitzinger and Giblin, 1996].

Overall, atmospheric deposition contributes approximately 1.82×10^9 kg N y^{-1} directly to the North Atlantic continental shelf, which accounts for about 20% of the combined inputs from rivers (6.44-8.82×10^9 kg N y^{-1}) and atmospheric deposition (Fig. 5). Nitrogen removal by burial is relatively small (2.8×10^8 - 1.26×10^9 kg N y^{-1}). However, denitrification in shelf sediments is estimated to remove an amount of N equivalent to approximately two times the terrestrial derived inputs to the shelf (rivers plus atmospheric deposition). The additional N needed to balance the shelf N budget is estimated to be from ocean inputs (e.g., onwelling, advection) [Seitzinger and Giblin, 1996].

An examination of the N budgets for various shelf regions throughout the North Atlantic suggests that the relative importance of atmospheric deposition differs geographically [Paerl and Whitall, 1999]. For example this analysis the continental shelf off the U.S. east coast and the Gulf of Mexico were divided into three regions:
1. the northeast coast of the U.S. (to Cape Hatteras),
2. the southeast coast of the U.S. (Cape Hatteras to Florida), and
3. the Gulf of Mexico.

The shelf area was operationally defined as extending to the 200 m contour [Pilson and Seitzinger, 1996]. The relative magnitude of atmospheric deposition is greatest on the continental shelf off the U.S. northeast coast where it is estimated to account for approximately 40-60% of the combined inputs from atmospheric deposition and rivers [Seitzinger and Giblin, 1996] (Fig. 6). In the shelf area off the southeast U.S. coast, the relative magnitude of atmospheric deposition is somewhat less but still accounts for approximately 30-40% of the N inputs. In the Gulf of Mexico, atmospheric deposition accounts for only approximately 12% of the combined inputs from atmospheric deposition and rivers. While these N inputs are clearly important sources of new N to these shelf regions, it appears that essentially all of the N inputs are eventually removed by denitrification in the shelf sediments (Fig. 6). In fact, as was concluded from the North Atlantic shelf budget as a whole, denitrification is estimated to remove even more N from each of these shelf regions than is entering from atmospheric and river sources combined.

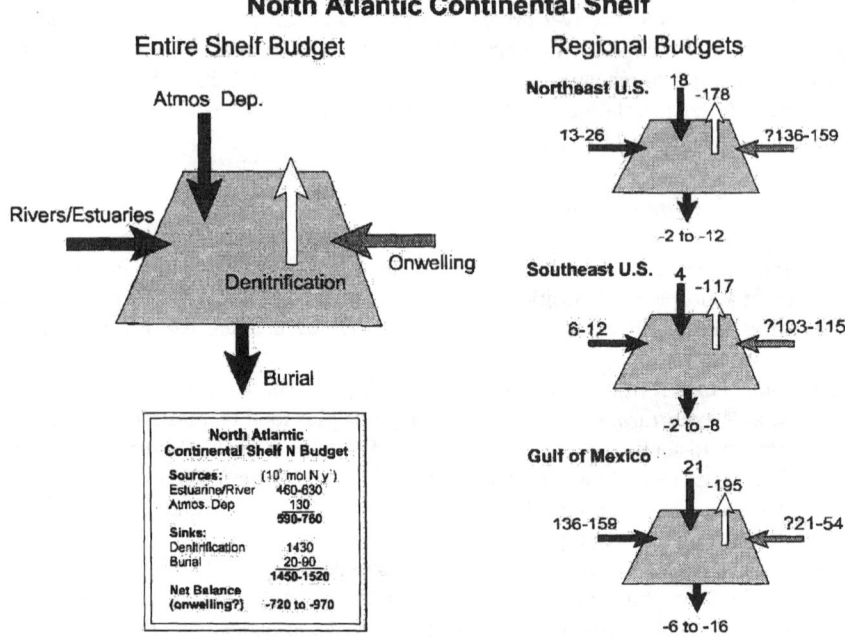

Figure 5. N budget for the North Atlantic continental shelf and for the three regions of shelf adjacent to North America. Numbers next to arrows indicate N inputs from estuarine export/large rivers and atmospheric deposition and N removal from burial and denitrification. Also indicated are N inputs attributed to ocean sources (e.g., onwelling/advection across the slope-shelf boundary) that are calculated by difference to balance the shelf N budget. Budgets derived from Seitzinger and Giblin [1996] that also used information contained in Howarth et al. [1996], Nixon et al. [1996], and Prospero et al. [1996].

This suggests that additional N is transported onto the shelf from ocean sources. In the northeast and southeast U.S. shelf regions the oceanic inputs are estimated to be four and eight times, respectively, greater than the combined atmospheric and river sources; however in the Gulf of Mexico, the ocean sources are estimated to be relatively smaller, contributing an amount of N equivalent to only about 20% of the atmospheric and river sources.

The relative importance of rivers as a source of new N likely decreases and the ocean sources likely increase as one moves from the land/sea boundary to the outer edge of the continental shelf. The distribution of atmospheric N deposition across that gradient is not well documented, but likely decreases away from land. The relative magnitude of atmospheric deposition compared to river and ocean inputs along that transect is expected to vary regionally. For example, in the continental shelf regions off the northeast and southeast U.S., AD-N is probably most important as a source of new N nearer to land and less important at the shelf/slope boundary. In the Gulf of Mexico the opposite pattern is predicted.

Modulation of Coastal Atmospheric N Deposition Impacts by Hydrology, Geology, and Geography

Terrestrial and aquatic ecosystems vary in size, shape, geology, climatic regimes, and hydrology. This variability, coupled with differing flora and fauna, can lead to distinct differences in ecosystem "sensitivity" in terms of biotic responses to and ecosystem assimilation of AD-N inputs. The size of the watershed and airshed relative to the receiving water body is likely to determine the contribution of AD-N to N budgets. Jointly, these physical features determine overall ecosystem biogeochemical and trophic responses to AD-N inputs. In addition, the quantity and quality of AD-N may vary geographically, depending on location of sources, transport in soils and streams and proximity to down stream ecosystems.

Once AD-N is deposited to a coastal watershed or water body, its residence time within that system and its transport to other systems may affect the potential impact on food web dynamics. The timing and form of this AD-N play large roles in determining the residence times of various AD-N forms in terrestrial systems. In watersheds the residence time of deposited N is on the order of days to years as regulated by hydrologic pathways and levels of ecosystem productivity. In areas where a few large storms contribute most of the yearly precipitation, or areas with a high proportion of precipitation as snow, followed by a rapid snowmelt, AD-N is more likely to be transported directly to stream ecosystems and not subjected to soil-system processing (e.g., denitrification, assimilation, mineralization). The average residence time can also be decreased significantly by individual storm events. Large storms tend to produce greater incidence of saturated overland flow, facilitating rapid transfer of AD-N components to aquatic systems. This same effect can be found in watersheds characterized by urbanized land use that tends to promote overland flow (i.e. over concrete). Once within the surface waters, their chemical form is altered by in-system processing [Kaplan, 1993], such that stream DIN and DON characteristics are probably defined more by watershed land use and disturbance, vegetation status and soil properties than by the original form of the AD-N.

These contrasting physiographic, hydrodynamic and biotic features lead to a range of positive (e.g., increased fertility and fisheries yields) and negative (e.g. harmful algal blooms, hypoxia, toxicity) responses to AD-N (as well as terrigenous organic N) loading scenarios along geographic gradients. This necessitates comparative regional studies of the role AD-N plays in determining ecosystem structure and function. Terrestrial and aquatic ecosystems varying in location, size, proximity to inputs (i.e., anthropogenic vs. natural), residence time, trophic state, and fertility should be included in a comparative evaluation.

Research and management recommendations

Regional and ecosystem-level studies reveal a broad spectrum of biogeochemical and trophic impacts accompanying anthropogenic new N enrichment of N-sensitive estuarine and coastal waters. The atmospheric fraction of this new N is substantial in magnitude and spatial scope and appears to be increasing in many coastal regions experiencing human population growth, resultant urbanization, and agricultural and industrial expansion. There is unprecedented pressure to characterize and quantify potentially detrimental ecological and economic impacts, including declines in water quality, fisheries and recreational resource and habitat losses. Relevant questions and information needs include:

1. What is the contribution of AD-N relative to other new and regenerated N sources in the N budgets of N sensitive water bodies along local, regional, national (i.e., regional comparisons) and international scales?

2. What are the biogeochemical and trophic roles and impacts of AD-N along the gradient spanning estuarine, coastal and open ocean waters? How do these impacts vary on seasonal and longer-term time scales?

3. What are the quantitative and qualitative ramifications of specific AD-N constituents (DIN, DON, PON) in terms of biogeochemical and trophic responses along spatio-temporal gradients?

4. There is a need to temporally and spatially couple specific land uses to sources (emission), routes of transport, deposition of and biological response to atmospheric N. Development of appropriately scaled models capable of coupling sources to fates is needed.

5. Utilizing combined process-based assessment and modeling approach, acute vs. chronic AD-N biogeochemical and trophic impacts should be evaluated in estuarine, coastal and oceanic waters.

6. Information from 1-5 should be incorporated into regional and global assessments of the roles and impacts of AD-N on marine productivity, trophodynamics and nutrient (N, C, P, etc.) cycling. In addition, AD-N impacts on air-sea exchange of N, C and other trace gases should be evaluated and factored into global fluxes and budgets.

Research efforts should be interdisciplinary, process-oriented, regional in scope and comparative. The thrust should be capable of coupling AD-N emission sources, transport and deposition (flux) dynamics to biogeochemical and trophic responses. This should be complemented by a modeling effort, aimed at linking AD-N generating, transport, and depositional and trophic utilization (from phytoplankton to fish) processes. Emphasis should be put on comparing sensitivities of and impacts in systems exhibiting various stages of eutrophication, ranging from advanced (e.g. shallow embayments in the Northeast, New York Bight, Chesapeake Bay), to moderate (e.g., Mississippi plume/Gulf of Mexico, Pamlico Sound, San Francisco Bay), to those exhibiting incipient but growing stages of eutrophication (e.g., Florida Bay, Puget Sound).

Assessment and research programs will, in all likelihood, rely on conventional monitoring complemented with uniquely suitable, novel approaches. Collection of atmospheric deposition in a "representative", non-biased manner continues to be a challenging aspect of any monitoring program (see other chapters). A growing arsenal of analytical methods is now available for separating and quantifying individual inorganic and organic constituents of AD-N. Tracing these constituents through surface and subsurface means of entry, into food webs and nutrient cycling steps (i.e., assimilation, denitrification) is now being pursued using stable N and O isotope techniques. Depending on the complexity of input sources, their isotopic signatures and points of entry to receiving waters, these techniques can be combined with mass flux measurements to develop source mixing models, of potential use to managers responsible for formulating, tracking and enforcing allowable N inputs to N sensitive waters.

Utilizing specific biomass and biodiversity indicators (e.g., HPLC-based diagnostic pigment analyses, nucleic acid-based probing techniques), the impacts of specific AD-N sources (types, quantities) on phytoplankton community production, composition and biogeochemical fates (i.e. waters column cycling, sedimentation, burial) can be examined. In addition, food web transfer and fate can be facilitated using these techniques.

In concert, researchers and managers now have available a wide array of techniques and approaches for quantifying, characterizing and evaluating AD-N in estuarine and coastal ecosystems under the influence of these growing and diversifying new N sources. Technological and conceptual advancements have facilitated interdisciplinary approaches which will be essential for understanding the ecological and biogeochemical impacts of AD-N on N-sensitive waters over a variety of relevant spatial and temporal scales. These interdisciplinary studies should incorporate air, water quality, fisheries managers, biogechemists, estuarine and marine ecologists, modelers, basic and applied researchers providing the linkage of atmospheric chemistry and physics (i.e. atmospheric forcing) with nutrient-production interactions, trophodynamics and water quality.

Acknowledgements: The authors are grateful to Richard Valigura and Robert Twilley for their reviews and editorial contributions.

References

Aber, J. D., K.N. Nadelhoffer, P. Steudler, J.M. Melillo, and R. Boone. 1989. Nitrogen saturation in northern forest ecosystems. BioScience 39:378-386.

Aber, J. D., and C.A. Federer. 1992. A generalized, lumped parameter model of photosynthesis, evapotranspiration and net primary production in temperate and boreal forest ecosystems. Oecologia 92:463-474.

Aber, J. D., A. Magill, S. G. McNulty, R. D. Boone, K. J. Nadelhoffer, M. Downs, and R. Hallett. 1995. Forest biogeochemistry and primary production altered by nitrogen saturation. Water, Air, and Soil Pollution 85:1665-1670.

Aber, J., W.McDowell, K. Nadelhoffer, A. Magill, G. Berntson, M. Kamakea, S. McNulty, W. Curie, L. Rustad and I. Fernandez. 1998. Nitrogen saturation in temperate forest ecosystems: Hypothesis revisited. BioScience 48:921-934.

Allott, T. E. H., C. J. Curtis, J. Hall, R. Harriman, and R. W. Battarbee. 1995. The impacts of nitrogen deposition on upland surface waters in Great Britain: A regional assessment of nitrate leaching. Water, Air, and Soil Pollution 85:297-302.

Ambio. 1990. Marine Eutrophication. Ambio 19:101-176.

Anderson, D. & D. Garrison (eds.). 1997. The Ecology and Oceanography of Harmful Algal Blooms. Limnol. Oceanogr. 42(2):1009-1305.

Andersson, L. and L. Rydberg. 1987. Trends in nutrient and oxygen conditions within the Kattegat: Effects of local nutrient supply. Estuarine, Coastal and Shelf Science 26: 559-579.

Antia, N, P. Harrison and L. Oliveira. 1991. The role of dissolved organic nitrogen in phytoplankton nutrition, cell biology and ecology. Phycologia 30:1-89.

Asman, W. 1994. Emission and deposition of ammonia and ammonium. Nova Acta Leopoldina 70:263-297.

Berg, G. M, P. M. Glibert, M. W. Lomas and M. A. Burford. 1997. Organic nitrogen uptake and growth by the chrysophyte *Aureococcus anophagefferens* during a brown tide event. Mar. Biol. 129:377-387.

Binkowski, F.S., and U. Shankar. 1995. The Regional Particulate Model, Part I: Model description and preliminary results J. Geophys. Res., 100:26191-26209.

Boicourt, W.C. 1992. Influences of circulation processes on dissolved oxygen in the Chesapeake Bay, p. 7-59. In: D.E. Smith, M. Leffler and G. Mackiernan. Oxygen Dynamics in the Chesapeake Bay-A Synthesis of Recent Results. A Maryland Sea Grant Book, College Park, MD.

Boler, R. 1998. Surface water quality, 1995-1997. Hillsborough County, Florida. Hillsborough County Environmental Protection Commission. Tampa, Florida.

Bowden, R. D., J. M. Melillo, P. A. Steudler, and J. D. Aber. 1991. Effects of nitrogen additions in annual nitrous oxide fluxes from temperate forest soils in the northeastern United States. Journal of Geophysical Research 96:9321-9328.

Boyer, J., D. Stanley, & R. Christian. 1994. Dynamics of NH_4^+ and NO_3^- uptake in the water column of the Neuse River estuary, North Carolina. Estuaries 17:361-371.

Boynton, W. R., J. H. Garber, R. Summers and W. M. Kemp. 1995. Inputs, transformations and transport of nitrogen and phosphorus in Chesapeake Bay and selected tributaries. Estuaries 18 (1B):285-314.

Bronk, D., P. Glibert and B Ward. 1994. Nitrogen uptake, dissolved organic nitrogen release, and new production. Science 265:1843-1846.

Buijsman, E., H.F.M. Maas and W.A.H. Asman. 1987. Anthropogenic ammonia emissions in Europe. Atmosph. Environ. 21:1009-1020.

Chang, J.S., R.A. Brost, I.S.A. Isaksen, S. Madronich, P. Middleton, W.R. Stockwell, and C.J. Walcek. 1987. A three-dimensional Eulerian acid deposition model: Physical concepts and formulation. J. Geophys. Res. 92:14681-14700.

Chesapeake Bay Program. 1996.

Chizmadia, P.A., M. J. Kennish and V. L. Ohori. 1984. Physical description of Barnegat Bay. In: Lecture Notes on Coastal and Estuarine Studies, edited by R.T. Barber, et al, pp. 1-28. Springer-Verlag, New York.

Christian, R. R, J. N Boyer and D. W. Stanley. 1991. Multi-year distribution patterns of nutrients within the Neuse River Estuary. Mar. Ecol. Progr. Ser. 71:259-274.

Cirmo, C. P. and C. T. Driscoll. 1993. Beaver pond biogeochemistry: Acid neutralizing capacity generation in a headwater wetland. Wetlands 13:277-292.

Clarke, J.F., E.S. Edgerton and B.E. Martin. 1997. Dry deposition calculations for the Clean Air Status and Trends Network. Atmos. Environ. 31:3667-3678.

Codispoti, L. A., J. Christensen, A. Devol, H. W. Paerl and T. Yoshinari. 1999. The influence of an inbalanced oceanic combined nitrogen budget on atmospheric carbon dioxide. Mar. Chem. (in press).

Cohn, R.D. and R.L. Dennis. 1994. The evaluation of acid deposition models using Principal Component spaces. Atmos. Environ. 28:2531-2543.

Collos, Y. 1989. A linear model of external interactions during uptake of different forms of inorganic nitrogen by microalgae. J. Plankt. Res. 11:521-533.

Cooper, S. and G. Brush. 1991. Long-term history of Chesapeake Bay anoxia. Science 254:992-996.

Copeland, B. J. and J. Gray. 1991. Status and Trends Report of the Albemarle-Pamlico Estuary (Ed. By Steel J). Albemarle-Pamlico Estuarine Study Report 90-01. NC Dept. of Environ. Health & Nat. Resources, Raleigh.

Cornell, S., A. Rendell and T. Jickells. 1995. Atmospheric inputs of dissolved organic nitrogen to the oceans. Nature 376:243-246.

Correll, D. and D. Ford. 1982. Comparison of precipitation and land runoff as sources of estuarine nitrogen. Estuar. Coast. Shelf Sci. 15:45-56.

D'Avanzo, C. and J. N. Kremer. 1994. Diel oxygen dynamics and anoxic events in an eutrophic estuary of Waquoit Bay, Massachusetts. Estuaries 17:131-139.

D'Elia, C. F. 1987. Nutrient enrichment of the Chesapeake Bay: too much of a good thing. Environment 29:2.

D'Elia, C. J., J. G. Sanders and W. R. Boynton. 1986. Nutrient enrichment studies in a coastal plain estuary: phytoplankton growth in large-scale, continuous cultures. Canadian Journal of Aquatic and Fisheries Sciences 43:397-406.

Deegan, L.A., J.T. Finn, S.G. Ayvazian, C.A. Ryder, J. Buonaccorsi. 1997. Development and validation of an estuarine biotic integrity index. Estuaries 20:601-617.

Dennis, R.L. 1997. Using the Regional Acid Deposition Model to determine the nitrogen deposition airshed of the Chesapeake Bay watershed, in J.E. Baker, editor, Atmospheric Deposition of Contaminants to the Great Lakes and Coastal Waters, SETAC Press, Pensacola, Florida, 393-413.

Diaz, R. J. and R. Rosenberg. 1995. Marine benthic hypoxia: A review of its ecological effects and the behavioural responses of benthic macrofauna. Oceanography and Marine Biology: an Annual Review 33:245-303.

Dodd, R. C., P. A. Cunningham, R. J. Curry and S. J. Stichter. 1993. Watershed planning in the Albemarle-Pamlico Estuarine System. Report No. 93-01, Research Triangle Institute, Research Triangle Park, NC. Dept of Environment, Health and Natural Resources.

Dortch, Q. 1990. The interaction between ammonium and nitrate uptake in phytoplankton. Mar. Ecol. Prog. Ser. 61:183-201

Dortch, Q. and T. E. Whitledge. 1992. Does nitrogen or silicon limit phytoplankton production in the Mississippi River plume and nearby regions? Continental Shelf Research 12:1293-1309.

Dortch, Q., R. Robichaux, S. Pool, D. Milsted, G. Mire, N. N. Rabalais, T. M. Soniat, G. A. Fryxell, R. E. Turner and M. L. Parsons. 1997. Abundance and vertical flux of *Pseudo-nitzschia* in the northern Gulf of Mexico. Marine Ecology Progress Series 146:249-264.

Dortch, Q. and H. Conway. 1984. Interactions between nitrate and ammonium uptake: variation with growth rate, nitrogen source, and species. Mar. Biol. 79:151-164.

Driscoll, C. T. and G. C. Schafran. 1984. Short-term changes in the base neutralizing capacity of an acid Adirondack Lake. Nature 310:208-310.

Driscoll, C. T., B. J. Wyskowski, C. C. Cosentini, and M. E. Smith. 1987. Processes regulating the temporal and longitudinal variations in chemistry of low-order woodland streams in the Adirondacks. Biogeochemistry 3:225-241.

Driscoll, C. T. and D. A. Schaefer. 1989. Overview of nitrogen processes. In: J.L. Malanchuk, and Jan Nilsson (Eds.), The Role of Nitrogen in the Acidification of Soils and Surface Waters, Nordic Council of Ministers, Denmark, pp. 4-1 to 4-12.

Duce, R. 1986. The impact of atmospheric nitrogen, phosphorus, and iron species on marine biological productivity, pp. 497-529. IN P. Buat-Menard (Ed.), The Role of Air-Sea Exchange in Geochemical Cycling, D. Reidel, Norwell, MA.

Duce, R. 1991. Chemical exchange at the air-coastal sea interface, pp. 91-110. IN R. Mantoura, J. Martin, and R. Wollast (Eds.), Ocean Margin Processes in Global Change, J. Wiley & Sons, Chichester.

Duce, R., et al. 1991. The atmospheric input of trace species to the world ocean. Global Biogeochem. Cycles 5:193-259.

Dugdale R and J Goering. 1967. Uptake of new and regenerated forms of nitrogen in primary productivity. Limnol. Oceanogr. 12:196-206.

Dunn, D. D. 1996. Trends in Nutrient Inflows to the Gulf of Mexico from Streams Draining the Conterminous United States 1972 – 1993. U.S. Geological Survey, Water-Resources Investigations Report 96—4113. Prepared in cooperation with the U.S. Environmental Protection Agency, Gulf of Mexico Program, Nutrient Enrichment Issue Committee, U.S. Geological Survey, Austin, Texas, 60 pp.

Duyzer, J. 1994. Dry deposition of ammonia and ammonium aerosols over heathland. J. Geophys. Res. 99:18,757-18,763.

Eadie, B. J., B. A. McKee, M. B. Lansing, J. A. Robbins, S. Metz and J. H. Trefry. 1994. Records of nutrient-enhanced coastal productivity in sediments from the Louisiana continental shelf. Estuaries 17:754-765.

Elmgren, R. 1989. Man's impact on the ecosystem of the Baltic Sea; enrgy flows today and at the turn of the century. Environmental Science and Tech. 9:635-638.

Eppley, R., J. Rogers, and J. McCarthy. 1969a. Half-saturation constants for uptake of nitrate and ammonium by marine phytoplankton. Limnol. Oceanogr. 14:912-920.

Eppley, R., J. Coatsworth, and L. Solorzano. 1969b. Studies on nitrate reductase in marine phytoplankton. Limnol. Oceanogr. 14:194-205.

Eshleman, K. N. 1988. Predicting regional episodic acidification of surfaces waters using emperical models. Water Resources Research 24:1118-1126.

Fisher, D. & M. Oppenheimer. 1991. Atmospheric nitrogen deposition and the Chesapeake Bay estuary. Ambio 20:102-108.

Forsberg, C. 1994. The large-scale flux of nutrients from land to water and the eutrophication of lakes and marine waters. Marine Pollution Bulletin 29:409-413.

Fowler, D., C.R. Flechard, M.A. Sutton and R.L. Storeton-West. 1998. Long term measurements of the land-atmosphere exchange of ammonia over moorland. Atmos. Environ.32:453-459.

Fu, J.-M. and J. W. Winchester. 1994. Sources of nitrogen in three watersheds of northern Florida, USA: Mainly atmospheric deposition. Geochimica et Cosmochimica Acta 58: 1581-1590.

Galloway, J.N., C. L. Schofield, G. R. Hendreu, N. E. Peters and A. H. Johannes. 1980. In: Ecological Impact of Acid Precipitation, SNSF-project, Oslo-Norway 264-265.

Galloway, J., H. Levy and P. Kasibhatia. 1994. Year 2020: Consequences of population growth and development on deposition of oxidized nitrogen. Ambio 23:120-123.

Galloway, J.N., W.H. Schlesinger, H. Lavy II, A. Michaels, and J.L. Schnoor. 1995. Nitrogen fixation: Anthropogenic enhancement-enviromantal response. Global Biogeochemical Cycles 9:235-252.

Geist, M.A. 1996. The Ecology of the Waquoit Bay National Estuarine Research Reserve. NOAA and MA DEM. Special Report. 119 p.

GESAMP. 1989. The Atmospheric Input of Trace Species to the World Ocean: Report and Studies No. 38. World Meteorological Association, Geneva.

Glibert, P. M., C. Garside, J. A. Fuhrman, and M. R. Roman. 1991. Time-dependent coupling of inorganic and organic nitrogen uptake and regeneration in the plume of the Chesapeake Bay estuary and its regulation by large heterotrophs. Limnol. Oceanog. 36:895-909.

Goldman, C.R. 1999. Four decades of change in two subalpine lakes. Baldi Lecture. Verh. Internat. Verein. Limnol. Vol. 27 (in press).

Goldman, J., J. McCarthy and D. Peavey. 1979. Growth rate influence on the chemical composition of phytoplankton in oceanic waters. Nature 279:210-215.

Goolsby, D. A., W. A. Battaglin, G. B. Lowrance, R. S. Artz, B. J. Aulenbach and R. P. Hooper. (in review). Gulf of Mexico Hypoxia Assessment, Topic #3, Flux and Sources of Nutrients in the Mississippi-Atchafalaya River Basin. Draft Report to White House Office of Science and Technology Policy, Committee on Environment and Natural Resources, Hypoxia Work Group, Mississippi River/Gulf of Mexico Watershed Nutrient Task Force, 90 pp.

Greening, H.S. (ed). 1999. Partnership for Progress: The Tampa Bay Nitrogen Management Consortium Action Plan 1995-1999. Prepared by the Tampa Bay Estuary Program. (in press).

Greening, H.S., G. Morrison, R.M. Eckenrod, and M.J. Perry. 1997. The Tampa Bay resource-based management approach, p. 349-355. In Treat, S.F. (ed.), Proceedings, Tampa Bay Area Scientific Information Symposium 3. Text, Tampa, Florida.

Hallegraeff, G. 1993. A review of harmful algal blooms and their apparent global increase. Phycologia 32:79-99.

Harrison, P. and D. Turpin. 1982. The manipulation of physical, chemical, and biological factors to select species from natural phytoplankton populations, pp. 275-289. IN G. Grice & M. Reeve (Eds.). Marine mesocosms: Biological and chemical research in experimental ecosystems. Springer-Verlag, New York.

Harrison, W., T. Platt, and M. Lewis. 1987. F-ratio and its relationship to ambient nitrate concentration in coastal waters. J. Plank. Res. 9:235-245.

Henriksen, A. and D. F. Brakke. 1988. Increasing contributions of nitrogen to the acidity of surface waters in Norway. Water, Air, and Soil Pollution 42:183-201.

Hersh, D.H. 1995. Abundance and Distribution of Intertidal and Subtidal Macrophytes in Cape Cod: the Role of Nutrient Supply and Other Controls. Ph.D. Thesis, Boston University.

Hinga, K.R., A.A. Keller and C.A. Oviatt. 1991. Atmospheric deposition and nitrogne inputs to coastal waters. Ambio 20:256-260.

Hobbie JE and NW Smith. 1975. Nutrients in the Neuse River Estuary, NC., Report No. UNC-SG-75-21, UNC Sea Grant Program, NC State Univ., 183p.

Holland E. A., F. J. Dentener, B. H. Braswell and J. M. Sulzman. 1999. Contemporary and pre-industrial global reactive nitrogen budgets. Biogeochem. (in press).

Hov, Ø., B.A. Hjøllo and A. Eliassen. 1994. Transport distance of ammonia and ammonium in Northern Europe 1. Model description J. Geophys. Res., 99:18,735-18,748.

Hov, Ø. and B.A. Hjøllo. 1994. Transport distance of ammonia and ammonium in Northern Europe 1. Its relation to emissions of SO_2 and NO_X, J. Geophys. Res., 99:18,749-18,755.

Howarth, R. W., G. Billen, D. Swaney, A. Townsend, N. Jaworski, K. Lajtha, J. A. Downing, R. Elmgren, N. Caraco, T. Jordan, F. Berendse, J. Freney, V. Kudeyarov, P. Murdoch and Z. Zhao-Liang. 1996. Regional nitrogen budgets and riverine N & P fluxes for the drainages to the North Atlantic Ocean: Natural and human influences. Biogeochemistry 35:75-79.

Huisman, J. and J. Weissing. 1995. Competition for nurients and light in a mixed water column: A theoretical analysis. Am. Nat. 146:536-564.

Jassby, A.D., J.E. Reuter, R.P. Axler, C.R. Goldman and S.H. Hackley. 1994. Atmospheric deposition of nitrogen and phosphorus in the annual nutrient load of Lake Tahoe (California-Nevada). Water Resources Res. 30(7):2207-2216.

Janicki A. and D. Wade. 1996. Estimating critical nitrogen loads for the Tampa Bay estuary: An empirically based approach to setting management targets. Technical Publication 06-96 of the Tampa Bay National Estuary Program. Coastal Environmental, Inc., St. Petersburg, Florida.

Jaworski, N., R. Howarth and L. Hetling. 1997. Atmospheric deposition of nitrogen oxides onto the landscape contributes to coastal eutrophication in the Northeast United States. Environment Sci Tech 31:1995-2004.

Johannes, A. H., E. R. Altwicker, and N. L. Clesceri. 1985. The Integrated Lake-Watershed Acidification Study: Atmospheric Inputs. Water, Air and Soil Pollution, 26, 339-353. Johannessen, M., A. Skartveit, R. F. Wright. In: Ecological Impact of Acid Precipitation, SNSF-project: Oslo, Norway, 1980, pp. 224-225.

Johansson, J.O.R. and T. Ries. 1997. Seagrass in Tampa Bay: Historic trends and future expectations, p. 139-150. In Treat, S.F. (ed.), Proceedings, Tampa Bay Area Scientific Information Symposium 3. Text, Tampa, Florida.

Johansson, J.O.R. and H.S. Greening. in press. Seagrass Restoration in Tampa Bay: A

Resource-Based Approach to Estuarine Management. CRC Publication.
Jonge, V. N. de. 1990. Response of the Dutch Wadden Sea ecosystem to phosphorus discharges from the Rhine River. Hydrobiologia 195:49-62.
Justic', D. 1991. Hypoxic conditions in the northern Adriatic Sea: historical development and ecological significance. Pages 95-105 in R. V. Tyson and T. H. Pearson, eds., Modern and Ancient Continental Shelf Anoxia, Geological Society Special Publication No. 58, The Geological Society, London, 470 pp.
Justic', D., T. Legovic and L. Rottini-Sandrini. 1987. Trends in oxygen content 1911-1984 and occurrence of benthic mortality in the northern Adriatic Sea. Estuarine and Coastal Shelf Science 24:435-445.
Justic', D., N. N. Rabalais, R. E. Turner, and W. J. Wiseman, Jr. 1993. Seasonal coupling between riverborne nutrients, net productivity and hypoxia. Marine Pollution Bulletin 26:184-189.
Justic', D., N. N. Rabalais and R. E. Turner. 1995a. Stoichiometric nutrient balance and origin of coastal eutrophication. Marine Pollution Bulletin 30:41-46.
Justic', D., N. N. Rabalais, R. E. Turner and Q. Dortch. 1995b. Changes in nutrient structure of river-dominated coastal waters: Stoichiometric nutrient balance and its consequences. Estuarine, Coastal and Shelf Science 40:339-356.
Justic', D., N. N. Rabalais and R. E. Turner. 1997. Impacts of climate change on net productivity of coastal waters: Implications for carbon budget and hypoxia. Climate Research 8:225-237.
Justic, D., N. N. Rabalais and R. E. Turner. 1996. Effects of climate change on hypoxia in coastal waters: A doubled CO_2 scenario for the northern Gulf of Mexico. Limnol. Oceanogr. 41:992-1003.
Kanciruk, P., J. M. Eilers, R. A. McCord, D. H. Landers, D. F. Brakke, and R.A. Linthurst. 1986. Characteristics of Lakes in the Eastern United States: Data Compendium of Site Characteristics and Chemical Variables. EPA/600/486.007c, Environmental Protection Agency, Washington DC.
Koch, E.W. and S. Beer. 1996. Tides, light and the distribution of *Zostera marina* in Long Island Sound, USA. Aquatic Botany 53:97-107
Langford, A.O. and F.C. Fehsenfeld. 1992. Natural vegetation as a source or sink for atmospheric ammonia: A case study. Science, 255:581-583.
Lenihan, H and C. H. Peterson. 1998. How habitat degradation through fishery disturbance enhances impacts of hypoxia on oyster reefs. Ecological Applic. 8:128-140.
Lewis, R.R., M.J. Durako, M.D. Moffler, and R.C. Phillips. 1985. Seagrass meadows of Tampa Bay - a review, p. 210-246. In S.F. Treat, J.L. Simon, R.R. Lewis, and R.L. Whitman, Jr. (eds.), Proceedings Tampa Bay Area Scientific Information Symposium. Burgess Publishing Company, Minneapolis, Minnesota.
Likens, G., F. Borman, and M. Johnson. 1974. Acid rain. Environment 14:33-40.
LMER Coordinating Committee. 1992. Understanding changes in coastal environments: The LMER Program. EOS 73(45):481-485.
Lohrenz, S. E., G. L. Fahnenstiel, D. G. Redalje, G. A. Lang, X. Chen, and M. J. Dagg. 1997. Variations in primary production of northern Gulf of Mexico continental shelf waters linked to nutrient inputs from the Mississippi River. Marine Ecology Progress Series 155:445-454.
Long Island Sound Study: report on Nitrogen and Organic Carbon Loads to Long Island Sound. 1996. Prepared by P. Stacey, Conn. Dept. of Environ. Protection. Bureaus of Water Management, Hartford, CT. 23 p.
Loye-Pilot, M. D., J. M. Martin and J. Morelli. 1990. Atmospheric input of inorganic nitrogen to the western Mediterranean. Biogeochem. 9:117-134.

Luke, W. and R. Dickerson. 1987. Flux of reactive nitrogen compounds from eastern North America to the western Atlantic Ocean. Global Biogeochem. Cycles 1:329-343.

Magnien, R., D. Boward and S. Bieber (eds). 1995. The State of the Chesapeake Bay. Environmental Protection Agency, Chesapeake Bay Program, Annapolis, MD ISBN 0-16-042627-8, p 1-45.

Mallin, M. A. and H. W. Paerl. 1994. Planktonic trophic transfer in an estuary: Seasonal, diel and community structure effects. Ecology 75:2168-2184.

Margalef, R. 1978. Life forms of phytoplankton as survival alternatives in an unstable environment. Oceanol. Acta 1:493-509

Martin, J. M., F. Elbaz-Poulichet, C. Gwue, M. D. Loye-Pilot and G. Han. 1989. River versus atmospheric input of material to the Mediterranean Sea: an overview. Mar. Chem. 28:159-182.

McCarthy, J. 1981. The kinetics of nutrient utilization. IN T. Platt (ed.). Physiological bases of phytoplankton ecology. Can. Bull. Fish. Aquat. Sci. 210:211-233.

Meyers, T.P., P. Finkelstein, J. Clarke, T.G. Ellestad, and P.F. Sims. 1998. A multilayer model for inferring dry deposition using standard meteorological measurements, J. Geophys. Res. 103:22,645-22,661.

Michaels, A. F., A. H. Knap, R. L. Dow, K. Gundersen, J. R. Johnson, J. Sorensen, A. Close and G. A. Knauer. 1994. Seasonal patterns of ocean biogeochemistry at the U.S. JGOFS Bermuda Atlantic Time Series Study site. Deep-Sea Res. 41:1013-1038.

Milliman, J. D. and R. H. Meade. 1983. World-wide delivery of river sediment to the ocean. Journal of Geology 91:1-21.

Molloy, C. and P. Syrett. 1988. Interrelationships between uptake of urea and uptake of ammonium by microalgae. J. Exp. Mar. Biol. Ecol. 118:85-95.

Mopper, K. and R. Zika. 1987. Free amino acids in marine rains: evidence for oxidation and potential role in nitrogen cycling. Nature 325:246-249.

Morris, I. 1974. Nitrogen assimilation and protein synthesis, pp. 583-609. IN W. Steward (ed.), Algal physiology and biochemistry. Blackwell, Oxford.

Moser, F. C. 1997. Sources and sinks of nitrogen and trace metals, and benthic macrofauna assemblages in Barnegat Bay, New Jersey. Pp. i-135. Ph.D. Dissertation, Rutgers, The State University of New Jersey.

Neilsen A and R Lewin. 1974. The uptake and utilization of organic carbon by algae: An essay in comparative biochemistry. Phycologia 13:227-264.

Nelsen, T. A., P. Blackwelder, T. Hood, B. McKee, N. Romer, C. Alvarez-Zarikian and S. Metz. 1994. Time-based correlation of biogenic, lithogenic and authigenic sediment components with anthropogenic inputs in the Gulf of Mexico NECOP study area. Estuaries 17:873-885.

Nelson, D. M. and Q. Dortch. 1996. Silicic acid depletion and silicon limitation in the plume of the Mississippi River: evidence from kinetic studies in spring and summer. Marine Ecology Progress Series 136:163-178.

Nixon, S. 1986. Nutrient dynamics and the productivity of marine coastal waters. Pp. 97-115, In: Halwagy, R.D., Clayton, B. & M. Behbehani (Eds.), Coastal Eutrophication. The Alden Press, Oxford.

Nixon, S. 1995. Coastal marine eutrophication: A definition, social causes, and future concerns. Ophelia 41:199-220.

Nixon, S.W., C.A. Oviatt, J. Frithsen and B. Sullivan. 1986. Nutrients and the productivity of estuarine and coastal marine systems. Journal of the Limnological Society of South Africa 12(1/2): 43-71.

Nixon, S.W., J. W. Ammerman, L. P. Atkinson, V. M. Berounski, G. Billen, W. C. Boicourt, W. R. Boynton, T. M. Church, D. M. Ditoro, R. Elmgren, J. H. Garber, A. E. Giblin, R. A.

Jahnke, N.J.P. Owens, M.E.Q Pilson and S.P. Seitzinger. 1996. The fate of nitrogen and phosphorus at the land-sea margin of the North Atlantic Ocean. Biogeochemistry 35:141-180.

Officer, C. B. and J. H. Ryther. 1980. The possible importance of silicon in marine eutrophication. Marine Ecology Progress Series 3:83-91.

Officer, C. B., R. B. Biggs, J. L. Taft, L. E. Cronin, M. A. Tyler and W. R. Boynton. 1984. Chesapeake Bay anoxia. Origin, development and significance. Science 223:22-27.

O'Halloran, I. 1993. Ammonia volatilization from liquid hog manure: influence of aerators and trapping systems. Soils Sci. Soc. Am. J. 57:1300-1303.

Paerl, H. W. 1983. Factors regulating nuisance blue-green algal bloom potentials in the lower Neuse River, N. C. Univ. North Carolina Water Resources Research Institute Report No. 188. 48 pp

Paerl, H. 1985. Enhancement of marine primary production by nitrogen-enriched acid rain. Nature 315:747-749

Paerl, H. W. 1987. Dynamics of blue-green algal (*Microcystis aeruginosa*) blooms in the lower Neuse River, North Carolina: causative factors and potential controls. Univ. North Carolina Water Resources Research Institute Report No. 229. 164 p.

Paerl, H. 1988. Nuisance phytoplankton blooms in coastal, estuarine, and inland waters. Limnol. Oceanogr. 33:823-847.

Paerl, H. 1991. Ecophysiological and trophic implications of light-stimulated amino acid utilization in marine picoplankton. Applied and Environmental Microbiology 57:473-479.

Paerl, H. 1993. Emerging role of atmospheric nitrogen deposition in coastal eutrophication: biogeochemical and trophic perspectives. Can. J. Fish. Aquat. Sci. 50:2254-2269.

Paerl, H. 1995. Coastal eutrophication in relation to atmospheric nitrogen deposition: current perspectives. Ophelia 41:237-259.

Paerl, H. 1997. Coastal eutrophication and harmful algal blooms: Importance of atmospheric deposition and groundwater as "new" nitrogen and other nutrient sources. Limnol. Oceanogr. 42:1154-1165.

Paerl, H. and M. Fogel. 1994. Isotopic characterization of atmospheric nitrogen inputs as sources of enhanced primary production in coastal Atlantic Ocean waters. Mar. Biol. 119:633-645.

Paerl, H., J. Rudek, and M. Mallin. 1990. Stimulation of phytoplankton production in coastal waters by natural rainfall inputs: nutritional and trophic implications. Mar. Biol.107:247-254.

Paerl, H., M. Mallin, C. Donahue, M. Go, and B. Peierls. 1995. Nitrogen loading sources and eutrophication of the Neuse River Estuary, NC: Direct and indirect roles of atmospheric deposition. UNC Water Res. Res. Inst. Rpt. 291. Raleigh, NC.

Paerl, H., J. Pinckney, J. Fear and B. Peierls. 1998. Ecosystem responses to internal and watershed organic matter loading: Consequences for hypoxia in the eutrophying Neuse River Estuary, NC, USA. Mar. Ecol. Progr. Ser. 166:17-25.

Paerl, H. W. and D. R. Whitall. 1999. Anthropogenically-derived atmospheric nitrogen deposition, marine eutrophication and harmful algal bloom expansion: Is there a link? Ambio 28:307-311.

Parker, C. and J. O'Reilly. 1991. Oxygen depletion in Long Island Sound: A historical perspective. Estuaries 14:248-265.

Peckol, P., B. DeMeo-Anderson, J. Rivers, I. Valiela, M. Maldonado, and J. Yates. 1994. Growth, nutrient uptake capacities and tissue constituents of the macroalgae *Cladophora vagabunda* and *Gracilaria tikvahiae* related to site-specific nitrogen loading rates. Mar. Biol. 121:175-185.

Peierls, B. L., N. F. Caraco, M. L. Pace, and J. J. Cole. 1991. Human influence on river nitrogen. Nature 350:386-387.

Peierls, B. L. and H. W. Paerl. 1997. The bioavailability of atmospheric organic nitrogen deposition to coastal phytoplankton. Limnology and Oceanography 42:1819-1823.

Pilson, M.E.Q. and S. P. Seitzinger. 1996. Areas of shallow water in the North Atlantic. Biogeochemistry 35:227-233.

Pinckney, J., D. Millie, B. Vineyard and H. Paerl. 1997. Environmental controls of phytoplankton bloom dynamics in the Neuse River Estuary (North Carolina, USA). Can. J. Fish. Aquat. Sci. 54:2491-2501.

Pinckney, J., H. Paerl, M. Harrington, and K. Howe. 1998. Annual cycles of phytoplankton community structure and bloom dynamics in the Neuse R. estuary, NC. Mar. Biol. 131:371-382.

Prado-Fiedler, R. R. 1990. Atmospheric input of inorganic nitrogen species to the Kiel Bight. Helgolander Meeresunt. 44:21-30.

Prospero, J. M. and D. L. Savoie. 1989. Effects of continental sources on nitrate concentrations over the Pacific Ocean. Nature 339:687-689.

Prospero, J., K. Barrett, T. Church, F. Dentener, R. Duce, J. Galloway, H. Levy, J. Moody and P. Quinn. 1996. Atmospheric deposition of nutrients to the North Atlantic Basin. Biogeochemistry 35:27-73.

Rabalais, N. N. 1998. on-line. Oxygen Depletion in Coastal Waters. In NOAA's State of the Coast Report. National Oceanic and Atmospheric Administration, Silver Spring, Maryland. http://state_of_coast.noaa.gov/bulletins/html/hyp_09/hyp.html.

Rabalais, N. N., R. E. Turner, D. Justic´, Q. Dortch, W. J. Wiseman, Jr. and B. K. Sen Gupta. 1996. Nutrient changes in the Mississippi River and system responses on the adjacent continental shelf. Estuaries 19:386-407.

Rabalais, N. N., R. E. Turner, W. J. Wiseman, Jr. and Q. Dortch. 1998. Consequences of the 1993 Mississippi River flood in the Gulf of Mexico. Regulated Rivers: Research & Management 14:161-177.

Rabalais, N. N., R. E. Turner, D. Justic, Q. Dortch and W. J. Wiseman, Jr. in review. Gulf of Mexico Hypoxia Assessment, Topic #1, Characterization of Hypoxia. Draft Report to White House Office of Science and Technology Policy, Committee on Environment and Natural Resources, Hypoxia Work Group, Mississippi River/Gulf of Mexico Watershed Nutrient Task Force, 219 pp.

Richardson, K. 1997. Harmful or exceptional phytoplankton blooms in the marine ecosystem. Adv. Mar. Biol. 31:302-385.

Riegman, R. 1992. *Phaeocystis* blooms and eutrophication of the continental coastal zones of the North Sea. Marine Biology 112: 479-484.

Ries, T. 1993. The Tampa Bay experience, p. 19-24. In Morriss, L.J. and D.A. Tomasko (eds.), Proceedings and conclusions of workshops on: Submerged aquatic vegetation and photosynthetically active radiation. Special publication SJ93-SP13. St. Johns River Water Management District, Palatka, Florida.

Rizzo, W. G., G. Lackey and R. R. Christian. 1992. Significance of euphotic, subtidal sediments to oxygen and nutrient cycling in a temperate estuary. Mar. Ecol. Progr. Ser. 86:51-61.

Rodhe, H.R., J. Soderlund and J. Ekstedt. 1980. Deposition of airborne pollutants on the Baltic. Ambio 9:168-173.

Rosenberg, R. 1980. Effect of oxygen deficiency on benthic macrofauna in fjords. Pages 499-514 in H. L. Freeland, D. M. Farmer and C. D. Levings (eds.), Fjord Oceanography, Plenum, New York.

Rudek, J., H. Paerl, M. Mallin, and P. Bates. 1991. Seasonal and hydrological control of phytoplankton nutrient limitation in the lower Neuse River Estuary, North Carolina. Mar. Ecol. Prog. Ser. 75:133-142.

Russell, K., J. Galloway, S. Macko, J. Moody and J. Skudlark. 1998. Sources of nitrogen in wet deposition to the Chesapeake Bay region. Atmosph. Environ. 32:2453-2465.

Ryther J. & W. Dunstan. 1971. Nitrogen, phosphorus, and eutrophication in the coastal marine environment. Science 171:1008-1112.

Seinfeld, J.H. and S.N. Pandis. 1998. Atmospheric Chemistry and Physics: From Air Pollution to Climate Change, John Wiley & Sons, Inc., New York.

Seitzinger, S.P., L.P. Nielson, J. Caffrey and P.B. Christensen. 1993. Denitrification measurements in aquatic sediments. A comparison of three methods. Biogeochemistry 23:147-167.

Seitzinger, S. P., I. E. Pilling and R. DeKorsey. 1993. Eutrophication and nutrient loading in Barnegat Bay: N or P limitation of primary production. Final Report No. 93-2F, New Jersey Department of Environmental Protection.

Seitzinger, S.P. and A.E. Giblin. 1996. Estimating denitrification in North Atlantic continental shelf sediments. Biogeochemistry 35:235-259.

Seitzinger, S and J Sanders. 1997. Contribution of dissolved organic nitrogen from rivers to estuarine eutrophication. Mar. Ecol. Prog. Ser. 159:1-12.

Seitzinger, S.P. and C. Kroeze. 1998. Global distribution of nitrous oxide production and N inputs in freshwater and coastal marine ecosystems. Global Biogeochem. Cycles, 12(1): 93-113.

Seitzinger, S.P. and R.W. Sanders. 1999. Atmospheric inputs of dissolved organic nitrogen stimulate estuarine bacteria and phytoplankton. Limnol. Oceanogr. 44: 721-730.

Sen Gupta, B. K., R. E. Turner and N. N. Rabalais. 1996. Seasonal oxygen depletion in continental-shelf waters of Louisiana: Historical record of benthic foraminifers. Geology 24: 227-230.

Short, F.T. and D.M. Burdick. 1996. Quantifying eelgrass habitat loss in relation ot housing development and nitrogen loading in Waquoit Bay, MA. Estuaries 19(3):730-739.

Smayda, T. J. 1990. Novel and nuisance phytoplankton blooms in the sea: Evidence for global epidemic. Pages 29-40 in E. Graneli, B. Sundstrom, R. Edler and D. M. Anderson (eds.), Toxic Marine Phytoplankton, Elsevier, New York.

Smetacek, V., U. Bathmann, E.-M. Nöthig and R. Scharek. 1991. Coastal eutrophication: Causes and consequences. Pages 251-279 in R. C. F. Mantoura, J.-M. Martin and R. Wollast (eds.), Ocean Margin Processes in Global Change. John Wiley & Sons, Chichester.

Sorteberg, A., Ø. Hov, S. Solber, K. Tørseth, H. Areskoug, M. Ferm, K. Granby, H. Lättilä, K. Persson and D. Simpson. 1998. Gaseous and particulate oxidized and reduced nitrogen species in the atmospheric boundary layer in Scandinavia in spring, J. Atmos. Chem., 30: 241-271.

Stanley, DW. 1983. Nitrogen cycling and phytoplankton growth in the Neuse River, NC WRRI Report No. 204. UNC Water Resources Research Instit., Raleigh, NC.

Steudler, P. A., R. D. Bowden, J. M. Melillo, and J. D. Aber. 1989. Influence of nitrogen fertilization on methane uptake in temperate forest soils. Nature 341:314-315.

Stoddard, J. L. 1994. Long term changes in watershed retention on nitrogen: Its causes and aquatic consequences. In: L. A. Baker (Ed.), Environmental Chemistry of Lakes and Reservoirs, Advances in Chemistry Series No. 237, American Chemcial Society, Washington DC, pp. 223-284.

Stolte, W., T. McCollin, A. Noordeloos and R. Riegman. 1994. Effects of nitrogen source on

the size distribution within marine phytoplankton populations. J. Exp. Mar. Biol. Ecol. 184:83-97.
Syrett, P. 1981. Nitrogen metabolism in microalgae. IN T. Platt (ed.). Physiological bases of phytoplankton ecology. Can. Bull. Fish. Aquat. Sci. 210:182-210.
Tampa Bay National Estuary Program. 1996. Charting the Course for Tampa Bay: Final Comprehensive Conservation and Management Plan. TBNEP, St. Petersburg, FL.
Tilman, D. 1977. Resource competition between planktonic algae: an experimental and theoretical approach. Ecology 58:338-348.
Timperley, M., R. Vigor-Brown, M. Kawashima and M. Ishigami. 1985. Organic nitrogen compounds in atmospheric precipitation: Their chemistry and availability to phytoplankton. Can. J. Fish. Aquat. Sci. 42:1171-1177.
Titman, D. 1976. Ecological competition between algae: experimental confirmation of resource-based competition theory. Science 192:463-465
Turner, R. E. and N. N. Rabalais. 1991. Changes in Mississippi River water quality this century. Implications for coastal food webs. BioScience 41: 140-148.
Turner, R. E. and N. N. Rabalais. 1994a. Coastal eutrophication near the Mississippi river delta. Nature 368: 619-621.
Turner, R. E. and N. N. Rabalais. 1994b. Changes in the Mississippi River nutrient supply and offshore silicate-based phytoplankton community responses. Pages 147-150 in K. R. Dyer and R. J. Orth (eds.), Changes in Fluxes in Estuaries: Implications from Science to Management. Proceedings of ECSA22/ERF Symposium, International Symposium Series, Olsen & Olsen, Fredensborg, Denmark.
Turner, R. E., N. Qureshi, N. N. Rabalais, Q. Dortch, D. Justic, R. F. Shaw and J. Cope. 1998. Fluctuating silicate:nitrate ratios and coastal plankton food webs. Proceedings of the National Academy of Science, USA 95:13048-13051.
Turpin, D.H. and P.J. Harrison. 1979. Limiting nutrient patchiness and its role in phytoplankton ecology. J. Exp. Mar. Biol. Ecol. 39:151-166.
U.S. Environmental Protection Agency. 1996. Air Quality Criteria for Ozone and Related Photochemical Oxidants, Report EPA/600/P-93/004aF, Volume I of III, Office of Research and Development, Washington, D.C.
Valiela, I., K. Foreman, M. LaMontagne, D. Hersh, J. Costa, P. Peckol, B. DeMeo-Anderson, C. D'Avanzo, M. Babione, C-H. Sham, J. Brawley, and K. Lajtha. 1992. Couplings of watersheds and coastal waters: Sources and consequences of nutrient enrichment in Waquoit Bay, Massachusetts. Estuaries. 15:443-457.
Valiela, I, G. Collins, J. Kremer, K. Lajtha, M. Geist, B. Seely, J. Brawley, and C.H. Sham. 1997. Nitrogen loading from coastal watersheds to receiving estuaries: new method and application. Ecological Applications 7:358-380.
Valigura, R., W. Luke, R. Artz and B. Hicks. 1996. Atmospheric Nutrient Inputs to Coastal Areas: Reducing the Uncertainties. US-NOAA Coastal Ocean Program Decision Analysis Series No. 9. Washington DC.
Van Dolah, P. and G. Anderson. 1991. Effects of Hurricane Hugo on salinity and dissolved oxygen conditions in the Charleston Harbor estuary. J. Coast. Res. 8:83-94.
Vitousek, P. M., J. D. Aber, R. W. Howarth, G. E. Likens, P. A. Matson, D. W. Schindler, W. H. Schesinger, and G. D. Tilman. 1997. Human alterations of the global nitrogen cycle: Causes and consequences, Issues in Ecology, 1, 1-15.
Vitousek P.M., H.A. Mooney, J. Lubchenko and J. M. Mellilo. 1997. Human domination of Earth's ecosystems. Science 277:494-499.
Vollenweider, R. A. 1976. Advances in defining critical loading levels of phosphorus in lake eutrophication. Memorie-Istituto Italiano de Idrobilogia 33:53-83.

Willey, J. D. and H. W. Paerl. 1993. Enhancement of chlorophyll *a* production in Gulf Stream seawater by synthetic vs. natural rain. Mar. Biol. 116:329-334.

Williams, M. W., and J. M. Melack. 1991. Precipitation chemistry in and ionic loading to an Alpine Basin, Sierra Nevada. Water Resources Research, 27, 1563-1574.

Wojcik, G.S. and J.S. Chang. 1997. A re-evaluation of sulfur budgets, lifetimes, and scavenging ratios for eastern North America, J. Atmos. Chem. 26:109-145.

Wyers, G.P. and J.W. Erisman. 1998. Ammonia exchange over coniferous forest. Atmos. Environ. 32:441-451.

3

Atmospheric Nitrogen Deposition to Coastal Estuaries and Their Watersheds

Tilden Meyers, Joseph Sickles, Robin Dennis, Kristina Russell, James Galloway, and Thomas Church

Abstract

As part of the overall process of evaluating the impacts of total atmospheric deposition of nitrogen on coastal ecosystems and estuaries, estimates of wet and dry deposition of nitrogen to both watershed and water bodies are required. Currently, most bays, estuaries, and watersheds lack onsite deposition monitoring data. In this first order attempt to assess the impact and fate of nitrogen delivered from atmospheric processes, estimates of wet and dry nitrogen deposition are derived using data from either regional monitoring activities or calibrated atmospheric deposition models. The components of estimated wet nitrogen deposition include, nitrate, ammonium and organic nitrogen. For dry deposition, the gas phase components include nitric acid vapor and ammonia, while the aerosol components consist of nitrate and ammonium. Total annual N deposition (wet plus dry) ranged from 3 to 14 kg N ha^{-1}, with wet deposition generally comprising about 60% of the total. Uncertainties in the deposition estimates arising from the extrapolation process are estimated to be no less than a factor of two, with generally more confidence in the wet deposition than the dry. Estimates derived for these watersheds are used in the process level models to evaluate the impact of atmospheric nutrient deposition on the health and quality of the coastal water body in question.

Introduction

This chapter presents a standardized approach to estimating nitrogen deposition to estuaries and their associated watersheds for selected regions of the coastal United States (excluding Hawaii and Alaska). After explaining the calculation methodology (including uncertainties), per unit area estimates of N deposition to the watershed and the estuary are tabulated together with total loadings. Prior to listing important findings and recommendations for future research, the chapter examines spatial and temporal variability in deposition and compares the findings of this chapter with previously published N deposition rates for specific watersheds/estuaries.

The N species considered are nitrate aerosol, nitric acid, ammonia, ammonium and total organic nitrogen. These species are removed by wet and dry deposition. The former process removes all the N species; removal rates are determined by the combination of the amount of precipitation with nitrate, ammonium and organic N concentrations. Dry deposition rates are determined for nitrate aerosols plus nitric acid, and for ammonia plus ammonium aerosols (NH_x). Dry deposition rates for organic nitrogen are not estimated due to a paucity of knowledge about organic N speciation in the atmosphere.

Methodologies for Wet Deposition of Nitrate

Ammonium and Nitrate

The average wet deposition of ammonium and nitrate to the 42 water bodies of interest were determined by using routine data collected by the National Atmospheric Deposition Program (NADP). The NADP database contains yearly and seasonal ammonium and nitrate deposition data based on weekly collections made at various sites throughout the United States. NADP sites within or near each of the watersheds of interest were identified (Table 1) and annual deposition averages for all of the years available were determined. The mean of the annual averages for each species was then taken to determine the overall average for the site. If only one site was contained within a watershed, then the mean for that site was used to represent the entire watershed. If multiple sites were contained in a watershed, then the means of the site averages were taken to provide one number to represent the deposition of a species to the watershed.

Because the NADP network has a weekly collection protocol, it imparts a potential artifact in the loss of nitrogen species due to biological activity in the collection buckets during the week long storage. This effect has been thoroughly evaluated using daily data from MAP3S (1981-1989) and AIRMoN (1992-1995) data sets [Butler et al., 1998]. The comparison used the daily data from both programs aggregated to a weekly basis to compare to the NADP weekly data sets. The results strongly suggest that the weekly and daily aggregated network concentrations for nitrate are comparable during both time periods. However, the results for ammonium show a statistically significant bias with the daily results having concentrations exceeding the weekly values by 15%.

At some locations, the AIRMoN (daily) and NADP(weekly) programs were co-located which permitted a direct side-by-side comparison of ammonium deposition. This comparison [Artz, unpublished] shows if somewhat higher aggregated daily versus weekly results (e.g., as much as 25% at some Midwest locations during the summer). Thus a uniform factor of 15% was taken to increase the NADP/NTN ammonium deposition data for all locations in making the assessment for wet ammonium deposition, with an estimated uncertainty of a factor of two.

Additional uncertainties in arriving at single numbers to represent annual wet depositions of ammonium and nitrate to each watershed arise from their calculations. Assuming that the mean of the annual averages is representative for a given site is reasonable, as most sites show no trend over time. Additionally, assuming that the average of the mean depositions of several sites within or near the watershed of interest is representative of the entire watershed and water body is also reasonable considering that the variation of average deposition is small within watersheds (e.g., for ammonium and nitrate annual averages at eight sites within the Long Island Sound Watershed, annual ammonium average = 1.83 ± 0.07 kg N ha^{-1} and annual nitrate average = 3.80 ± 0.16 kg N ha^{-1}).

TABLE 1. Nitrogen Wet Deposition to Water Bodies of Interest.

Water body	NADP Sites	Years of Data	Avg. NH$_4^+$ kg N ha^{-1}	Avg. NO$_3^-$ kg N ha^{-1}
Casco Bay	ME02	1981-1996	*1.21*	**2.22**
Great Bay	MA13	1982-1996	1.14	2.69
	ME02	1981-1996	1.21	2.22
	Average		*1.18*	*2.46*
Merrimack River	MA13	1982-1996	1.47	2.69
	NH02	1979-1995	2.04	3.67
	Average		*1.76*	*3.18*
Waquoit Bay Buzzards Bay	MA01	1982-1995	*1.16*	*2.74*
Narragansett Bay	MA08	1984-1996	1.77	3.90
	MA13	1982-1996	1.14	2.69
	Average		*1.46*	*3.30*
Long Island Sound	NY12	1980-1983	1.57	3.33
	NY20	1979-1996	1.66	3.31
	NY51	1980-1984	1.64	4.18
	NY99	1984-1996	2.01	4.59
	MA08	1984-1996	1.77	3.90
	NH02	1979-1995	2.04	3.67
	VT01	1982-1996	1.92	3.74
	VT99	1985-1996	1.99	3.65
	Average		*1.83*	*3.80*
Gardiners Bay	NY51	1980-1984	1.64	4.18
	NY99	1984-1996	2.01	4.59
	Average		*1.83*	*4.39*
Hudson River/Raritan Bay	NJ99	1982-1996	2.03	3.79
	NY12	1980-1983	1.57	3.33
	NY20	1979-1996	1.66	3.31
	NY51	1980-1984	1.64	4.18
	NY68	1986-1996	2.15	4.53
	NY99	1984-1996	2.01	4.59
	VT01	1982-1996	1.92	3.74
	VT99	1985-1996	1.99	3.65
	Average		*1.87*	*3.89*
Barnegat Bay New Jersey Inland Bays	NJ99	1982-1996	*2.03*	3.79
Delaware Bay	NJ99	1982-1996	2.03	3.79
	PA72	1986-1996	1.94	4.44
	Average		*1.99*	*4.12*
Delaware Inland Bays	MD13	1983-1996	**1.93**	3.24
Chesapeake Bay	MD03	1985-1996	2.14*	3.78
	MD13	1983-1996	1.93	3.24
	VA00	1985-1996	2.10	3.69
	VA28	1989-1996	2.33	2.93

TABLE 1. Continued

Water body	NADP Sites	Years of Data	Avg. NH$_4^+$ kg N ha^{-1}	Avg. NO$_3^-$ kg N ha^{-1}
	WV18	1979-1996	2.33	4.74
	PA15	1984-1996	1.99	3.88
	PA29	1983-1996	2.56	4.95
	NY08	1979-1996	2.45	3.82
	NY52	1983-1996	3.66	6.45
	NY65	1980-1996	1.52	2.81
	Average		*2.30*	*4.03*
Maryland Inland Bays	MD13	1983-1996	*1.93*	*3.24*
Pamlico Sound	NC03	1979-1996	1.55*	2.34
	NC41	1979-1996	2.32	2.46
	Average		*1.94*	*2.40*
Ace Basin/ St. Helena	SC06	1985-1996	**1.00***	*1.94*
North Inlet/ Wynah Bay	NC34	1979-1996	2.42	2.71
	NC36	1984-1996	1.45	2.52
	SC06	1985-1996	1.00*	1.94
	Average		*1.62*	*2.39*
Charleston Harbor	SC06	1985-1996	**1.00***	**1.94**
St. Catherine's/ Sapelo Sound	GA20	1984-1996	**1.48**	**1.88**
Altamaha River	GA20	1984-1996	1.48	1.88
	GA41	1982-1988	1.52	1.96
	Average		*1.50*	*1.92*
Ogeechee Coastal	GA20	1984-1996	*1.48*	*1.88*
Indian River	FL11	1982-1996	1.03	1.61
	FL99	1984-1996	0.91*	1.96*
	Average		*0.97*	*1.79*
Charlotte Harbor Sarasota Bay Tampa Bay	FL41	1984-1996	**1.08***	**2.15***
Apalachicola Bay	FL14	1985-1996	1.08	1.85
	GA41	1982-1988	1.52	1.96
	Average		*1.30*	*1.91*
Mobile Bay	AL10	1984-1996	1.22	1.94
	AL99	1984-1996	2.22*	2.27*
	GA41	1982-1988	1.52	1.96
	MS14	1981-1989	1.85	2.44
	MS19	1989-1996	1.84	2.16
	TN98	1990-1995	2.23	2.69
	Average		*1.81*	*2.24*
Barataria-Terrebonne Bays	LA12	1983-1996	2.43	2.55
	LA30	1984-996	2.46	2.70
	Average		*2.45*	*2.63*
Calcasieu Lake	LA12	1983-1996	*2.43*	*2.55*
Sabine Lake	LA06	1982-1988	1.58	2.11
	TX21	1983-1996	1.84	2.26

TABLE 1. Continued

Water body	NADP Sites	Years of Data	Avg. NH_4^+ kg N ha^{-1}	Avg. NO_3^- kg N ha^{-1}
	TX38	1982-1996	1.94	1.98
	Average		*1.79*	*2.12*
Laguna Madre	TX03	1984-1990	**1.56**	**1.24**
West Mississippi Sound	LA30	1984-1996	**2.46**	**2.7**
Apalachee Bay	FL14	1985-1996	1.08	1.85
	GA99	1994-1996	1.26	1.66
	Average		*1.17*	*1.76*
Corpus Christi Bay	TX03	1984-1990	1.56	1.24
	TX16	1984-1990	1.06	0.91
	Average		*1.31*	*1.08*
Matagorda Bay	TX16	1984-1990	1.06	0.91
	TX53	1981-1988	1.32	1.53
	Average		*1.19*	*1.22*
Galveston Bay	TX10	1985-1996	1.59	1.73
	TX38	1982-1996	1.94	1.98
	TX56	1985-1996	1.92	1.92
	Average		*1.82*	*1.88*
San Francisco Bay Watershed	CA45	1979-1997	0.33	0.46
	CA88	1779-1997	1.56	0.71
	CA99	1982-1997	1.05	1.10
	Average		*0.98*	*0.76*
Puget Sound Watershed	WA14	1981-1997	0.44	0.63
	WA19	1984-1997	0.51	1.37
	WA21	1984-1997	0.37	0.70
	Average		*0.44*	*0.90*

*indicates significantly increasing deposition over the time indicated
NOTE: Ammonium depositions should be increased by 15% in order to account for losses associated with weekly collection

Organic Nitrogen

Organic nitrogen in wet deposition has been measured in a limited number of locations over extended time periods throughout the eastern United States in conjunction with ongoing inorganic nitrogen collections [see Gorzelska et al., 1997 for review]. However, because many of the measurements have been made on samples collected in a variety of materials, stored under various conditions, and using a variety of methodologies, the accuracy of the results may be questionable. In all cases, the organic nitrogen is calculated as the difference between a total inorganic nitrogen assay and one conducted after strong oxidation of the sample. Thus, errors in the two analyses and uncertainties in the extent of oxidation for all organic species can propagate to a large extent in the calculated difference. Attempts to

TABLE 2: Organic Nitrogen in Wet Deposition Events to Eastern U.S. Sites (expressed as percent of total nitrogen).

Location	Time Period	Technique	Wet Organic-N Deposition (%)	Source
Narragansett, RI	1993-1996 (summer)	Persulfate Oxidation	28	1
New Jersey	1995-1997	High temperature oxidation	19	2
Newark, DE	March 1997 Sept. 1997	High temperature oxidation, frozen collection	10	3
Lewes, DE	Oct. 1993 - Dec. 1994	Persulfate Oxidation	23	4
Rhode River, MD	1978-1994	Kjeldahl Method	31 (wet+dry)	5
Charlottesville, VA	Apr. 1996 - Sept. 1997	High temperature oxidation	7	3
Morehead City, NC	July 1994 - Oct. 1996	High temperature oxidation	18	6
Galveston, TX	Apr. 1993 - Jan. 1994	High temperature oxidation	17	7

1) Nixon and Buckley, pers. com.; 2) Seitzinger and Styles, pers. com.; 3) Keene et al., pers. com.; 4) Scudlark et al., 1998; 5) Correll et al., 1994; 6) Peierls and Paerl, 1997; 7) Shon, 1994

speciate the organic nitrogen in precipitation has shown that primary amino acids and urea are minor (10%) components at continental locations [Gorzelska et al., 1992; Cornell et al., 1998].

Recent studies [Scudlark et al., 1998; Russell et al., 1999] evaluated different collection materials and two wet oxidation techniques (UV and persulfate). Glass or stainless steel and persulfate oxidation are the preferred materials and oxidizing reagent respectively, yielding results with uncertainties of 10-20%. However, storage experiments suggest that unless immediate freezing of the sample occurs some samples can lose up to half of their organic nitrogen within hours.

More recent studies [Keene et al., pers. comm.] show major losses of precipitation organic nitrogen can occur even during collection, so freezing during collection also appears to be requisite. Further, using instruments for high temperature oxidation requires additional care in assuring the quantitative introduction of the sample and conversion of both inorganic and organic analytes. The latter effect may cause the organic nitrogen to be over estimated because it is calculated as a difference between total and inorganic nitrogen.

A compilation of organic nitrogen results (both published and unpublished) in precipitation at a number of geographically diverse east coast locations is reported in Table 2. Noted are the locations, years of results, method of analyses, and references. The results are reported only as the average percentage organic nitrogen deposition to total wet nitrogen over the period of collection. The results range from 7-31% with no apparent trend in location latitude. Most center around 20%, and are close to the mean (19%). A few urban locations are around 50%, apparently from local pollutants (e.g., PAN) and as a result, are not included. Thus, the organic nitrogen wet deposition was calculated for the assessment using a uniform quantity of 20% of the total nitrogen wet deposition with an apparent uncertainty of a factor of two.

Methodologies for Dry Deposition of Nitrate

Estimates of dry nitrogen (N) deposition were developed from field measurements and from models. Model estimates came from the US EPA's Regional Acid Deposition Model (RADM) (see Section "*Extended RADM*," below). Measurement data came largely from the US EPA's Clean Air Status and Trends Network (CASTNet) data archive [Clarke et al., US E.P.A., 1998]. There are 45 CASTNet sites in the eastern U.S., and most of these are at inland locations. As a result, dry deposition monitoring data that correspond directly with the watersheds of interest are not available from CASTNet. In those cases where one or more monitoring sites are located in the regional vicinity of the watershed, data from these sites were selected and averaged. This approach assumes that site specific deposition is representative of the watershed under consideration. Since separations on the order of up to 200 km may exist between the monitoring sites and the subject water body, this approach provides only a crude first order approximation of N deposition for a specific watershed. The monitoring data employed are the averages of annual dry deposition values (see Table 3); usually for the 1989 - 1995 period, but in some cases where sites had been recently added or deleted from the monitoring network, shorter periods were employed. In one instance (*i.e.*, Tampa Bay), monitoring data from outside of CASTNet were used [Pribble and Janicki, 1998].

CASTNet was designed to be a rural monitoring network directed toward providing data to describe broad spatial and site-specific temporal patterns of total deposition [Clarke et al., 1997; US E.P.A., 1998]. CASTNet siting criteria were designed to avoid undue influence from point sources, area sources, and local activities. Atmospheric concentration data for nitric acid (HNO_3), particulate nitrate (NO_3^-), and particulate ammonium (NH_4^+) along with other non-nitrogen pollutant species are obtained from weekly filter pack (FP) measurements at all sites. The FP consists of three types of filters in series operating with a flow rate of 1.5 l min^{-1}. The first, a Teflon filter, collects aerosols (*i.e.*, SO_4^{2-}, NO_3^-, and NH_4^+); next, a nylon filter collects HNO_3; and finally, two carbonate-impregnated cellulose filters collect SO_2. Continuous measurements of meteorological variables as required for the estimation of deposition velocities (V_d), are also made at each site.

Based on collocated FP sampling at various CASTNet sites, estimates of precision for network measurements of NO_3^- are ± 8%, while corresponding estimates for HNO_3 and NH_4^+ are within ± 4% [US E.P.A, 1998]. FP measurements are subject to various sampling biases [Sickles, 1987; Sickles et al., 1990; 1999]. The FP offers some degree of size selectivity in sampling aerosols, probably capturing fine aerosols (perhaps less than 5 μm) adequately but missing the larger sizes. This may result in a low bias for aerosol N deposition, especially in coastal regions where gaseous HNO_3 may become attached to the relatively large sea salt aerosols [Pio and Lopez, 1998]. Most of the CASTNet sites are located sufficiently inland (>80 km) to minimize these and other marine influences. Sampling biases for oxidized N occur largely as a result of volatilization or displacement reactions involving $_{an}$ NH_4NO_3 aerosol captured on the first filter and tend to be more important at higher temperatures (*e.g.*, summer) and less important in winter. Based on comparisons with annular denuders

TABLE 3. Nitrogen Dry Deposition to Watersheds of Interest Expressed in kg N ha^{-1} yr^{-1}.

Watershed	CASTNet Site(s)	Years of Data	Average Dry NH$_4^+$	Average Dry NO$_3^-$
Casco Bay	HOW132	1993 - 1995	0.15	0.95
	WST109	1989 - 1995	0.11	0.62
		Average	*0.13*	*0.79*
Great Bay	HOW132	1993 - 1995	0.15	0.95
	WST109	1989 - 1995	0.11	0.62
	LYE145	1994 - 1995	0.31	3.25
		Average	*0.19*	*1.61*
Merrimack River	WST109	1989 - 1995	0.11	0.62
	LYE145	1994 - 1995	0.31	3.25
	ABT147	1994 - 1995	0.39	3.20
		Average	*0.27*	*2.36*
Massachusetts Bay	ABT147	1994 - 1995	0.39	3.20
Waquoit Bay	WSP144	1989 - 1995	0.51	3.62
Buzzard's Bay				
Narragansett Bay				
Gardiners Bay		*Average*	*0.45*	*3.41*
	WST109	1989 - 1995	0.11	0.62
Long Island Sound	LYE145	1994 - 1995	0.31	3.25
	WPB104	1989 - 1993	0.42	3.05
	CAT175	1994 - 1995	0.42	4.18
	ABT147	1994 - 1995	0.39	3.20
		Average	*0.33*	*2.86*
Hudson River/	WFM105	1989 - 1992	0.26	2.13
Raritan Bay	CTH110	1989 - 1995	0.49	4.80
	CAT175	1994 - 1995	0.42	4.18
	WPB104	1989 - 1993	0.42	3.05
	WSP144	1989 - 1995	0.51	3.62
		Average	*0.42*	*3.56*
Barnegat Bay	WSP144	1989 - 1995	0.51	3.62
NJ Inland Bays				
Delaware Bay				
DE Inland Bays	WSP144	1989 - 1995	0.51	3.62
MD Inland Bays	BEL116	1989 - 1995	0.60	2.94
		Average	*0.56*	*3.28*
Chesapeake Bay	BEL116	1989 - 1995	0.60	2.94
	ARE128	1989 - 1995	0.68	4.34
	PSU106	1989 - 1995	0.64	3.07
	LRL117	1989 - 1995	0.39	3.11

TABLE 3. Continued.

Watershed	CASTNet Site(s)	Years of Data	Average Dry NH_4^+	Average Dry NO_3^-
	PAR107	1989 - 1995	0.46	2.36
	VPI120	1989 - 1995	0.47	4.32
	PED108	1989 - 1995	0.46	2.77
	SHN118	1989 - 1995	0.59	4.87
		Average	*0.54*	*3.47*
Pamlico Sound	BFT142	1994 - 1995	0.28	1.43
Winyah Bay	CND125	1990 - 1995	0.50	2.52
		Average	*0.39*	*1.98*
Altamaha River	GAS153	1989 - 1995	0.44	2.24
Tampa Bay	Pribble and Janicki [1998]	1996 -1998	na[1]	na[1]
Apalachee Bay	SUM156	1989 - 1995	0.24	1.3
Apalachicola Bay	GAS153	1989 - 1995	0.44	2.24
	SUM156	1989 - 1995	0.24	1.30
		Average	*0.34*	*1.77*
Mobile Bay	CVL151	1989 - 1995	0.38	2.07
	SND152	1989 - 1995	0.71	2.36
	SUM156	1989 - 1995	0.24	1.30
		Average	*0.44*	*1.91*

[1] Reported as Total Dry N Deposition. [See Tables 2 and 4 of Pribble and Janicki, 1998].

conducted in the summer and early fall, FP NO_3^- concentrations may be biased low by 20%, and HNO_3 may be high by 10 to15 % [Sickles et al., 1990; 1999]. Concentrations of NH_4^+ may be biased high by 15 to 30%, the extent to which NH_3 is captured by acid aerosols during sampling. Since NH_3 is not monitored in CASTNet, consideration of NH_4^+ without NH_3 acts to underestimate reduced airborne N concentration. In addition, these biases are likely to depend on the atmospheric composition, temperature, and relative humidity and thus are difficult to predict since they will be influenced strongly by site location and season of the year.

Dry deposition (*i.e.*, fluxes) of gas phase and aerosol N species are determined by combining their measured concentrations with deposition velocities estimated from a model. Deposition velocities for gases and aerosols were estimated at an hourly temporal resolution using an earlier version of the Multi-Layer Model (MLM), similar to that described by Meyers et al. [1998]. The MLM is based on the resistance analogue with resistance to deposition grouped into three terms: R_a, aerodynamic resistance from the planetary boundary layer downward through the turbulent atmospheric surface layer; R_b, resistance through the (approximately 1-mm thick) laminar boundary layer at the leaf surface; and R_c, net surface resistance including contributions of various vegetative (*e.g.*, stomatal, cuticular), soil, rock, and water surfaces. The species and site dependent deposition velocity is the reciprocal of the overall resistance to deposition and is computed according to Eq. (1).

$$V_d = 1/(R_a + R_b + R_c) \qquad (1)$$

During both the day and night, V_d values for HNO_3 and aerosols are strongly influenced by R_a, while vegetative and other surface resistances may dominate V_d for other species (*e.g.*, O_3, SO_2). In the MLM, V_d for fine aerosols (<2 μm) is computed with no distinctions for different size or chemical composition [Meyers et al., 1998], and as a result, estimates of V_d for NO_3^- and NH_4^+, as well as SO_4^{2-}, are identical.

Inputs required to compute V_d using the MLM include: wind speed, wind direction, the standard deviation of wind direction, temperature, relative humidity, solar radiation, precipitation, surface wetness, leaf area index, plant species, and percent green leafout, among others. The MLM was designed to compute the V_d for a small spatial scale (*i.e.*, approximately a 1 km radius surrounding the site with consistent terrain and surface cover). Uncertainties in the determination of the input parameters result in irreducible uncertainties of ± 20 to 25% in hourly estimates of V_d for HNO_3 [Cooter and Schwede, 1999], which likely become negligible for annual periods. Limited comparisons with the modified Bowen ratio method show the model to underestimate the deposition velocity for HNO_3 by 10%, while estimates for aerosols have not been evaluated [Meyers et al., 1998]. Differences in annual V_d for HNO_3 of ± 40% has been observed at four sites located within 0.5 km and attributed to spatial inhomogeneities in meteorology and terrain [Brook et al., 1997]. This is consistent with the findings of Clarke et al. [1997] who found the HNO_3 V_d on a ridge to average 1.7 times higher than at a nearby (*i.e.*, within 1 km) site located in a valley. Thus, estimates of V_d for HNO_3 appear to be biased low by 10% with an uncertainty of ± 50%.

Dry deposition (*i.e.*, flux) of an airborne species is represented by Eq. (2), where D is weekly dry deposition, $\overline{C}$ is the weekly average concentration of the gas or aerosol species of interest, and $\overline{V_d}$ is weekly average deposition velocity for the gas or aerosol of interest.

$$D = \overline{V_d}(z) \, \overline{C}(z) + \overline{V_d' C'} = \overline{V_d}(z) \, \overline{C}(z) + \sigma_v \sigma_c r_{vc} \qquad (2)$$

Here, the primes indicate short-term variations from long-term weekly averages denoted by the overbars. The second term on the right-hand side of Eq. (2) represents the correlation between the hourly concentrations and hourly deposition velocities. In a preliminary analysis for SO_2 deposition, Meyers and Yuen [1987] found the correlation to be significant for summer periods. In a more detailed analyses, Matt and Meyers [1991] found that neglecting the term in Eq. (2) resulted in underestimates of SO_2 deposition by nearly 40% for summer periods for a site in Eastern Tennessee. No significant correlations were found for other times during the year. Corresponding information about N deposition is scarce. With little evidence to be used for guidance, the cross correlation term is frequently neglected (as it is in the current study), and Eq. (3) results.

$$D = \overline{V_d}(z) \, \overline{C}(z) \qquad (3)$$

Weekly estimates of dry deposition are made by aggregating hourly V_d to the weekly time scale and then applying Eq. (3). With regard to nitrogen, this approach can result in a low bias in the flux estimates in those instances where there is appreciable covariance between C and V_d values (*e.g.*, 20% for HNO_3 in the summer [Clarke et al., 1997]).

Estimates of deposition will be influenced by the bias and uncertainty associated with concentration, deposition velocity, and their product (see Eq. 2 and 3). Measured concentrations in the current study have not been adjusted for biases and thus probably represent lower bound estimates for NO_3^-, while the HNO_3 concentration may be biased high. As noted earlier, V_d for HNO_3 may be biased low by 10% (due to model shortcomings). Although opposing biases in C and V_d may tend to balance, neglecting the cross-correlation term of Eq. (2) may result in a 10 to 20% low bias in annual HNO_3 deposition. Spatial inhomogeneities may contribute $\pm$ 50% uncertainties to local annual estimates of V_d for HNO_3, and the uncertainty is likely to be larger on a regional scale. Since no adjustments have been made, the resulting oxidized N deposition likely reflects lower bound estimates, and since oxidized N is dominated by HNO_3, uncertainty of a factor of two is likely for regional oxidized N deposition.

Dry deposition of NH_3 is expected to be relatively rapid and similar to HNO_3. Although gaseous NH_3 is not monitored in CASTNet, in areas distant from major sources its ambient concentration is expected to be less than half that of aerosol NH_4^+ [Sickles et al., 1990; 1999; Langford et al., 1992]. Some gaseous NH_3 may be captured by the CASTNet FP sampler and reported as NH_4^+ aerosol. Although this may result in a high bias in NH_4^+ deposition results, the overall deposition of reduced N will be biased low to the extent that NH_3 is not captured, but this bias is likely less than 30%. In addition, because of the large difference in gas and aerosol deposition velocities, the spatial location of reduced N deposition may be misrepresented. Thus, the uncertainty of regional reduced N deposition is expected to be similar to that for oxidized N deposition (*i.e.*, a factor of two).

Extended RADM

Dry deposition estimates based on the CASTNet data were developed for only a portion of the watersheds of the estuaries in this assessment. For most of the remaining watersheds, a model was used to provide deposition estimates. The extended (Regional Acid Deposition Model (RADM)) was used for this task because it has a number of critical capabilities that make it a reasonable choice. The objective was to combine the data and model predictions together in a manner that would make the mix of estimates as consistent as possible. The basis for consistency was determined to be the CASTNet data. The model was used as a sophisticated, "calibrated" extrapolator/interpolator constrained to approximate closely the CASTNet deposition data.

The Extended RADM (ExtRADM) is an extension of the Regional Acid Deposition Model [Chang et al., 1987] developed under the National Acid Precipitation Program, NAPAP, to include particle physics relevant to N deposition from the Regional Particulate Model [Binkowski and Shankar, 1995]. The model was designed for regional analyses of acidic and nutrient deposition. It includes full oxidant chemistry to predict oxidized forms of N, wet deposition to account for the full budget of the air pollutants involved in N deposition, particle physics of the interaction among sulfuric acids/sulfate, HNO_3, and NH_3, and uses national emissions inventories of sulfur, nitrogen oxides, volatile organic compounds, and NH_3 emissions to predict both oxidized and reduced forms of N. It predicts the partitioning of HNO_3 into particulate NO_3^- and HNO_3 gas and the partitioning of NH_3 into particulate NH_4^+ and NH_3 gas which is critical to the estimation of dry deposition. The meteorological fields used to drive the ExtRADM are from version 4 of the NCAR/Penn State Mesoscale Meteorological Model (MM4), where the MM4 is run using 4-dimensional data assimilation (4DDA) to produce the most accurate recreation of past weather for the chemical transport model [Seaman and Stauffer, 1989].

Since ExtRADM predicts chemistry on a synoptic or episodic time scale, an aggregation technique is used to calculate annual estimates of acidic deposition. The development and evaluation of the aggregation simulation set is described by Brook et al. [1995a; 1995b]. Meteorological cases with similar 850-mb wind flow patterns were grouped by applying cluster analysis to classify the wind flow patterns from 1982 to 1985, resulting in 19 sampling groups, or strata. Meteorological cases were randomly selected from each stratum; the number selected was based on the number of wind flow patterns in that stratum relative to the number of patterns in each of the other strata, to approximate proportionate sampling. A total of thirty cases are used in the current aggregation approach. Each case is run for five days, using a separate initial condition for each season specific to the emissions scenario being run. Only the last three days are used to avoid the influence of initial conditions. The aggregation method recreates average meteorology representative of at least a decade. Deposition results for the 30 selected 3-day cases are weighted according to the strata sampling frequencies to form annual averages. Application of the aggregation technique is described in Dennis et al. [1990].

The1990 base inventory for all emissions, except NH_3, was used as input to the model. Estimates of NH_3 emissions are much more uncertain than emissions estimates for sulfur and nitrogen oxides. Since NH_3 and particulate NO_3^- concentrations are believed to be closely associated, the 1985 NAPAP NH_3 emissions inventory was adjusted upwards on a seasonal basis in a simple model inversion exercise until model predictions of the seasonal and annual patterns of the ratio of particulate to total nitrate (*i.e.*, $NO_3^-/(NO_3^- + HNO_3)$) across the eastern United States were similar to the CASTNet data. Essentially, the model physics was deemed to be more reliable than the emissions inventory.

Annual concentrations and deposition computed by the ExtRADM for each model grid cell (80 x 80 km) containing one of the CASTNet sites from Table 3, and data from the sites were compared for the same time periods to develop a calibration between the Extended RADM and CASTNet. The development of "calibrated" predictions was broken into three parts: first, a calibrated prediction of total nitrate ($NO_3^- + HNO_3$) dry deposition; second, a calibrated prediction of dry NH_4^+ deposition; and third, a model prediction of total dry NH_3 plus NH_4^+ deposition. This approach was found to maximize the ability to constrain the model results to be as CASTNet-like as possible. For brevity, we use the term "prediction" for "calibrated predictions." For the total nitrate predictions (first part), a multi-step procedure was used to constrain the ExtRADM to CASTNet-like conditions. First, ambient concentrations of total nitrate ($NO_3^- + HNO_3$) from the ExtRADM grid cells were associated with ambient concentrations of total nitrate for the CASTNet sites. Then CASTNet ambient concentrations of total nitrate were associated with CASTNet deposition of total nitrate. Total nitrate concentration was the association of choice because it avoided the added uncertainty introduced by deposition velocity differences due to differences in the ExtRADM and CASTNet deposition velocity models and meteorology. Linear relationships were found to represent very well the associations in each of the individual steps. For ambient concentration of total nitrate, there was good correlation between ExtRADM and

TABLE 4. Dry N Deposition (oxidized plus reduced, excluding NH_3) predicted by the constrained ExtRADM and estimated from CASTNet data.

Watershed	ExtRADM Predicted, kg N ha^{-1}	CASTNet Based, kg N ha^{-1}	Percent Difference %
Casco Bay	1.45	0.92	58
Great Bay	2.06	1.8	15
Merrimack River	1.99	2.63	-24
Massachusetts Bay	3.3	3.86	-15
Waquoit Bay	3.19	3.86	-17
Buzzards Bay	3.19	3.86	-17
Narragansett Bay	3.42	3.86	-11
Long Island Sound	2.45	3.19	-23
Gardiners Bay	4.54	3.86	18
Hudson River/Raritan Bay	2.34	3.98	-41
Barnegat Bay	5.3	4.13	28
New Jersey Inland Bays	5.07	4.13	23
Delaware Bay	3.89	4.13	-6
Delaware Inland Bays	3.84	4.18	-8
Chesapeake Bay	3.46	4.03	-14
Maryland Inland Bays	3.65	4.18	-13
Pamlico Sound	3.19	2.37	35
Winyah Bay	2.85	2.37	20
Altamaha River	3	2.68	12
Apalachee Bay	2.12	1.54	38
Apalachicola Bay	2.75	2.11	30
Mobile Bay	3.18	2.35	35

CASTNet (*i.e.*, $R^2 = 0.57$). CASTNet total nitrate ambient concentration and CASTNet total nitrate dry deposition were similarly correlated (*i.e.*, $R^2 = 0.56$). The association between ExtRADM NH_4^+ dry deposition results and CASTNet results also had a similar correlation (*i.e.*, $R^2 = 0.57$) and was used to develop a predictive relationship for NH_4^+ dry deposition.

Comparison of ExtRADM and CASTNet Total Dry N Deposition (excluding NH_3)

The predictions of the constrained ExtRADM for the deposition by watershed were compared with the CASTNet estimates for the same watersheds to develop a sense of the uncertainty in the ExtRADM "predictions." These results are given in Table 4, along with the percent difference. Ninety percent of the ExtRADM predictions are within ± 40% of the CASTNet estimates, and these differences are relatively equally distributed around

0%. Comparison of ExtRADM predictions with CASTNet results, in essence compares a modeled regional (80 x 80 km) estimate with a regional estimate based on the average of measurements at specific sites. Although is it likely that the ExtRADM predictions have greater uncertainty than the averaged CASTNet measurements, the uncertainty from the model predictions appears consistent with the factor of two estimates for regional N deposition described earlier for the CASTNet measurements.

Estimates of Dry NH_3 and NH_4^+ Deposition

ExtRADM results can also be used to provide estimates of the additional dry deposition associated with NH_3 that is not included in the above results. ExtRADM results for the ratio of gaseous NH_3 to gaseous plus aerosol NH_X concentrations (*i.e.*, $NH_3/(NH_3 + NH_4^+)$) range from 9% to 47% for the grid cells containing CASTNet sites, have a median of 21%, and upper and lower quartiles of 31% and 15%, respectively. The ExtRADM NH_3 ratios for the watersheds range from 8% to 83% with a median of 26% and upper and lower quartiles of 32% and 20%, respectively. Independent measurements of both NH_3 and NH_4^+ at several sites in the Eastern U.S. suggest that NH_3 ratios range between 0.1 and 0.3 [Sickles et al., 1990; 1999; Lanford, 1992] and suggest the model results are of the correct magnitude. The central linear tendency of the ExtRADM results is that dry NH_X deposition is a factor of two larger than the aerosol NH_4^+ deposition. Total dry N deposition including NH_3, therefore, is estimated by doubling the aerosol NH_4^+ dry deposition and summing with NO dry deposition. We estimate of a factor of two for the uncertainty associated with NH_X deposition.

Total N Deposition to Watershed

Estimates of reduced and oxidized wet N deposition were obtained from NADP (see Table 1) and combined with estimates of reduced and oxidized dry N deposition, taken from Table 3 to arrive at the estimates of total (wet plus dry) N deposition shown in Table 5. The wet NH_4^+ data were adjusted upward by 15% to account for losses during sample collection, and the sum of wet NO_3^- and wet NH_4^+ was adjusted upward by 25% to account for DON. There is a factor of two uncertainty associated with the wet estimates.

In those cases where monitoring data is available, dry N deposition was estimated and is summarized in Table 5. The uncertainty associated with these dry deposition estimates is likely to be on the order of a factor of two. In one test case for a rural site in Harvard Forest near Petersham, MA, we found agreement within 10% between N deposition estimated using the current approach and recently reported direct deposition measurements [Munger et al., 1998; Munger, 1998]. In 17 of the remaining 19 cases, ExtRADM was used to estimate total dry N deposition using the approach described earlier. The uncertainty associated with these modeled dry deposition estimates is likely to be similar to those of the CASTNet results described earlier. In the two cases of San Francisco Bay and Puget Sound, where monitoring data were unavailable and the model could not be applied, very crude estimates were made. In these cases, the wet nitrogen deposition data from Table 1 for the watersheds associated with San Francisco Bay and Puget Sound were corrected as noted, arriving at 2.36 and 1.76 kg N ha^{-1} yr^{1}, respectively. Using NADP wet N deposition, corrected as above, and the corresponding CASTNet and ExtRADM dry deposition estimates for the 40 eastern watersheds, a linear regression between wet and dry N deposition was determined (*i.e.*, Dry

TABLE 5. Nitrogen Deposition to Watersheds of Interest Expressed in kg N ha^{-1} yr^{-1}.

Watershed	Wet NH$_4$[1]	Corrected Wet NH$_4$[2]	Wet NO$_3$[1]	Wet Total N[3]	Dry NH$_4$[4]	Dry NO$_3$[4]	Dry Total N[5]	Wet+Dry Total N[6]
Casco Bay	1.21	1.3915	2.22	4.51	0.13	0.79	1.05	5.56
Great Bay	1.18	1.357	2.46	4.77	0.19	1.61	1.99	6.76
Merrimack River	1.76	2.02	3.18	6.5	0.27	2.36	2.9	9.4
MA Bay	1.14	1.311	2.69	5	0.45	3.41	4.31	9.31
Waquoit Bay	1.16	1.334	2.74	5.09	0.45	3.41	4.31	9.4
Buzzards Bay	1.16	1.334	2.74	5.09	0.45	3.41	4.31	9.4
Narragansett Bay	1.46	1.679	3.3	6.22	0.45	3.41	4.31	10.53
Gardiners Bay	1.83	2.1045	4.39	8.12	0.45	3.41	4.31	12.43
Long Island Sound	1.83	2.1045	3.8	7.38	0.33	2.86	3.52	10.9
Hudson/Raritan	1.87	2.1505	3.89	7.55	0.42	3.56	4.4	11.95
Barnegat Bay	2.03	2.3345	3.79	7.66	0.51	3.62	4.64	12.3
NJ Inland Bay	2.03	2.3345	3.79	7.66	0.51	3.62	4.64	12.3
DE. Bay	1.99	2.2885	4.12	8.01	0.51	3.62	4.64	12.65
DE Inland Bays	1.93	2.2195	3.24	6.82	0.56	3.62	4.74	11.56
MD Inland Bays	1.93	2.2195	3.24	6.82	0.56	3.62	4.74	11.56
Chesapeake Bay	2.3	2.645	4.03	8.34	0.54	3.47	4.57	12.91
Pamlico Sound	1.94	2.231	2.4	5.79	0.39	1.98	2.76	8.55
Winyah Bay	1.62	1.863	2.39	5.32	0.39	1.98	2.76	8.08
Charleston Harbor	1	1.15	1.94	3.86	0.26[7]	2.02[7]	2.54[7]	6.4
St. Helena Sound	1	1.15	1.94	3.86	0.35[7]	2.39[7]	3.09[7]	6.95
Ogeechee Coastal	1.48	1.7	1.88	4.48	0.27[7]	2.99[7]	3.53[7]	8.01
St.Cath./Sapelo S.	1.48	1.7	1.88	4.48	0.22[7]	2.19[7]	2.63[7]	7.11
Altamaha River	1.5	1.725	1.92	4.56	0.44	2.24	3.12	7.68
Indian River	0.97	1.12	1.79	3.63	0.19[7]	1.79[7]	2.17[7]	5.8
Charlotte Harbor	1.08	1.24	2.15	4.24	0.22[7]	1.70[7]	2.14[7]	6.38
Sarasota Bay	1.08	1.24	2.15	4.24	0.20[7]	2.19[7]	2.59[7]	6.83
Tampa Bay	1.08[1]	1.24[2]	2.15[1]	4.24[3]	na	na	3.87[8]	8.11
Apalachee Bay	1.17	1.3455	1.76	3.88	0.24	1.3	1.78	5.66
Apalachicola Bay	1.3	1.495	1.91	4.26	0.34	1.77	2.45	6.71
Mobile Bay	1.81	2.0815	2.24	5.4	0.44	1.91	2.79	8.19
West Miss. Sound	2.46	2.83	2.7	6.91	0.48[7]	2.58[7]	3.54[7]	10.45
Barataria Bay	2.45	2.82	2.63	6.81	0.49[7]	5.71[7]	6.69[7]	13.5
Terrebonne Bay	2.45	2.82	2.63	6.81	0.36[7]	5.11[7]	5.83[7]	12.64
Calcasieu Lake	2.43	2.79	2.55	6.68	0.40[7]	3.17[7]	3.97[7]	10.65
Sabine Lake	1.79	2.06	2.12	5.22	0.37[7]	2.72[7]	3.46[7]	8.68

TABLE 5. Continued.

Watershed	Wet NH_4[1]	Corrected Wet NH_4[2]	Wet NO_3[1]	Wet Total N[3]	Dry NH_4[4]	Dry NO_3[4]	Dry Total N[5]	Wet+Dry Total N[6]
Galveston Bay	1.82	2.09	1.88	4.97	0.33[7]	3.28[7]	3.94[7]	8.91
Matagorda Bay	1.19	1.37	1.22	3.24	0.25[7]	1.78[7]	2.28[7]	5.51
Corpus Christi Bay	1.31	1.51	1.08	3.23	0.18[7]	1.30[7]	1.66[7]	4.89
Upper Lag. Madre	1.56	1.79	1.24	3.79	0.14[7]	1.25[7]	1.53[7]	5.32
Lower Lag. Madre	1.56	1.79	1.24	3.79	0.10[7]	1.03[7]	1.23[7]	5.02
San Francisco Bay	0.98	1.13	0.76	2.36	na	na	1.48[9]	3.84
Puget Sound	0.44	0.51	0.9	1.76	na	na	1.11[9]	2.87

[1] NADP data (see Table 1).
[2] Column 2 is multiplied by 1.15 to correct for NH_4^+ losses.
[3] Sum of columns 3 plus 4, multiplied by 1.25 to correct for unmeasured DON.
[4] From CASTNet monitoring data in Table 3, except where footnoted.
[5] Column 6, multiplied by a factor of 2 to account for NH_4 deposition, plus column 7.
[6] Sum of columns 5 plus 8.
[7] Model estimate from ExtRADM (see text).
[8] See Pribble and Janicki [1998].
[9] Estimated from a linear regression of wet and dry deposition (see text)

Total N = m[Wet Total] + b, where m= 0.61, b=0.027, and r^2=0.54). The resulting predictions yield estimates of wet and dry total N and 95% CI of 3.84 [3.30-4.38] and 2.87 [2.25-3.49] kg N ha^{-1} yr^{-1}. There is more uncertainty associated with them than with the remaining 40 estimates. They were estimated with no direct data or modeling input; they were estimated from a relationship that may apply only for eastern sites; even if the relationship applies in the west, it fails to account for 46% of the variability between wet and dry deposition at eastern sites; and the very nature of the upwind air for both of the western watersheds is much different from those in the east (*i.e.*, marine, very clean, and probably subject to an appreciable influence by fog).

Total deposition of nitrogen (wet + dry) to the various watersheds (Table 5) ranged from around 3 kg N/ha/year for the Western U.S. locations to nearly 14 kg N ha^{-1} yr^{-1} for bay locations in Louisiana. Bay locations from North Carolina to Long Island Sound also had relatively high atmospheric inputs with total deposition rates exceeding 10 kg N ha^{-1} yr^{-1}. On average, dry deposition comprised about 40% of the total nitrogen deposition to the watersheds (Table 5), and provided no less than about 20%. Of the nitrogen deposited to the watershed, approximately 70% was delivered in the form of NO_3^-, with values ranging from 50% to 70%. Loadings to the watershed were much more variable since watershed areas vary in size. For example, total loading to the Sarasota Bay watershed was less than one million kg N per year while Chesapeake Bay, which has largest watershed area, is estimated to have about 230 million kg N deposited via atmospheric processes (Table 6).

Total Deposition to Water

Total deposition to the estuaries and water bodies (on a per unit area basis) was less than corresponding deposition rates to the associated watershed since the dry deposition is

TABLE 6. Total nitrogen deposition and loadings to watersheds and water bodies.

	Estuary	Deposition to the Watershed kg ha^{-1} yr^{-1}	Deposition to the Water Surface kg ha^{-1} yr^{-1}	Total Load to the Watershed 10^6 kg yr^{-1}	Total Load to the Water Surface 10^6 kg yr^{-1}
1	Casco Bay	5.56	4.74	1.42	0.20
2	Great Bay	6.76	5.22	1.73	0.02
3	Merrimack River	9.4	7.15	12.22	0.01
4	Massachusetts Bay	9.31	5.96	2.27	0.57
5	Waquoit Bay	9.4	6.06	0.04	0.00
6	Buzzards Bay	9.4	6.06	1.18	0.39
7	Narragansett Bay	10.53	7.19	4.53	0.30
8	Gardiners Bay	12.43	9.08	0.88	0.47
9	Long Island Sound	10.9	8.18	45.41	2.70
10	Hudson River/ Raritan Bay	11.95	8.55	49.71	0.68
11	Barnegat Bay	12.3	8.69	1.72	0.16
12	New Jersey Inland Bays	12.3	8.69	4.22	0.24
13	Delaware Bay	12.65	9.04	42.07	1.87
14	Delaware Inland Bays	11.56	7.87	0.65	0.06
15	Maryland Inland Bays	11.56	7.87	0.33	0.04
16	Chesapeake Bay	12.91	9.35	215.34	10.53
17	Pamlico Sound	8.55	6.38	22.85	3.57
18	Winyah Bay	8.08	5.91	37.94	0.05
19	Charleston Harbor	6.4	4.43	26.32	0.04
20	St. Helena Sound	6.95	4.55	8.52	0.09
21	Ogeechee Coastal	8.01	5.29	0.19	0.35
22	St. Catherines/ Sapelo Sounds	7.11	5.08	1.60	0.10
23	Altamaha River	7.68	5.23	28.39	0.02
24	Indian River	5.8	4.13	1.79	0.36
25	Charlotte Harbor	6.38	4.72	7.46	0.24
26	Sarasota Bay	6.83	4.84	0.45	0.06
27	Tampa Bay	8.11	5.13	4.63	0.46
28	Apalachee Bay	5.66	4.27	8.09	0.76
29	Apalachicola Bay	6.71	4.78	35.04	0.28
30	Mobile Bay	8.19	5.99	93.71	0.65
31	West Mississippi Sound	10.45	7.66	40.14	3.22
32	Barataria Bay	13.5	8.36	6.46	0.71
33	Terrebonne/Timbalier Bays	12.64	8.12	3.33	1.02

TABLE 6. Continued.

Estuary	Deposition to the Watershed kg ha^{-1} yr^{-1}	Deposition to the Water Surface kg ha^{-1} yr^{-1}	Total Load to the Watershed 10^6 kg yr^{-1}	Total Load to the Water Surface 10^6 kg yr^{-1}
34 Calcasieu Lake	10.65	7.57	11.57	0.20
35 Sabine Lake	8.68	6	46.59	0.16
36 Galveston Bay	8.91	5.87	55.09	0.85
37 Matagorda Bay	5.51	3.74	67.09	0.42
38 Corpus Christi Bay	4.89	3.6	21.77	0.21
39 Upper Laguna Madre	5.32	4.14	5.63	0.34
40 Lower Laguna Madre	5.02	4.07	6.61	0.53
41 San Francisco Bay	3.84	2.69	45.77	0.36
42 Puget Sound	2.87	2.01	8.05	0.53

expected to be less. Few if any direct measurements of nitrogen dry deposition to water bodies have been made. However, since the deposition process for gas and aerosol phase nitrate and ammonia is largely dominated by turbulence (surface uptake resistances are small), to a first order, the overall deposition rate to the water should follow the similar reductions in the friction velocity (u_*) from watersheds to water bodies. Water surfaces are generally much smoother aerodynamically than their terrestrial counterparts. Typical drag coefficients (u_*^2/u^2) for forest ecosystems are on the order of 0.016 [Meyers and Baldocchi, 1991] while measurements over coastal regions are on the order of 0.0018 [Crawford et al., 1993]. This translates into a deposition velocity for gas phase nitrate that is 3 to 6 times less over water than over surrounding aerodynamically rougher forested watersheds. As an approximation to estimate the dry input of nitrogen to the water bodies, estimates from the corresponding watersheds were reduced by a factor of four. This resulted in a reduction of total deposition that ranged from 15% to 40% from what was deposited to the corresponding watershed. The average reduction in the deposition from the watershed to the water body was about 30%. Total loading to the water was much less that observed for the watershed mainly because of the reduction in area. Total nitrogen loadings ranged from around 0.01 million kg N yr^{-1} to nearly 10 million kg N yr^1 for Chesapeake bay which is the largest water body considered in this study. Since the estimated dry deposition to the water bodies (per unit area basis) was only 25% of the input to the watershed, nitrogen deposition directly to the water via dry deposition processes constitutes roughly 10% of the total nitrogen input per unit area basis.

Annual and Temporal Variability in Wet Deposition

Temporal trends in annual wet deposition of ammonium and nitrate were investigated. Analysis of variance tests were performed for each of the sites to determine whether deposition was significantly changing over time. If a p-value for the trendline was less than 0.05, then the trend was considered significant. Most of the sites in this analyses showed not

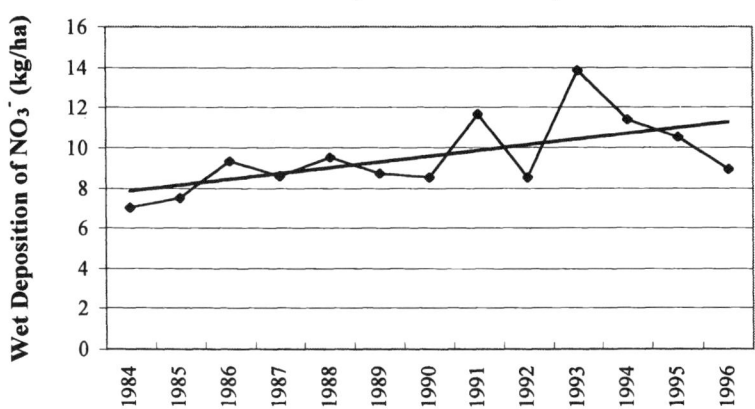
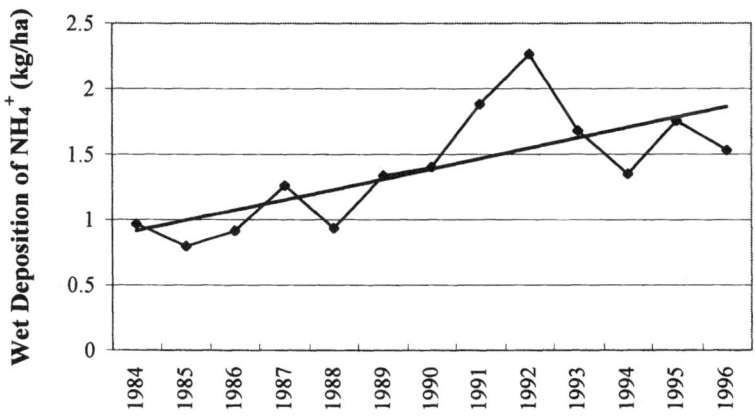

Figure 1. Annual wet deposition of nitrate and ammonium at the Verna Well Field site (FL41).

trends, which is consistent with the assumptions used in this study. A few sites showed increasing deposition with time. AL99, Sand Mountain Experimental Station, which is located in the Mobile Bay watershed, showed both significantly increasing ammonium and nitrate annual wet depositions (Figures 1). FL41, Verna Well Field, located within or near the Charlottes Harbor, Sarasota Bay, and Tampa Bay watersheds, also showed significantly

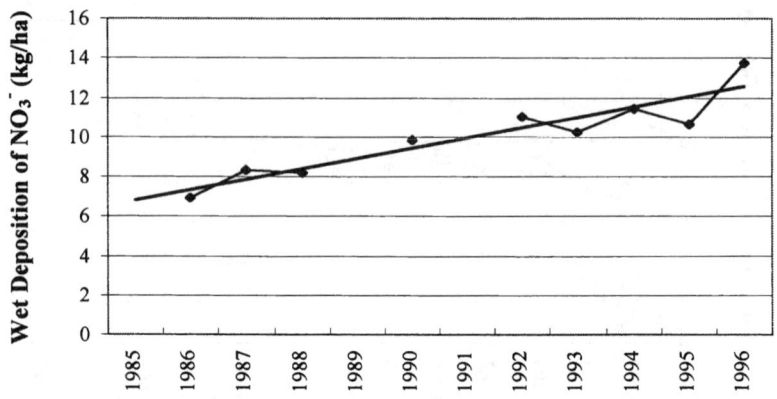

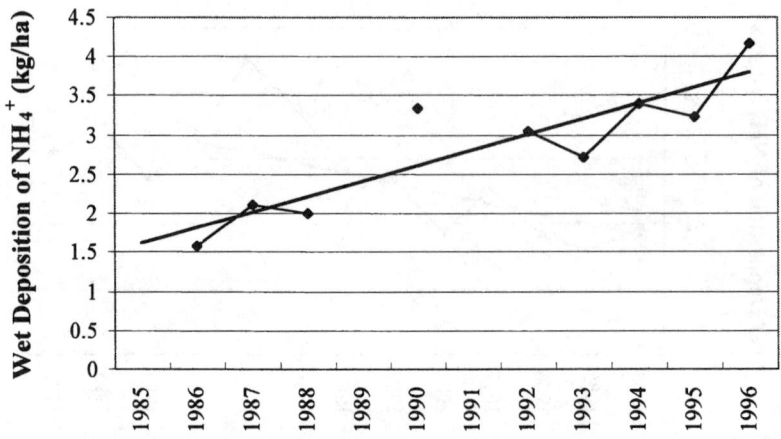

Figure 2. Annual wet deposition of nitrate and ammonium at the Sand Mountain Experiment Station site (AL99).

increasing ammonium and nitrate annual wet depositions (Figures 2), as did FL99, Kennedy Space Center, in the Indian River watershed (Figures 3). MD03, White Rock, (Chesapeake Bay Watershed), NC03, Lewiston, (Pamlico Sound watershed), NC35, Clinton Crops Research Station (Neuse River watershed), and SC06, Santee National Wildlife Refuge

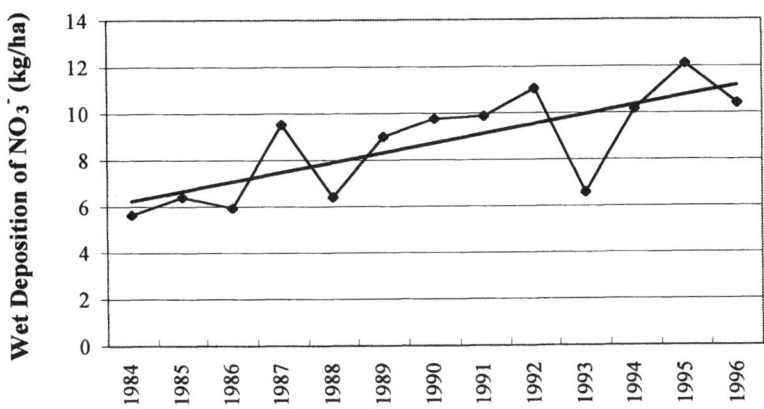
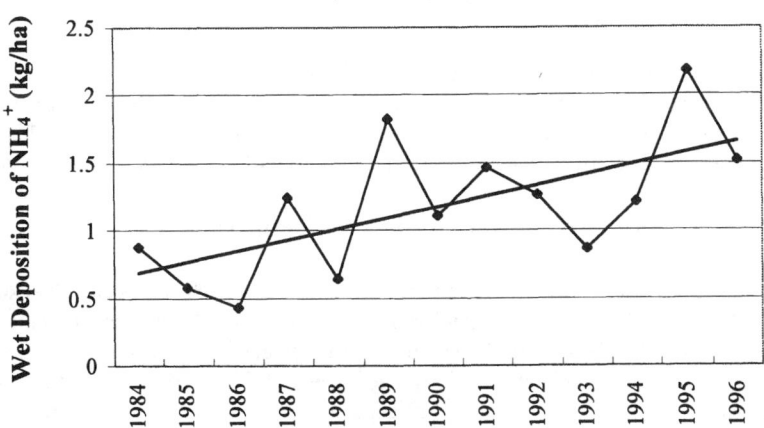

Figure 3. Annual wet deposition of nitrate and ammonium at the Kennedy Space Center site (FL99).

watershed (Ace Basin/St. Helena, North Inlet/Wynah Bay, and Charleston Harbor watersheds), all showed significantly increasing ammonium depositions. Other sites also showed increasing or decreasing trends, although not significant ones. Most notable was FL11, Everglades National Park, which had p-values of 0.16 for increasing ammonium and

TABLE 7. Variability of Dry Deposition for Nitrogenous Species at Single and Multiple CASTNet Monitoring Sites.

Deposition	Inter-annual Variability (CV, %) of Annual Means 1989-1995[1]						Regional Variability
	ASH 135[2]	WSP 144[2]	ARE 128[2]	BEL 116[2]	SUM 156[2]	NE US[3]	NE US[4]
HNO_3	15	7	13	9	12	6	42
NO_3^-	12	13	16	16	12	11	63
$HNO_3 + NO_3^-$	14	7	12	8	12	3	45
NH_4^+	10	11	5	11	11	7	39
Total N	13	7	11	8	11	5	40

[1] Variability expressed as coefficient of variation (CV) in percent, where $CV = 100 SD/\bar{x}$.
[2] Computed using the mean of the six annual average deposition values at this site (n = 6).
[3] Computed using the mean of the six annual average deposition values for 10 sites in the Northeastern U.S. (n = 6) [Sickles, 1999].
[4] Computed using the mean of the site-specific six-year average deposition values for the 10 sites in the Northeastern U.S. (n = 10) [Sickles, 1999].

0.09 for increasing nitrate and is also located in Florida where deposition is also increasing at FL41, Verna Well Field and FL99, Kennedy Space Center. Although some of the increasing trends were significant at a few sights, the uncertainties in these trends lie within the estimated uncertainty for the total deposition to each of the watersheds/water bodies.

Annual and Temporal Variability in Dry Deposition

Nationally, NO_x emissions have been relatively constant (± 1.4%) between 1989 and 1995 [US E.P.A., 1998]. A recent trends analysis of total airborne oxidized N concentration (i.e., the sum of HNO_3 and NO_3^-) has shown only small changes to have occurred at 34 eastern CASTNet sites between 1989 and 1995 [Holland et al., 1999]. The median change over this period was -8%. It is likely that dry N deposition to the watersheds of interest also experienced similar small changes.

Year-to-year (interannual) variability is fairly small for dry deposition of N species. As seen in Table 7, interannual variability (expressed as CV's) for five different CASTNet sites pertinent to the current study ranges between 7 and 16%. This range narrows slightly to between 3 and 11% when multiple sites in a region are aggregated [Sickles, 1999]. In the current study; however, spatial variability is more important than the temporal variability. As an example, the variability of six-year means, determined across the ten sites considered in the Northeastern U.S. ranges between 40 and 60% [Sickles, 1999]. Coefficients of variation for dry deposition computed for the 11 areas in Table 3 with multiple CASTNet sites range between 15 and 89% for nitrate and between 11 and 56% for NH_4^+, with respective median values of 30 and 40%. These results are consistent with earlier estimates of a factor of two uncertainty in current estimates of annual dry atmospheric N deposition for watersheds under consideration.

Summary and Recommendation

Existing wet and dry deposition databases were used to estimate the total N deposition to various coastal watersheds in the U.S. to evaluate the relative importance of atmospherically delivered N on the these watersheds. Since extrapolations were made on the existing data to estimate the N input, factors of uncertainty of at least a factor of two are placed on these estimates, especially for dry deposition. To reduce the uncertainty of these estimates, several recommendations are suggested. Monitoring systems need to be located within the watersheds of interest. At least two sites (which contain wet and dry monitoring gear) are necessary to evaluate the uncertainty with respect to spatial sampling. For wet deposition, the input of organics appears to play a significant role in the wet deposition input. Care must be taken to either provide daily sampling, or samples must be preserved to maintain sample integrity. For dry deposition monitoring in coastal regions, current sampling protocols may not be adequate to provide reliable concentration data in a marine environment where significant amounts of aerosol sea salt exist. Although aerosol nitrate comprised only a small fraction of the total dry input, coarse sea salt aerosol (which will readily absorb HNO_3) could play an important role. Sampling methodologies may also need to be evaluated for operations in marine environments to determine sampling artifacts.

References

Brook, J.R.,F. Di-Giovanni, S. Cakmak, T.P. Meyers. 1997. Estimation of dry deposition velocity using inferential models and site-specific meteorology: uncertainty due to siting of meteorological towers. *Atmospheric Environment*, 31: 3911-3919.

Butler, T.J. and G.E. Likens. Weekly and daily precipitation chemistry network comparisons in the eastern U.S.: NADP/NTN vs. MAP3S/AIRMoN. *Atmospheric Environment* 32 (21): 3749-3765. 1998.

Clarke, J.F., E.S. Edgerton, B.E. Martin. 1997. Dry deposition calculations for the clean air status and trends network. *Atmospheric Environment*, 31: 3667-3678.

Cooter, E.J., D. Schwede. 1999. Sensitivity of the NOAA multilayer model to instrument error and parameterization uncertainty. *Journal of Geophysical Research*, submitted.

Cornell, S.E., T.D. Jickells, and C.A. Thornton. Urea in Rainwater and Atmospheric Aerosol. *Atmospheric Environment* 32 (11): 1903-1910. 1998.

Correll, D. L., T. E. Jordan and D. E. Weller. Long-term nitrogen deposition on the Rhode River watershed. pp. 508-518. In P. Hill and S. Nelson (eds.) *Toward a Sustainable Watershed: The Chesapeake Experiment, Proceedings of the 1994 Chesapeake Research Conference, Chesapeake Research Consortium Publication No. 149.* 1994.

Crawford, T., R. T. McMillen, T. P. Meyers, and B. B. Hicks, 1993. Spatial and temporal variability of heat, water vapor, carbon dioxide exchange and momentum air-sea exchange in a coastal environment, *J. Geophysical Research*, 98, 12,869-12,880.

Gorzelska, K., J.N. Galloway, K. Watterson, and W.C. Keene. Water-Soluble Primary Amine Compounds in Rural Continental Precipitation. *Atmospheric Environment* 26A (6): 1005-1018. 1992.

Gorzelska, K., J.R. Scudlark, and W.C. Keene. Dissolved organic nitrogen in the atmospheric environment. In: *Atmospheric Deposition of Contaminants to the Great Lakes and Coastal Waters*. Ch. 20. Pp. 379-392. Society of Environmental Toxicology and Chemistry. Pensacola, FL. 1997.

Matt, D. R. and T. P. Meyers, 1991. On the use of the inferential technique to estimate dry deposition of SO_2, *Atmospheric Environment*, 27, 493-501.

Meyers, T., and D. D. Baldocchi, 1993; The budgets of turbulent kinetic energy and Reynolds stress within and above a deciduous forest, *Agricultural and Forest Meteorology*, 53, 207-222.

Meyers, T.P., P. Finkelstein, J. Clarke, T.G. Ellestad, P.F. Sims. 1998. A multi-layer model for inferring dry deposition using standard meteorological measurements. *Journal of Geophysical Research*, 103: 22645-22661.

Meyers, T. P. And T. S. Yuen, 1987. An assessment of averaging strategies associated with day/night sampling of dry deposition fluxes of SO_2 and O_3, *J. Geophysical Research*, 92, 6705-6712.

Munger, J.W., S-M. Fan, P.S. Bakwin, M.L. Goulden, A.H. Goldstein, A.S. Colman, S.C. Wofsy. 1998. Regional budgets for nitrogen oxides from continental sources: variations of rates for oxidation and deposition with season and distance from source regions. *Journal of Geophysical Research*, 103: 8355-8368.

Munger, J.W. 1998. Personal communication, Harvard University, Boston, MA, December 21, 1998.

Pio, C.A., D.A. Lopez. 1998. Chlorine loss from marine aerosol in a coastal atmosphere. *Journal of Geophysical Research*, 103: 25263-25272.

Peierls, B.L. and H.W. Paerl. Bioavailability of atmospheric organic nitrogen deposition to coastal phytoplankton. Limnology and Oceanography 42 (8): 1819-1823. 1997.

Pribble, J.R. and A.J. Janicki. 1998. "Atmospheric deposition contributions to nitrogen and phosphorus loadings in Tampa Bay: Intensive wet and dry deposition data collection and analysis," PBS&J Report prepared for the Tampa Bay National Estuary Program, St. Petersburg, FL.

Scudlark, J.R., K.M. Russell, J.N. Galloway, T.M. Church, and W.C. Keene. Organic nitrogen in precipitation at the mid-Atlantic U.S. coast: Methods evaluation and preliminary measurements. *Atmospheric Environment* 32 (10): 1719-1728. 1998.

Sickles, J.E., II. 1987. "Sampling and analytical methods development for dry deposition monitoring," Research Triangle Institute Report RTI/2823/00-15F, 215 pp.

Sickles, J.E., II, L.L. Hodson, W.A. McClenny, R.J. Paur, T.G. Ellestad, J.D. Mulik, K.G. Anlauf, H.A. Wiebe, G.I. Mackay, H.I. Schiff, D.K. Bubacz. 1990. Field comparison of methods for the measurement of gaseous and particulate contributors to acidic dry deposition, *Atmospheric Environment*, 24A: 155-165.

Sickles, J.E., II. 1999. Airborne concentrations of sulfur- and nitrogen-containing pollutants in the northeastern United States, *Journal of Air and Waste Management Association*, 49: 882-893.

Sickles, J.E., II, L.L. Hodson, L.M. Vorburger. 1999. Evaluation of the filter pack for long duration sampling of ambient air. *Atmospheric Environment*,33: 21187-2202.

Shon, Z.-H. Atmospheric Input of N to the Coastal Region of Southeastern Texas. M.S. Thesis. Texas A&M. 1994

4

Contribution of Atmospheric Deposition to the Total Nitrogen Loads to Thirty-four Estuaries on the Atlantic and Gulf Coasts of the United States

Mark S. Castro, Charles T. Driscoll, Thomas E. Jordan, William G. Reay, Walter R. Boynton, Sybil P. Seitzinger, Renée V. Styles and Jaye E. Cable

Abstract

In this chapter, we quantify the net anthropogenic nitrogen (N) inputs (N fixation by crops and pastures, N fertilization, atmospheric nitrate deposition, net food and net feed import) to 34 watershed/estuary systems on the Atlantic and Gulf coasts of the United States (U.S.), identify the dominant watershed N sources contributing to the total N loads to these 34 estuaries, and determine the contribution made by atmospheric N deposition to the total N loads to each estuary. We quantified the net anthropogenic N inputs to the watersheds of these 34 watershed/estuary systems using the approach of Jordan and Weller [1996]. We developed land-use (upland forests, urban and agricultural lands) specific approaches to estimate the N available for transport to the estuary from different watershed sources (runoff from agriculture, urban areas, upland forests, and point sources) above and below the fall line. These approaches were used to estimate the contribution made by atmospheric N to the total N runoff from these 3 different land-uses. We then compared the contribution made by these watershed N sources to the total N input to each estuary (including atmospheric N deposition to the surface of the estuary) to determine the relative contribution made by each N source to the total N input to the estuary.

Introduction

Nitrogen (N) has been identified as a key element controlling the productivity and eutrophication of coastal ecosystems [Ryther and Dunstan 1971; Nixon 1986; D'Elia et al. 1992; Fisher and Oppenheimer 1991]. Symptoms of eutrophication include increased frequency of harmful algal blooms, hypoxic and anoxic bottom waters, loss of sea grasses, and reduced fish stocks [Paerl 1988; Hallegraeff 1993; Valiela and Costa 1988; Valiela et al. 1990; Boynton et al. 1995]. Coastal eutrophication is typically related to human-induced increases of N inputs to coastal waters and is likely to persist or expand in the future as a consequence of increased population growth in coastal regions [Lee and Olsen 1985; Nixon 1995; Lapointe and Matzie 1996; Vitousek et al. 1997].

There are many sources of N to coastal waters, which makes identification of the N sources causing eutrophication difficult. Nitrogen enters coastal waters directly from point sources, nitrogen fixation and atmospheric deposition to the surface of the estuary, and indirectly as surface and subsurface runoff from the watershed. A significant portion of the N runoff from non-point sources may be derived from atmospheric deposition. The quantity of atmospheric N deposition supplied to coastal waters is not well established. However, in coastal ecosystems that have been characterized, atmospheric N deposition is a significant fraction of the total N input [Valigura et al. 1996; Jaworski et al. 1997; Paerl 1995; 1997]. For example, the contribution made by atmospheric N deposition to the total N load to the Chesapeake Bay and its tributaries ranges from 25 to 85% of the total N load [Fisher and Oppenheimer 1991; Hinga et al. 1991; Boynton et al. 1995; Jaworski et al. 1997]. This large uncertainty, however, demonstrates that considerable work is needed to better quantify the contribution made by atmospheric deposition to the total N loads of coastal ecosystems.

The objectives of the analyses described in this chapter were to: (1) quantify the net anthropogenic N inputs (N fixation, N fertilization, atmospheric nitrate deposition, net food and net feed import) to 34 watershed/estuary systems on the Atlantic and Gulf coasts of the U.S., (2) identify the dominant N sources contributing to the total N inputs to these estuaries, and (3) determine the contribution made by atmospheric N deposition to the total N inputs to each of the 34 estuaries.

Characteristics of the Watershed/Estuary Systems

We conducted our analyses on 34 watershed/estuary systems on the Atlantic and Gulf coasts of the U.S. (Figure 1). Watershed land-uses and population densities varied widely across systems (Table 1). Watersheds in the Northeast are largely forested (45 to 77%), except for Massachusetts Bay where forests represent only 22% of the total watershed area. With the exception of the Hudson River-Raritan Bay watershed, agriculture is a relatively small fraction of the total watershed area (≤11%) in the Northeast. Agriculture accounts for 23% of the total land area in the Hudson River-Raritan Bay watershed. In the Mid-Atlantic region, the importance of agricultural lands increases, largely at the expense of upland forests. Agricultural and upland forests account for 32 to 37% and 46 to 56% of the total watershed areas, respectively. In the Southeast region, upland forests (42 to 63% of the total watershed area) and agriculture (24 to 36% of total watershed area) dominate all watersheds except for St. Catherines - Sapelo, GA and Indian River, FL. The St. Catherine-Sapelo watershed has little agriculture (1 %), but considerable upland forests (63%) and wetlands (30%). Land-use in the Indian River watershed is dominated by urban areas (37%) and agriculture (27%).

Watersheds on the Gulf Coast have the most diverse characteristics of any region (Table 1). On the East Gulf Coast, 2 Florida watersheds, Charlotte Harbor and Tampa Bay, have considerable rangeland (30 and 21%, respectively) and agriculture (40 and

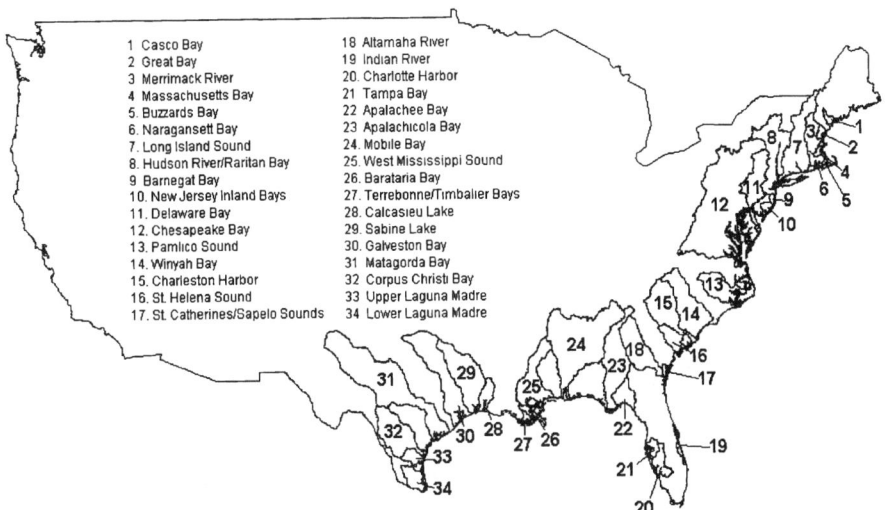

Figure 1. Locations and watershed boundaries of the 34-watershed/estuary systems examined in this chapter.

35%, respectively). In addition, 28% of the Tampa Bay watershed is urban. Four watersheds (Apalachee Bay, FL, Apalachicola Bay, FL, Mobile Bay, LA and West Miss. Sound, LA) are clearly dominated by agriculture (22 to 28%) and upland forests (59 to 70%). Two watersheds (Barataria Bay, LA and Terrebonne-Timbalier, LA) are dominated by wetlands, which account for approximately 75% of the total land area. On the West Gulf Coast, 2 watersheds, Calcasieu Lake and Sabine Lake, contain substantial upland forests (56% and 64% of total, respectively) and agriculture (26 and 27% of total, respectively). Galveston Bay watershed has significant agriculture (49%), moderate upland forests (27% of total) and urban areas (14% of total). Four watersheds have considerable (18 to 36%) agriculture and rangelands; rangeland accounts for 48% or more of the Matagorda, Corpus Christi, Upper and Lower Laguna Madre watersheds (Table 1).

There were strong regional trends in the population densities. Watersheds in the Northeast have the highest population densities that range from 0.9 to 8.1 persons ha^{-1}; 8 of 12 watersheds had population densities greater than 1.5 persons ha^{-1} (Table 1). In the Mid-Atlantic region, the Delaware Bay watershed has an elevated population density (2.3 persons ha^{-1}), while the other two Mid-Atlantic watersheds have population densities less than 1 persons ha^{-1}. With the exception of the Indian River watershed (1.2 persons ha^{-1}), watersheds in the Southeast region have relatively low population densities of less than 0.8 persons ha^{-1}. Similarly, with the exception of Tampa Bay (2.8 persons ha^{-1}), Barataria Bay (1.4 persons ha^{-1}) and Galveston Bay (1.2 persons ha^{-1}) watersheds, watersheds on the Gulf Coast have population densities less than 0.6 persons ha^{-1}.

Because water is a primary transport medium for N, knowledge of water budgets is essential to understanding contaminant movement within a given watershed. Water budgets are dependent upon numerous factors, such as climate, topography, geologic framework, vegetation, and human alterations. Due to these interacting factors, water

Table 1. Population density, total land area, and the percent of the total land area occupied by different land-uses.

	Population Density (persons ha⁻¹)	Total Watershed Area (km²)	Percent Urban	Percent Agriculture	Percent Rangeland	Percent Upland Forests	Percent Wetlands	Percent Barren
Northeast Region								
Casco Bay	0.9	2188	15.5	11.1	0.05	71.4	1.2	1.2
Great Bay	0.9	2491	18.7	8.1	0.10	70.1	2.6	2.6
Merrimack River	1.6	12458	16.5	5.4	0.01	76.5	1.1	1.1
Massachusetts Bay	8.1	2089	75.3	1.2	0	22.3	0.7	0.7
Buzzards Bay	3.2	1021	21.0	8.5	0.37	63.9	5.0	5.0
Narragansett Bay	3.7	4018	40.2	4.9	0.19	52.2	1.7	1.7
Long Island Sound	1.8	40774	16.0	11.2	0.08	70.8	1.6	1.6
Hudson River-Raritan Bay	3.6	36114	17.5	23.2	0.64	57.2	1.1	1.1
Barnegat Bay	2.8	1365	34.5	5.4	0	44.9	12.4	12.4
New Jersey Inland Bays	2.3	3215	16.1	9.2	0	49.1	23.8	23.8
Mid-Atlantic Region								
Delaware Bay	2.3	30792	16.6	32.0	0.27	46.6	3.4	3.4
Chesapeake Bay	0.8	160765	8.4	33.3	0.16	56.4	1.0	1.0
Pamlico Sound, NC	0.5	25090	7.1	36.6	0.09	45.5	10.3	10.3
Southeast Region								
Wynah Bay	0.5	43492	4.7	36.0	0.04	48.4	10.5	10.5
Charleston Harbor	0.7	41143	10.6	24.1	0.02	58.8	2.9	2.9
St. Helena Sound	0.3	11970	1.8	35.4	0.09	42.0	20.2	20.2
St. Catherines-Sapelo	0.2	1973	4.7	1.3	0	63.3	30.2	30.2
Altamaha River	0.5	36711	5.5	26.0	0.01	63.0	5.1	5.1
Indian River	1.2	2441	37.1	27.2	19.90	4.6	8.5	8.5
East Gulf Coast								
Charlotte Harbor	0.6	7610	6.5	40.0	29.50	5.8	12.9	12.9
Tampa Bay	2.8	5005	27.7	34.5	21.19	4.8	9.2	9.2
Apalachee Bay	0.2	14215	3.3	21.8	0.08	60.5	13.8	13.8
Apalachicola Bay	0.5	48216	6.8	28.0	0.04	59.0	5.7	5.7
Mobile Bay	0.3	112665	4.2	23.4	0.19	69.6	1.9	1.9
West Miss. Sound	0.5	38407	6.4	26.8	0.03	59.7	6.2	6.2
Barataria Bay	1.4	4156	8.8	16.9	0	0.3	73.7	73.7
Terrebonne-Timbalier Bays	0.3	2097	8.6	12.5	0	0.2	78.5	78.5
West Gulf Coast								
Calcasieu Lake	0.3	9820	3.2	25.8	0.97	55.7	12.5	12.5
Sabine Lake	0.2	51657	3.5	27.3	0.14	64.4	3.2	3.2
Galveston Bay	1.2	60322	14.3	49.0	7.36	26.9	1.7	0.4
Matagorda Bay	0.1	114981	3.8	29.5	47.56	18.5	0.3	0.1
Corpus Christi Bay	0.1	44256	1.6	17.5	64.32	16.2	0.1	2.3
Upper Laguna Madre	0.2	10395	1.1	26.6	65.22	3.6	2.3	3.8
Lower Laguna Madre	0.5	12916	5.1	36.1	52.06	5.1	1.0	0.5

budgets generally display a large degree of spatial and temporal variations. Estimates of watershed precipitation, potential evapotranspiration (PET), mean temperature, and riverine discharges are presented in Table 2. In general, there is a north to south increase in rainfall and percent of rainfall lost to evapotranspiration along the Atlantic coast. PET rates approximate theoretical rates assuming that water in the root zone is not limiting to the PET process (PET is a near maximum value based on Thornewaite's Equation). Rainfall within the Northeast region is on the order of 110 cm (range: 92 to 116 cm), compared to 115 (range: 104 to 135 cm) in the Mid-Atlantic, and 130 cm (range: 122 to

Table 2. Characteristics of the watershed/estuary systems.

	Watershed to Surface Water Area	Mean Annual Temperature[a] (°C)	Total Annual Precipitation[b] (cm)	Potential Evapotranspiration[c] (cm)	Annual Discharge[d] (cm)
Northeast Region					
Casco Bay	5	7.4	113	54	63
Great Bay	52	7.3	92	54	44
Merrimack River	805	10.7	105	61	57
Massachusetts Bay	2	10.7	105	61	47
Buzzards Bay	2	10.7	105	61	44
Narragansett Bay	10	10.2	116	70	49
Long Island Sound	12	10.9	106	70	59
Hudson River-Raritan Bay	45	12.6	112	70	62
Barnegat Bay	7	12.6	112	70	48
New Jersey Inland Bays	12	11.7	102	70	45
Mid-Atlantic Region					
Delaware Bay	15	12.3	104	75	61
Chesapeake Bay	14	14.1	109	66	41
Pamlico Sound, NC	4	16.7	135	81	38
Southeast Region					
Wynah Bay	491	24.2	131	89	40
Charleston Harbor	484	24.2	131	89	35
St. Helena Sound	59	24.7	125	89	30
St. Catherines-Sapelo	11	24.7	125	91	43
Altamaha River	939	22.4	128	91	38
Indian River	3	21.4	122	113	24
East Gulf Coast Region					
Charlotte Harbor	15	28.9	135	113	38
Tampa Bay	6	22.4	119	113	46
Apalachee Bay	8	26.7	167	113	33
Apalachicola Bay	81	25.7	162	113	50
Mobile Bay	104	25.2	162	90	52
West Mis. Sound	9	23.8	149	98	50
Barataria Bay	5	25.3	157	108	51
Terrebonne-Timbalier	2	25.3	157	108	51
West Gulf Coast Region					
Calcasieu Lake	38	19.9	139	108	50
Sabine Lake	195	21.7	123	108	30
Galveston Bay	41	23.5	107	109	22
Matagorda Bay	103	22.7	92	112	6
Corpus Christi Bay	78	22.0	76	128	2
Upper Laguna Madre	13	28.3	68	127	4
Lower Laguna Madre	10	28.3	68	126	4

[a] Mean Annual Temperatures are based on data from 1961 to 1992 for each watershed from NOAA National Climatic Center, Asheville, NC.
[b] Total Annual Precipitation was computed from the same NADP sites used in Chapter 3 to compute the wet deposition rates of N.
[c] Potential Evaportranspiration was calculated using Thronewaite's Method (Thornewaite and Mather 1957)
[d] Annual Discharge was provided by Richard Alexander (Chapter 7, this book).

131 cm) in the Southeast. Evapotranspiration comprises approximately 60, 65, and 75 percent of the rainfall within the Northeast, Mid-Atlantic and Southeast regions, respectively. Rainfall and PET are more variable on the East and West Gulf Coasts. Except for Tampa Bay (119 cm), watersheds in the East Gulf Coast receive a large amount of rain (range: 135 to 167 cm) with PET rates on the order of 70 percent (range: 55 to 95%) of the rainfall. There is a rather dramatic decrease in rainfall on the West Gulf Coast from Calcasieu Bay (139 cm) to the more southern Upper and Lower Laguna

Madre watersheds (68 cm). PET shows an inverse pattern, with values increasing from the more northern West Gulf Coast watersheds (Calcasieu PET ~78% of rainfall) to the southern Upper and Lower Laguna Madre watersheds where PET values can approach greater than 180% of rainfall rates; elevated PET rates can lead to water budget deficits.

The amount of water available for storage and runoff (both surface runoff and subsurface drainage) can be estimated by adjusting rainfall/snowfall rates for evapotranspiration losses. Much of the subsurface drainage remains within relatively shallow aquifer systems and flows towards local discharge sites such as springs, streams, lakes, wetlands, and nearby coastal waters. Groundwater discharge is an important component of the hydrologic cycle in many estuarine systems such as Cape Cod embayments [Valiela and Costa 1988; Giblin and Gaines 1990; Weiskel and Howes 1991], Rhode Island embayments [Lee and Olsen 1985], Great South Bay, NY [Bokuniewicz 1980], Chesapeake Bay [Simmons et al. 1989, U.S. EPA 1993], and Apalachicola Bay [Winchester and Fu 1992]. No attempt was made to separate total runoff into surface and subsurface components due to the complexity of the problem and lack of site specific studies. Watershed discharge rates, based on USGS stream discharge data and adjusted for watershed area, are presented in Table 2. As with precipitation and PET, watershed discharge rates displayed regional patterns. Watershed discharge rates decreased from the Northeast to the Southeast region. Watershed discharge on the East Gulf Coast was comparable to the Northeast and Mid-Atlantic regions, whereas the West Gulf Coast displayed more variable discharge rates. The northern portion of the West Gulf Coast was comparable to East Gulf Coast while the more southern watersheds exhibited the lowest discharge rates.

Net Anthropogenic Nitrogen Inputs to Watersheds

We calculated the net anthropogenic N inputs to the watersheds of each of the 34 watershed/estuary systems. The anthropogenic N sources in our analysis included: (1) application of fertilizer N to crops, (2) biotic N fixation by crops and pastures, (3) atmospheric deposition of nitrate, (4) net import of food for humans and (5) net import of feed for livestock (Figure 2) These N inputs represent allochthonous (external) sources of anthropogenic N to watersheds. Inputs to whole watersheds were calculated by summing inputs to the counties and portions of counties that are included in the watersheds. We briefly summarize our methods used to make these calculations in this section of our chapter, but details of these calculations are described in Jordan and Weller (1996). Data sources used for these calculations and others are identified in Table 3.

Nitrogen fertilization

Our estimates of the fertilizer N inputs to crops were based on the 1987 agricultural census county-level fertilizer sales [Alexander and Smith 1990]. We assumed that all fertilizer sold in a county was applied to agricultural lands in that county.

Nitrogen fixation

We estimated both non-symbiotic and symbiotic N fixation for crops, pastures, hay fields and upland forests. Rates of non-symbiotic N fixation were obtained from field studies reported in the literature (Table 4). Symbiotic N fixation was assumed to be a

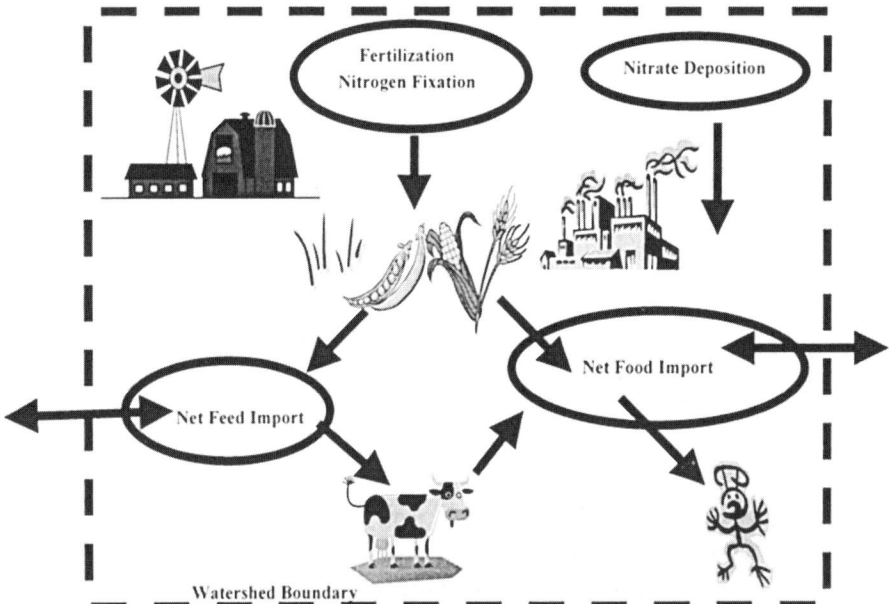

Figure 2. Conceptual diagram of the net anthropogenic N inputs to the watershed. The circles contain the 5 anthropogenic N inputs to the watersheds and the arrows represent the N fluxes within and between watersheds. The square box (dashed line) represents the watershed boundaries.

Table 3. Data sources for variables or characteristics used in this chapter.

Variable or Characteristic	Data Sources
Watershed and Estuary Characteristics	Pacheco 1999
Human Population	1990 Census
N Fertilizer	1987 Agricultural Census
Crop Harvest	1987 Agricultural Census
Atmospheric Deposition	Chapter 3, this book
Livestock Populations	Bureau of Census 1993
Nitrogen Discharged from Wastewater Treatment Plants	Pacheco 1999
1991 NASQAN N Fluxes to Fall Line	Pacheco 1999

function of the type of legume, crop N harvest, N in the unharvested portion of the crops, soil N availability, and fertilization rate [Meisinger and Randall 1991]. Crop harvest yields were obtained from the 1987 agricultural census data [Jordan and Weller 1996]. Nitrogen removed in crop harvest was estimated by multiplying crop harvest by the percent N in each crop [Meisinger and Randall 1991]. Soil N was estimated based on the assumption that the soils contained 2% organic matter and the legumes received the recommended fertilization rates of 0 kg N ha^{-1} for soybeans and peanuts, 17 kg N ha^{-1} for cowpeas, 50 kg N ha^{-1} for snap beans and green peas and 78 kg N ha^{-1} for lima beans. Equations used to calculate symbiotic N fixation are summarized in Table 4.

Table 4. Nitrogen fixation rates and equations used to estimate N fixation for different crops.

Plant Type	N Fixation Rate (kg N ha^{-1} yr^{-1})	Reference
Symbiotic N Fixation		
Crops		
Soybean and Peanuts	1.11 x harvested crop	Meisinger and Randall (1991)
Green Peas and Snap Peas	0.90 x harvested crop	Meisinger and Randall (1991)
Cow Peas	0.98 x harvested crop	Meisinger and Randall (1991)
Lima Beans	0.75 x harvested crop	Meisinger and Randall (1991)
Hays		
Alfalfa	0.98 x harvested crop	Meisinger and Randall (1991)
Non-Alfalfa	0.10 x harvested crop	Meisinger and Randall (1991)
Non-Symbiotic N Fixation		
Crops, vegetables and orchards	5	Stevenson (1982)
Rice Fields	52	Evans and Barber (1977)
Eastern non-wooded pastures	15	Keeney (1979)
Western non-wooded pastures	1	Woodmansee (1978)
Upland Forests	1	Roskoski (1980); Harvey et al. (1987); Hendrickson (1990)

Atmospheric deposition of nitrate

Nitrate and NH_4^+ are the dominant forms of inorganic N in atmospheric deposition. Nitrate in atmospheric deposition is mostly derived from high-temperature combustion processes and is largely anthropogenic [Berner and Berner 1987; Jordan and Weller 1996]. In contrast, NH_4^+ deposition represents the return of ammonia gas that is released from the land surface via volatilization from fertilizer and animal wastes. Ammonia gas is highly reactive in the atmosphere and is not transported great distances from the emission source [Schlesinger and Hartley 1992; Jordan and Weller 1996]. As a result, we assumed that NH_4^+ deposition originates in the watershed and is not an external anthropogenic N source. Similarly, we assumed that organic N is not an external anthropogenic N source. Thus, NO_3^- was assumed to be the only new atmospheric N source deposited to the selected watersheds. Estimates of the wet and dry NO_3^- deposition rates (kg N ha^{-1} yr^{-1}) to both the watershed and surface of the estuary were provided by the atmospheric deposition work group (Chapter 3).

Net import of food and feed

Net exchange of N in shipments of food and feed were calculated as the difference between the production and consumption of human food and animal feed in each watershed [Jordan and Weller 1996]. If consumption exceeded production, we assumed that this demand was met by N import from outside the watershed. Conversely, if production exceeded consumption, we assumed that the excess production was exported from the watershed.

Sources of Uncertainty: Anthropogenic N Input to Watersheds

Among the anthropogenic N inputs to watersheds, our estimate of fertilizer N input is probably the most certain because fertilizer sales records and the N content of fertilizers are well documented. In contrast, our estimate of biotic N fixation associated with agriculture is not well known. Nitrogen fixation by legumes can vary greatly

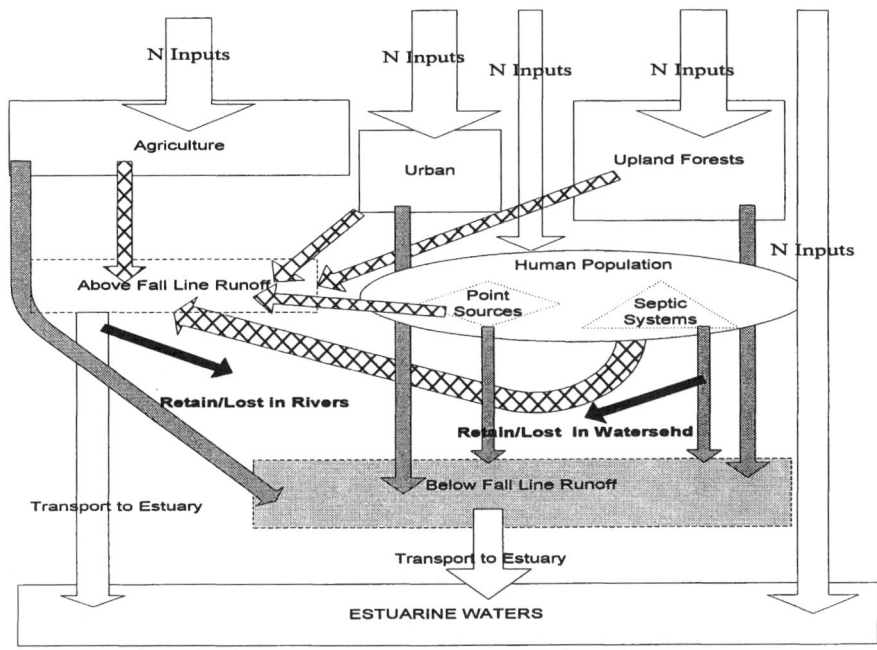

Figure 3. This schematic diagram shows the pathways used to move N from the watersheds to the estuaries.

depending on N availability in soil, crop type and productivity [Meisinger and Randall 1991]. However, for most budgets, biotic N fixation was not a large source of anthropogenic N. Therefore, uncertainties about N fixation did not have a significant effect on our N budgets. Estimates of net import of N in food and feed are also uncertain because they depend on a number of assumptions about the production and consumption of crops and animal products [Jordan and Weller 1996]. Uncertainties in wet and dry deposition (no less than a factor of 2) rates of NO_3^- are described in Chapter 3 [Meyers et al., this book]. In general, uncertainties about N budgets decrease as the size of the watershed increases because N budgets for larger watersheds are based on data from more counties and are less sensitive to uncertainties in data from a particular county.

Nitrogen Export from Watershed to Estuaries

We used land-use specific approaches to estimate N export from the watersheds to the estuaries (Figure 3). We estimated the amount of N available for water-borne transport from agricultural lands (crops and pastures), urban areas, and upland forests. With the exception of Barataria Bay and Terrebonne-Timbalier Bay, these 3 land-uses accounted for between 69 and 99 % of the total land area in the 34 watersheds. Watersheds of Barataria Bay and Terrebonne-Timbalier Bay were dominated (75% of total area) by wetlands. When a watershed included a fall line, we made estimates of N export above and below the fall line. Eighteen of the 34 watershed/estuary systems had above fall line N inputs. Nitrogen exported from above the fall line was attenuated by

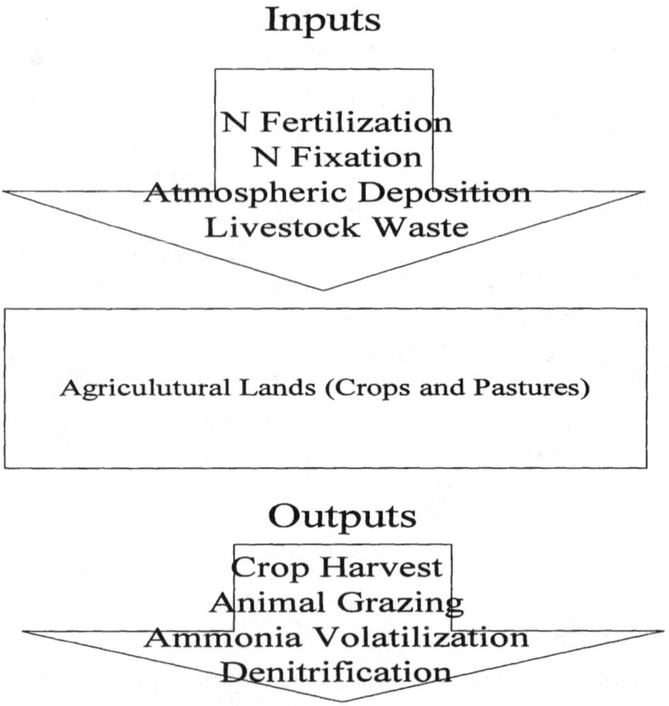

Inputs

N Fertilization
N Fixation
Atmospheric Deposition
Livestock Waste

Agriculutural Lands (Crops and Pastures)

Outputs

Crop Harvest
Animal Grazing
Ammonia Volatilization
Denitrification

Figure 4. Components of our N budget for agricultural lands that were used to estimate N available for transport to the estuary and the contribution made by atmospheric deposition to the total N exported from agricultural lands.

30% to account for N retention and losses (e.g. denitrification) during riverine transport to the estuary (discussed below). Nitrogen exported from below the fall line of the watersheds was not attenuated, but was directly transported to the estuary.

Agricultural lands

For agricultural lands (crops and pastures), N available for water-borne transport to the estuary was estimated as the difference between N inputs and N outputs (Figure 4). We assumed that the atmospherically-derived component of the N exported from agricultural lands was equal to the proportion made by atmospheric N deposition to the total N input to agricultural lands.

Nitrogen inputs for our agricultural input-output budgets included N fertilization, N fixation, atmospheric deposition of NH_4^+ and NO_3^- and livestock waste (Figure 4). Atmospheric deposition, fertilization and N fixation were computed as discussed above. Livestock wastes were estimated as the difference between livestock consumption of N in feed and livestock production of N in meat, milk, and eggs for human consumption [Jordan and Weller 1996].

Outputs from agricultural lands included crop harvest, pasture grazing, volatilization of ammonia and denitrification. County data on crop harvest was obtained from the 1987 agricultural census [Bureau of Census 1993]. Nitrogen removed in crop harvest was estimated by multiplying crop harvest by the percent N in each crop [Meisinger and Randall 1991]. Grazing was estimated based on population of cattle, sheep and horses [Bureau of Census 1993]; their total dietary N requirements and the proportions of their dietary N obtained from grazing [Jordan and Weller 1996]. Ammonia volatilization was assumed to equal 10% of the N inputs from fertilizer and atmospheric deposition, and 20% of the livestock manure N inputs [Schlesinger and Hartley 1992]. Denitrification losses of N_2 and N_2O, except from rice fields, were estimated to be 10% of the N inputs from fertilizer and atmospheric deposition and 20% of livestock N waste [Meisinger and Randall 1991]. This percentage of N lost from denitrification is a median from several reported values [Ryden 1983; Colbourn 1992; Ruz-Jerez et al. 1994; Schnabel and Stout 1994; Jarvis et al. 1991; Paul and Zebarth 1997a, 1997b; Howarth et al. 1998]. Denitrification losses from rice fields were assumed to be 30% of the inorganic N inputs which is characteristic of poorly drained soils with 2 to 5% organic matter [Meisinger and Randall 1991].

Urban areas

Nitrogen export from urban areas included effluent from wastewater treatment plants (point sources), leachate from septic systems and non-point source (NPS) runoff from pervious and impervious surfaces in urban areas (Figure 5). We assumed that both point and septic system N inputs did not contain any atmospheric N. Atmospheric N inputs to urban areas were calculated by multiplying the areal rates of total (wet and dry) deposition of NH_4^+ and NO_3^- to the watershed (from Chapter 3) by the total urban area in the watershed (Table 1). We assumed that 75% of the atmospheric N input to urban areas was exported from the urban area to the estuary [Fisher and Oppenheimer 1991].

Data on point source total N (organic and inorganic forms) released from wastewater treatment plants were only available for the region below the fall line of each watershed [Pacheco 1993]. We found a strong linear relationship between the below fall line total N discharged from wastewater treatment plants and the human population below the fall line that use municipal sewage treatment facilities (Figure 5b). This relationship was used with the above fall line population to estimate the total N discharged from the wastewater treatment plants above the fall line.

The slope of the relationship between total N discharged from wastewater treatment plants and the sewered population suggests that the per capita N excretion rate is 4.2 kg N person^{-1} yr^{-1} (Figure 5). This rate is similar to previous estimates of human N excretion [4.4 –5.2 kg N person^{-1} yr^{-1} in Howarth et al. 1996 and references therein] and per capita N load from wastewater effluent [3.3 kg N person^{-1} yr^{-1} in Meybeck et al. 1989; 2.0 to 7.23 kg N person^{-1} yr^{-1} in U.S. EPA 1980].

To estimate N export from septic systems into the estuaries, we multiplied our human per capita N excretion rate (4.2 kg N person^{-1} yr^{-1}) by the above and below fall line human population on septic systems from the 1990 census data. We assumed that only 25% of the total N in septic systems is retained or removed in the septic tank or unsaturated zone. The other 75% of the N in septic systems is exported to the estuary.

The soil water assessment model (SWAT) was used to estimate non-point source total N (organic and inorganic forms) runoff from pervious and impervious land in urban areas [R. Srinivasan, personnel communication]. SWAT is a watershed scale hydrologic, distributed parameter, continuous time model developed at the Grassland Laboratory,

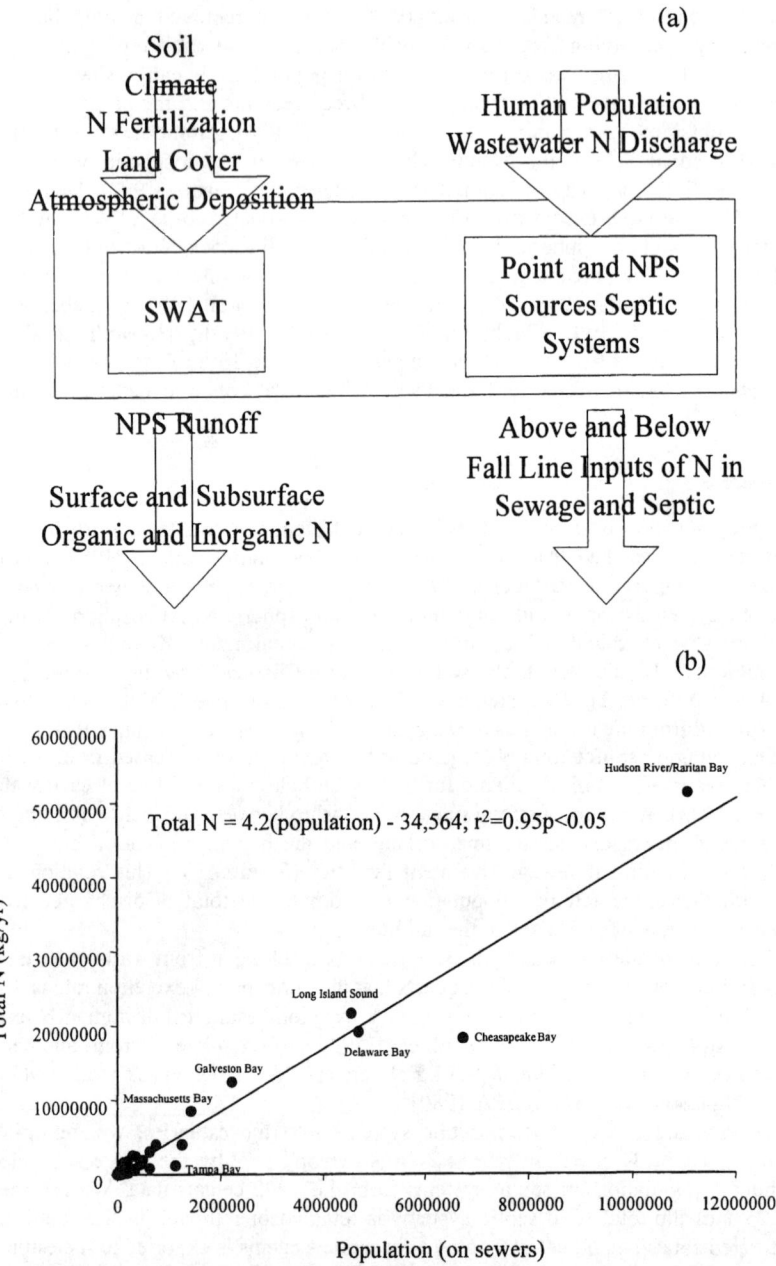

Figure 5. Schematic diagram of the methods used to estimate N export from urban areas (a) and the relationship between total N discharged from wastewater treatment plants below the fall line and 1990 human population on sewers below the fall line of each watershed (b).

Temple, Texas (http://www.brc.tamus.edu/swat/). SWAT is designed to predict the impact of various types of land management practices on water, sediment and agricultural chemical yields in large complex watersheds with varying soils, land use, and management conditions. It can also estimate the loadings from point sources. SWAT requires specific information about weather, soil properties, topography, vegetation and land management (pervious and impervious land uses).

Upland forests

Atmospheric deposition (NH_4^+ and NO_3^-) and non-symbiotic N fixation were assumed to be the only N inputs to forests (Figure 6a). The contribution made by atmospheric N deposition to the total N runoff from upland forests was the same proportion that atmospheric N deposition made to the total N inputs. We estimated N export from upland forests using a non-linear regression relationship between wet deposition of NH_4^+ and NO_3^- and stream water N export of dissolved inorganic N (NH_4^+ and NO_3^-; Figure 6b). This relationship (stream water N export = $0.0482 e^{(0.481 \times \text{wet deposition N input})}$; $p<0.05$) was developed using results from many forested watershed studies in the U.S., representing a broad range of wet N deposition and forest types [Castro and Morgan 1999 and references therein; Driscoll unpublished data]. We realize that dissolved organic nitrogen (DON) can also be exported from upland forests, but did not consider this N source because few studies have measured DON export from upland forests.

Nitrogen Losses in Surface Waters

The following equation was developed to estimate N losses during transport in surface waters ($r^2 = 0.79$, $\rho = 0.01$; Seitzinger et al., in prep):

$$\% \text{ N removed} = 84.9 * (\text{depth/time of travel})^{-0.37} \qquad (1)$$

(units: depth (m), time of travel (yr)). The "% N removed" is the percent of N inputs to the river/stream reach that is removed, not the percent of N inputs to the watershed. Equation (1) was developed using published studies in individual river/stream reaches and lakes from various locations including the Northeast US, Canada, Europe and New Zealand. The rivers ranged from first order headwater streams to the tidal freshwater portions of major rivers. The correct application of this equation requires information on depth and time of travel for all river reaches throughout a watershed, as well as the stream/river network configuration so that N can be routed properly through the stream/river reaches.

Depth, time of travel and stream/river network configuration data were not available for the 34 watershed/estuary systems included in this chapter. Thus, it was not possible to apply equation (1) to estimate surface water N removal for these 34 watershed/estuary systems. However, depth, time of travel and stream/river network configuration data have been compiled, and estimates of N removal have been made for several Northeastern U.S. rivers (Kennebec, Androscoggin, Saco, Charles, Upper Hudson, Mohawk and Delaware; Seitzinger et al., in prep). Calculated riverine N losses were similar across these 7 rivers and ranged from 25 to 35% of the N inputs to these rivers. Therefore, to estimate N removal in the above fall line portion of rivers in the 34

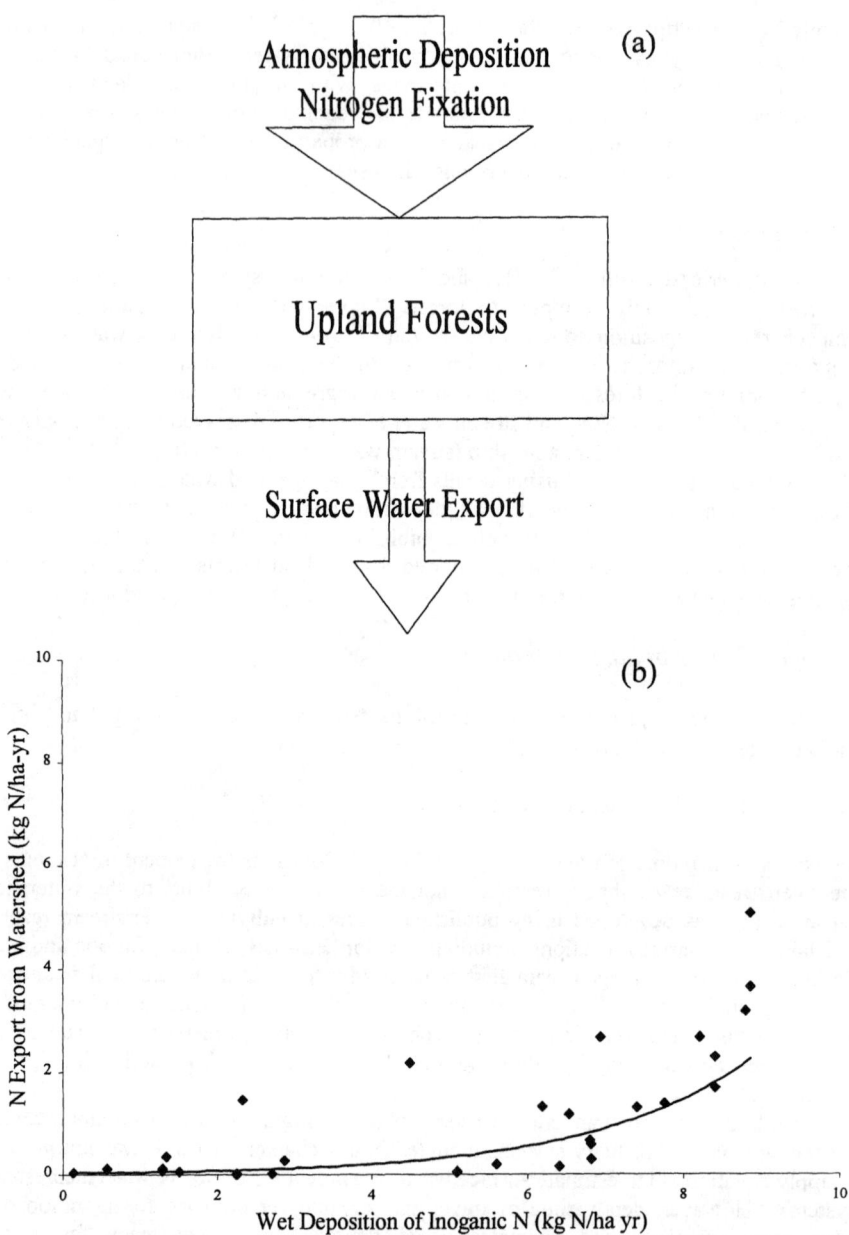

Figure 6. Schematic diagram of the method used to estimate N export from upland forests (a) and the relationship between wet deposition inputs to forested watersheds and inorganic N (NH_4^+ and NO_3^-) exported from these watersheds in surface waters (b).

watershed/estuary systems in this chapter, we chose to use a midpoint value of 30%. This value is within the range of river retention estimates reported by Howarth et al. [1996].

Results and Discussion

Net Anthropogenic N Inputs to the Watershed

Net anthropogenic N inputs to the 34 watersheds ranged from 5.8 kg N ha^{-1} yr^{-1} in St. Catherines-Sapelo, GA to 70.8 kg N ha^{-1} yr^{-1} in Tampa Bay, FL (Figure 7). Net anthropogenic N inputs averaged over all 34 watersheds (study watershed average) was 28 kg N ha^{-1} yr^{-1}. In the Northeast, 6 of 10 watersheds had net anthropogenic N inputs that were above the study watershed average. Massachusetts Bay had the second highest (62 kg N ha^{-1} yr^{-1}) net anthropogenic N input compared to all 34 watersheds. In the Mid-Atlantic region, all 3 watersheds had net anthropogenic N inputs above the study watershed average. In contrast, 4 of the 6 watersheds in the Southeast region and 11 of the 15 watersheds on the Gulf Coast had net anthropogenic N inputs that were below the study watershed average. These two regions (Southeast and Gulf Coast), however, contain 3 watersheds, all of which are located in Florida, that had some of the highest (> 50 kg N ha^{-1} yr^{-1}) net anthropogenic N inputs in the study.

There were strong regional differences in the dominant anthropogenic N inputs. In the Northeast, total net anthropogenic N inputs were dominated by net import of food and atmospheric NO_3^- deposition, with minor inputs from agricultural activities (Figure 7a) On average, net food import and atmospheric NO_3^- deposition account for 49% and 22% of the total N inputs, respectively. High net food import is required to support the large populations (Table 1).

Unlike the Northeast, the dominant anthropogenic N sources to the 3 watersheds in the Mid-Atlantic region varied considerably. (Figure 7b). Net food import (24% of total) and N fertilization (29% of total) dominated the total net anthropogenic N inputs to the watershed of Delaware Bay. Net feed import (31%) and N fertilization (26%) were the dominant inputs to Chesapeake Bay. Nitrogen fertilization was the dominant (56%) anthropogenic N input to Pamlico-Pungo Sound. Atmospheric NO_3^- deposition accounted for 11 to 21% of the total anthropogenic N inputs to these 3 watersheds.

In the Southeast, total net anthropogenic N inputs to 2 watersheds (St. Helena, SC and Indian River, FL) were clearly dominated by N fertilization (51 and 79%, respectively). The Indian River watershed had the highest fertilizer N inputs (40.1 kg N ha^{-1} yr^{-1}) of any of the watersheds (Figure 7b). Three Southeast watersheds (Winyah Bay, SC, Charleston, SC and Altamaha River, GA) were dominated by both net feed import (21%, 27% and 30%) and N fertilization (46%, 29% and 38%), and one watershed (St. Catherines-Sapelo, GA) was dominated by atmospheric NO_3^- deposition (62% of total). With the exception of St. Catherines-Sapelo, atmospheric NO_3^- deposition accounted for 7 to 24% of the total anthropogenic N inputs to the other watersheds in this region.

Watersheds on the East Gulf Coast were dominated by either N fertilization (Charlotte Harbor, FL, Tampa Bay, FL, Apalachee Bay, FL and Appalachicola Bay FL), net feed import (Mobile Bay, LA and West Mississippi Bay) or atmospheric nitrate deposition (Barataria Bay, LA and Terrebonne-Timbalier Bays, LA; Figure 7c). Much like the Indian River watershed in the Southeast region, 2 of the Florida watersheds (Charlotte and Tampa Bay) on the East Gulf Coast had high N fertilization rates (35 and 39 kg N ha^{-1} yr^{-1}, respectively). These high fertilization rates are needed to support crops

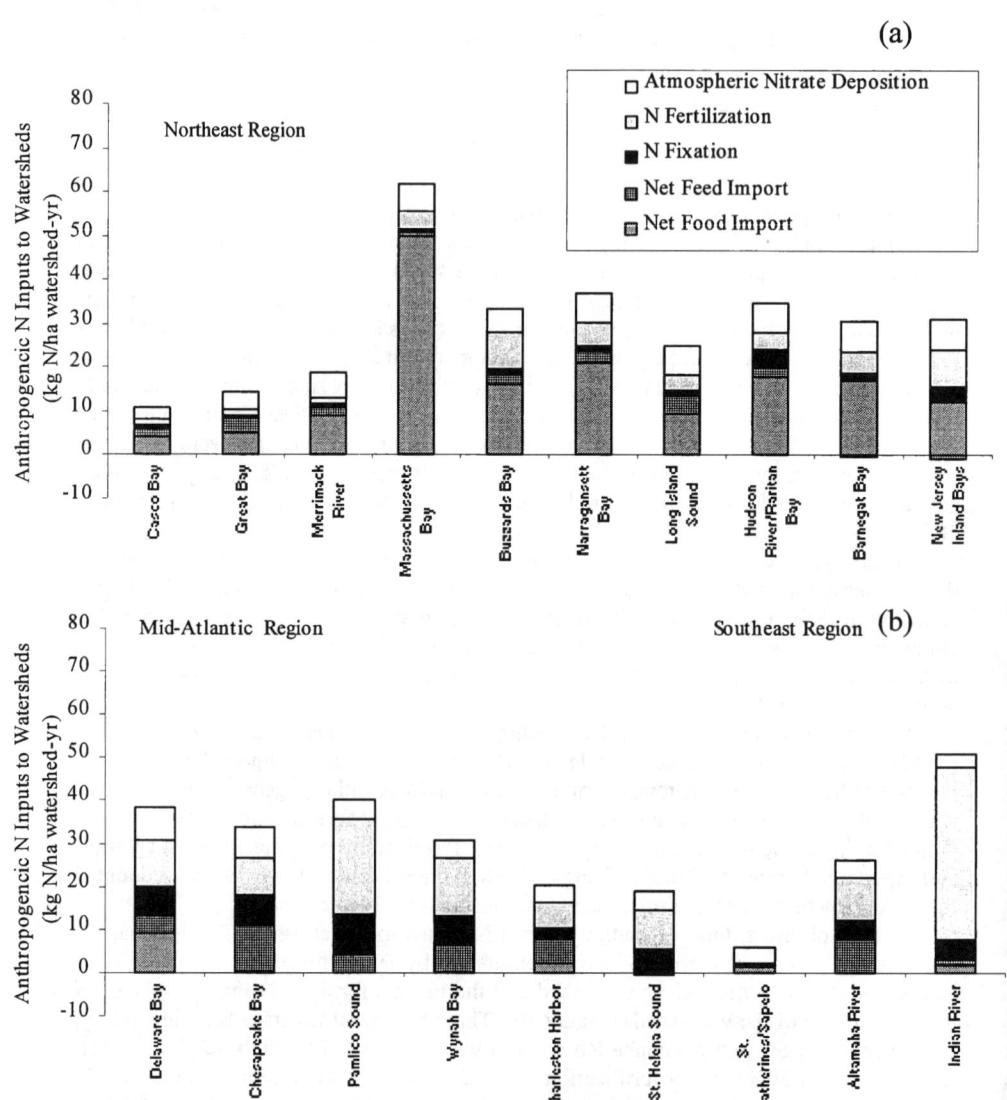

Figure 7. Net anthropogenic N inputs to watersheds.

of fruits and vegetables with multiple annual cropping made possible by the long growing season. Atmospheric NO_3^- accounted for 34 and 53% of the total anthropogenic N inputs to Barataria Bay and Terrebonne-Timbalier, respectively. In contrast, atmospheric NO_3^- deposition accounted for 6 and 26% of the total anthropogenic N inputs to the other watersheds in this region. In the West Gulf Coast region, total anthropogenic N inputs were dominated by N fertilization (37 and 95%) for 6 of the 7 watersheds, and 1

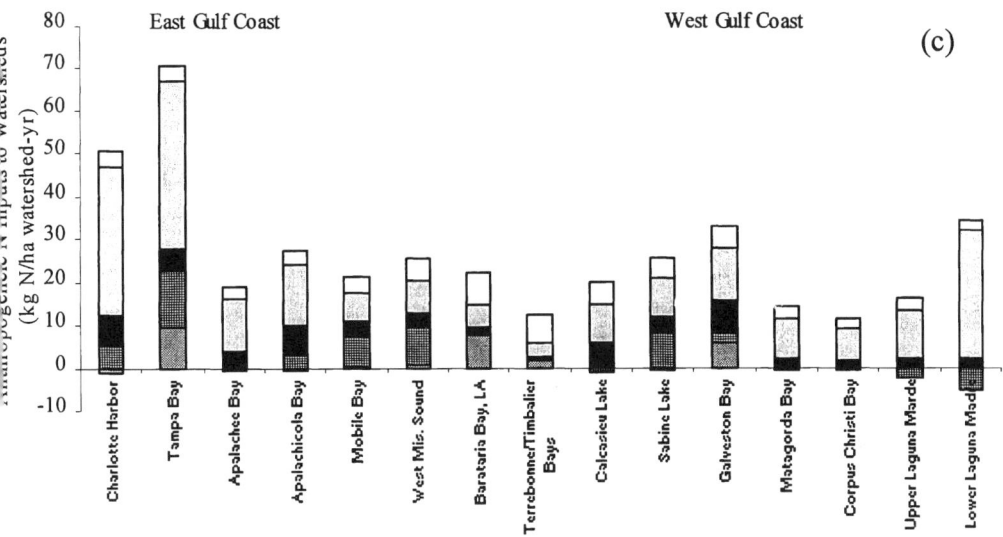

Figure 7. Continued.

watershed had considerable net feed import (Sabine Lake, LA). Atmospheric deposition contributed between 7 and 40% of the total N inputs to these watersheds.

Sources of Uncertainty: Predicted and Measured N Fluxes From Watersheds

Before estimating the contribution made by atmospheric N deposition to the total N load to the Atlantic and Gulf coast estuaries in the U.S., we used the assumptions discussed above (Figure 3) to estimate total riverine N fluxes to the fall line. Then we compared our estimates to measurements of riverine N fluxes to the fall line (for 1991) from the USGS NASQAN program available for 18 watersheds [Pacheco 1999]. Nitrogen loads at the NASQAN stations, most of which are at or near the fall line, were estimated using the Load Estimation regression model created by the USGS [Pacheco 1999].

In general, the agreement between calculated and measured N fluxes was not satisfactory because 8 of the watersheds had predicted riverine N fluxes which were 2 to 11 times greater than the measured fluxes (Figure 8a). Some discrepancy was expected because of the many potential sources of error in our approach. For example, we are comparing the results of calculations that were derived from a variety of data sources obtained from the late 1980s to mid 1990s (Table 3). Year-to-year variability in these and the NASQAN data undoubtedly contribute to differences between predicted and measured riverine N fluxes.

The amount of disagreement between predicted and measured N fluxes differed according to the dominant N sources in the watershed. There was reasonably good agreement for watersheds with greater than 50% of the riverine N from urban sources (e.g. Merrimack River, Hudson River-Raritan Bay, Long Island Sound, Narragansett Bay and Delaware Bay; Figure 8a). This good agreement suggests that our budgets are

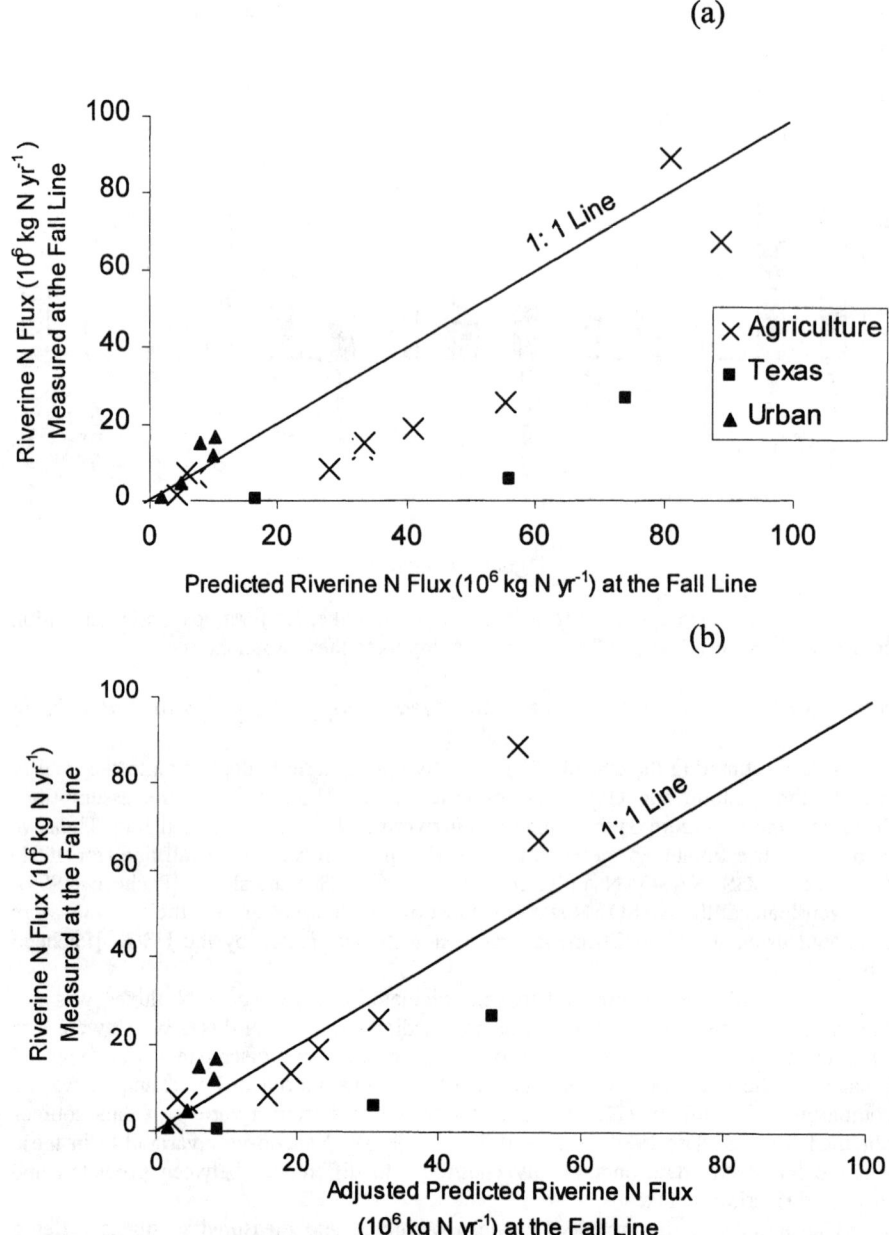

Figure 8. Comparison of predicted N export to the fall line with measurements made by the NASQAN program in 1991 (a) and our adjusted predictions (with the 50% reduction in N available for export from agricultural land) compared to the NASQAN measurements at the fall line (b).

accurate and that our estimate of in-river N loss (30% of the above fall line inputs) is reasonable. In contrast, 8 of 11 watersheds with greater than 50% of their riverine N flux from agricultural sources (including the 3 watersheds from Texas) had much higher predicted riverine N fluxes than was measured (Figure 8a). Similarly, David et al. [1997] reported that only about 50% of the excess N in an agricultural watershed in Illinois was exported from the watershed. Other studies also report that N draining from agricultural lands can be retained in wetlands, riparian zones or aquifers and subsequently stored or converted to gaseous forms [Howarth et al. 1996]. For example, processes in riparian forests can take up 70 to 90% of N released by adjacent agricultural lands [Lowrance et al. 1984; Peterjohn and Correll 1984; Jacobs and Gilliam 1985; Jordan et al. 1993]. Denitrification was estimated to remove 27% of NO_3^- infiltrating glaciofluvial sands and loamy sands of an aquifer under a Rhode Island riparian forest [Groffman et al. 1996]. Additional *in situ* NO_3^- removal within an aquifer may occur by ground water mixing and dilution, microbial immobilization and plant uptake [Nelson et al. 1995]. We did not consider N uptake by adjacent land-uses in our agricultural budget calculations, because we do not know the spatial arrangement of land-uses, and the fluxes of N between adjacent land-uses. We have, however, adjusted our estimates to account for this discrepancy.

To adjust our calculations for the apparent overestimate in leaching of N from agricultural lands, we considered scenarios where some of the excess N from agriculture was retained or denitrified in soil or groundwater. When greater than 50% of the excess agricultural N was allowed to be retained in soil or groundwater, our estimates of riverine N flux were less than the measured fluxes for the NASQAN sites, except for the watersheds in Texas. Therefore, we assumed that 50% of the excess N from agriculture lands is retained or denitrified in soil or groundwater (Figure 8b). This assumption resulted in better agreement between predicted and measured N fluxes ($r^2 = 0.67$; and predicted N flux = 1.1* (NASQAN − 1.95); the slope of this relationship was very close to 1. This adjustment was applied to our agricultural budgets and all subsequent calculations.

Watershed N Retention

The percentage of the net anthropogenic N input to the watershed that was retained by the watershed (inputs - outputs = retained) ranged from 18 to 80%, and the overall average retention was 62% (Table 5). The percentage of anthropogenic N retained by upland forests and agricultural lands was high. For upland forests, the average retention was 95%. Upland forests in the Chesapeake Bay watershed had the lowest N retention (90%). Retention of anthropogenic N on agricultural lands was somewhat less than the retention by upland forests. Average retention of anthropogenic N inputs by agricultural lands was 80%, with Indian River exhibiting the lowest retention of 69%. Our estimates of N retention by agricultural lands assumes that at least 50% of N available for export from agricultural lands is retained or denitrified in adjacent landscapes prior to discharge into surface waters (discussed above). Our retention estimates are consistent with those reported by Jordan et al. [1997a, 1997b]. They reported that watersheds on the Piedmont [Jordan et al. 1997a] and Coastal Plain [Jordan et al. 1997b] of Chesapeake Bay retained between 50 and 70% of the net anthropogenic N input to the watershed.

Urban-dominated watersheds had the lowest watershed retention of N. For example, retention of anthropogenic N by the Massachusetts Bay watershed was only 18% and in

Table 5. Total net anthropogenic N inputs to each watershed and the retention of this N by different land-uses in the watersheds.

Watershed/Estuary Systems	Total Anthropogenic Nitrogen Input (kg N ha^{-1} yr^{-1})	Watershed Retention (% of Input)	Retention by Upland Forests (% of N input)	Retention by Agriculture (% of N)
Northeast Region				
Casco Bay	10.97	62.7	95.1	92.3
Great Bay	14.34	54.8	95.4	83.6
Merrimack River	18.60	49.3	93.8	88.1
Massachusetts Bay	61.62	18.2	96.3	76.4
Buzzards Bay	33.24	48.2	96.2	73.0
Narragansett Bay	36.79	38.9	94.6	77.5
Long Island Sound	24.85	53.9	91.8	87.1
Hudson River/Raritan Bay	34.91	47.4	92.0	89.2
Barnegat Bay	30.38	79.5	91.9	75.2
Great Bay	30.33	70.9	91.9	73.7
Mid-Atlantic Region				
Delaware Bay	38.18	60.9	90.9	85.7
Chesapeake Bay	34.05	72.6	89.8	85.2
Pamlico Sound	40.03	66.1	94.4	76.7
Southeast Region				
Wynah Bay	30.28	67.3	95.1	78.6
Charleston Harbor	20.28	47.9	96.7	80.8
St. Helena Sound	18.36	75.1	96.9	81.3
St Catherines-Sapelo	5.83	79.3	96.1	78.8
Altamaha Sound	26.13	68.9	96.2	78.4
Indian River	50.98	60.3	96.7	68.5
East Gulf Coast Region				
Charlotte Harbor	49.70	65.1	96.1	74.4
Tampa Bay	70.82	68.2	96.4	73.9
Apalachee Bay	18.44	70.9	96.2	78.9
Apalachicola Bay	26.76	66.7	96.2	78.3
Mobile Bay	21.47	65.8	95.0	79.4
West Mis. Sound	25.44	49.5	92.8	80.6
Barataria Bay	22.12	65.9	94.8	79.7
Terrebonne-Timbalier Bay	12.19	67.7	94.4	81.4
West Gulf Coast Region				
Calcasieu River	18.82	59.2	93.7	76.4
Sabine River	24.90	68.5	95.6	79.7
Galveston Bay	32.85	54.5	96.2	79.2
Matagorda Bay	14.02	65.8	97.0	82.6
Corpus Christi Bay	10.93	67.1	96.7	85.1
Upper Laguna Madre	13.55	75.5	96.2	88.9
Lower Laguna Madre	29.09	66.5	96.0	80.2

Narragansett Bay, watershed retention of anthropogenic N was 38%. For these urban-dominated watersheds, net food import is the most important source of anthropogenic N. This N input is consumed by humans and discharged to surface waters with little attenuation in wastewater treatment plants. This pathway of N transport bypasses many of the mechanisms of N retention and loss that application of N to the land surface are subjected to, thereby decreasing watershed retention.

Watershed N Export to Estuaries

Sources of N to Estuaries

Our analysis showed considerable variability in N export (kg N ha^{-1} of watershed yr^{-1}) to the Atlantic and Gulf coast estuaries from their watersheds (Table 6). Watershed export of N was relatively low (< 7 kg N ha^{-1} of watershed yr^{1}) in Casco Bay, Great Bay and Barnegat Bay in northern New England. Nitrogen export increased in the watersheds of southern New England and New York, with particularly high inputs to Massachusetts Bay (51 kg N ha^{-1} of watershed yr^{-1}). With the exception of Barnegat Bay, N inputs to these northern estuaries were strongly influenced by urban sources. Point sources accounted for 48 to 88% of the total N exported from these northern watersheds to their estuaries (Table 6). For Barnegat Bay, point sources of N from wastewater treatment plants are discharged offshore and do not contribute N to this estuary, but septic systems (1 kg ha^{-1} of watershed yr^{-1}) account for 16% of the total N input to this estuary. Atmospheric deposition (50% of total N input) and agriculture (32% of the total N inputs) are the dominant sources of N to the Barnegat Bay estuary (Tables 7). Immediately south of Hudson River-Raritan Bay, watershed N export to estuaries remained high, but the source of the N inputs shifted from largely urban-dominated to a combination of urban and agricultural sources, or predominately agricultural sources (Table 7).

In the Mid-Atlantic region, watershed N export ranged from 8 to 15 kg N ha^{-1} of watershed yr^{-1} (Table 6). Nitrogen inputs to the Delaware Bay estuary were dominated (56% of total) by urban point sources (Table 7). Nitrogen inputs to the other two Mid-Atlantic estuaries (Chesapeake Bay and Pamlico-Pungo) were clearly dominated by agricultural sources (>55% of total N loads) and point sources (~ 20% of total N load).

Nitrogen inputs to 4 of the 6 estuaries in the Southeast region were dominated by agricultural N inputs ($\geq$ 70% of total N inputs). Exception to this occurred in the St. Catherines-Sapelo, GA and Charleston Harbor, SC (Tables 7) estuaries. St. Catherines-Sapelo, GA estuary had about equal inputs of N from agricultural runoff (22%), urban non-point sources (26%), point sources (23%) and atmospheric deposition (28%). Charleston Harbor, SC had very high point source inputs (60%) and agricultural inputs (31%).

Watershed N export to estuaries on the East Gulf coast varied considerably, ranging from 5 to 25 kg N ha^{-1} of watershed yr^{-1}; 6 out of 8 of these estuaries had export rates less than 10 kg N ha^{-1} of watershed yr^{-1} (Table 6). Two of the Florida estuaries (Tampa Bay and Charlotte Harbor) had N watershed export rates of approximately 20 kg N ha^{-1} of watershed yr^{-1} or greater. Nitrogen export to all of the East Gulf coast estuaries, except Barataria Bay, Terrebonne-Timbalier Bays and West Miss. Sound, were dominated by agriculture (> 60%). Watershed N export to Barataria Bay, Terrebonne-Timbalier Bays and West Miss. Sound, were dominated by urban point sources (>50% of total N input).

Watershed export to estuaries on the West Gulf coast ranged from 3 to 12 kg N ha^{-1} of watershed yr^{-1} (Table 6). Nitrogen inputs to all estuaries in this region, except for Galveston Bay, were dominated by agricultural sources (> 50% of total N input). Galveston Bay estuary had significant N inputs from both point sources (43% of total) and agricultural runoff (38% of total N input, Table 7).

Table 6. Nitrogen export (kg N ha^{-1} yr^{-1}) to the estuaries from different watershed nitrogen sources.

Watershed/Estuary Systems	Agriculture Runoff	Urban NPS Runoff	Upland Forest Runoff	Human Sewage	Atmospheric Deposition	Total Runoff From Watershed	Adjusted[a] Total Runoff from Watershed
Northeast Region							
Casco Bay	0.43	0.80	0.035	2.58	0.71	4.55	4.55
Great Bay	0.96	1.36	0.032	3.13	1.01	6.50	6.50
Merrimack River	0.59	0.49	0.047	7.02	1.35	9.49	7.98
Massachusetts Bay	1.57	0.14	0.008	44.42	4.59	50.73	50.73
Buzzards Bay	4.15	0.25	0.024	14.20	1.62	20.25	20.25
Narragansett Bay	2.46	0.73	0.028	17.06	3.05	23.33	21.64
Long Island Sound	1.42	0.53	0.058	7.77	1.70	11.49	10.59
Hudson River/Raritan Bay	1.76	0.82	0.046	16.06	1.93	20.62	19.91
Barnegat Bay[b]	1.99	0.15	0.036	1.00	3.15	6.33	6.33
Great Bay	3.67	0.12	0.040	3.68	1.88	9.38	9.38
Mid-Atlantic Region							
Delaware Bay	4.54	0.43	0.042	8.82	2.08	15.91	14.79
Chesapeake Bay	5.26	0.20	0.058	2.37	1.70	9.58	8.22
Pamlico Sound	10.55	0.05	0.025	2.57	1.04	14.23	11.51
Southeast Region							
Wynah Bay	7.48	0.08	0.024	2.27	0.73	10.58	8.69
Charleston Harbor	3.23	0.20	0.020	6.18	0.69	10.32	7.41
St Helena Sound	3.66	0.01	0.013	0.59	0.36	4.63	3.51
St Catherines-Sapelo	0.29	0.35	0.024	0.30	0.38	1.35	1.35
Altamaha Sound	5.66	0.06	0.024	1.70	0.68	8.12	5.84
Indian River	18.86	0.31	0.002	2.93	2.18	24.28	24.28
East Gulf Coast							
Charlotte Harbor	15.26	0.03	0.002	1.63	1.10	18.02	18.02
Tampa Bay	18.87	0.19	0.002	3.93	1.90	24.89	24.89
Apalachee Bay	4.39	0.05	0.023	0.47	0.39	5.33	4.70
Apalachicola Bay	6.90	0.13	0.022	1.75	0.70	9.51	6.71
Mobile Bay	4.43	0.11	0.035	2.09	0.70	7.37	5.32
West Mis. Sound	4.72	0.12	0.043	7.15	1.08	13.11	6.49
Barataria Bay	1.89	0.17	0.000	5.17	1.09	8.33	8.33
Terrebonne-Timbalier Bays	1.11	0.17	0.000	2.55	0.88	4.71	4.71
West Gulf Coast Region							
Calcasieu River	4.84	0.06	0.035	2.42	1.11	8.46	6.51
Sabine River	5.36	0.07	0.028	1.58	0.83	7.87	5.97
Galveston Bay	5.68	0.22	0.010	6.46	2.62	14.99	11.61
Matagorda Bay	2.81	0.04	0.006	0.58	1.64	5.06	3.91
Corpus Christi Bay	1.95	0.01	0.005	0.58	1.07	3.61	2.76
Laguna Madre	2.52	0.01	0.001	0.28	0.49	3.31	3.31
Laguna Madre	7.48	0.03	0.002	1.47	0.76	9.74	9.74

[a] For watersheds with above fall line areas, N export from above the fall line region were reduced by 30% to account for N losses and retention during river transport to the fall line.

[b] Effluent from sewage treatment plants in the Barnegat Bay watershed are discharged offshore. Sewage inputs to the Barnegat Bay estuary are from septic systems in the watershed.

Atmospheric N Deposition

In the Northeastern region, the contribution made by atmospheric N deposition to the total N inputs ranged from 10 % for Hudson River-Raritan Bay to 58 % for Barnegat Bay, NJ (Table 8). Direct deposition to the surface of the estuary was most important for the Casco Bay and Buzzards Bay estuaries where it accounted for 57% and 70% of the total atmospheric N input, respectively (Figure 9). Direct deposition also accounted for

Table 7. Watershed sources of N to the estuaries. Percent contribution (% of total) made by each watershed N source to the total watershed N inputs to each estuary.

Watershed/Estuary Systems	Watershed Type[a]	Agriculture Runoff	Urban NPS Runoff	Upland Forest Runoff	Human Sewage	Atmospheric Deposition
Northeast Region						
Casco Bay	Urban	9.4	17.6	0.8	56.7	15.5
Great Bay	Urban	14.8	21.0	0.5	48.2	15.5
Merrimack River	Urban	6.2	5.1	0.5	74.0	14.2
Massachusetts Bay	Urban	3.1	0.3	0.0	87.6	9.1
Buzzards Bay	Urban	20.5	1.2	0.1	70.1	8.0
Narragansett Bay	Urban	10.5	3.1	0.1	73.1	13.1
Long Island Sound	Urban	12.3	4.6	0.5	67.7	14.8
Hudson River/Raritan Bay	Urban	8.5	4.0	0.2	77.9	9.4
Barnegat Bay[b]	Atmospheric	31.5	2.4	0.6	15.8	49.8
Great Bay	Urban	39.1	1.2	0.4	39.2	20.0
Mid-Atlantic Region						
Delaware Bay	Urban	28.5	2.7	0.3	55.5	13.1
Chesapeake Bay	Agriculture	54.8	2.1	0.6	24.7	17.7
Pamlico Sound	Agriculture	74.1	0.4	0.2	18.0	7.3
Southeast Region						
Wynah Bay	Agriculture	70.7	0.8	0.2	21.4	6.9
Charleston Harbor	Urban	31.3	1.9	0.2	59.9	6.6
St. Helena Sound	Agriculture	79.1	0.2	0.3	12.6	7.7
St. Catherines-Sapelo	Mixed	21.8	26.1	1.8	22.5	27.9
Altamaha Sound	Agriculture	69.7	0.7	0.3	21.0	8.3
Indian River	Agriculture	77.7	1.3	0.0	12.1	9.0
East Gulf Coast Region						
Charlotte Harbor	Agriculture	84.7	0.2	0.0	9.1	6.1
Tampa Bay	Agriculture	75.8	0.8	0.0	15.8	7.6
Apalachee Bay	Agriculture	82.3	1.0	0.4	8.9	7.4
Apalachicola Bay	Agriculture	72.6	1.4	0.2	18.4	7.4
Mobile Bay	Agriculture	60.2	1.5	0.5	28.4	9.5
West Mis. Sound	Urban	36.0	0.9	0.3	54.6	8.2
Barataria Bay	Urban	22.7	2.1	0.0	62.1	13.1
Terrebonne-Timbalier Bays	Urban	23.7	3.6	0.0	54.1	18.6
West Gulf Coast Region						
Calcasieu River	Agriculture	57.2	0.8	0.4	28.6	13.1
Sabine River	Agriculture	68.1	0.9	0.4	20.1	10.5
Galveston Bay	Mixed	37.9	1.5	0.1	43.1	17.4
Matagorda Bay	Agriculture	55.4	0.7	0.1	11.4	32.3
Corpus Christi Bay	Agriculture	54.0	0.2	0.1	16.0	29.6
Upper Laguna Madre	Agriculture	76.2	0.3	0.0	8.6	14.9
Lower Laguna Madre	Agriculture	76.8	0.3	0.0	15.1	7.8

[a] For watersheds with above fall line areas, N export from above the fall line region were reduced by 30% to account for N losses and retention during river transport to the fall line.

[b] Effluent from sewage treatment plants in the Barnegat Bay watershed are discharged offshore. Sewage inputs to Barnegat Bay estuary are from septic systems in the watershed.

37% of the total atmospheric N input to Massachusetts Bay. In contrast, direct deposition contributed less than 30% of the total atmospheric N input to the other 8 estuaries in the northeast.

In the Mid-Atlantic region, atmospheric N deposition accounted for 16%, 23% and 18% of the total N loads to Delaware Bay, Chesapeake Bay and Pamlico Sound estuaries, respectively. For the Delaware and Chesapeake Bay estuaries, transmission through the

Table 8. Atmospheric nitrogen inputs to each of the selected watershed/estuary systems.

Watershed/Estuary Systems	Atmospheric Nitrogen Deposition to Surface of Estuary (10^6 kg N yr^{-1})	Atmospheric Nitrogen Export from Watershed to Estuary (10^6 kg N yr^{-1})	Total Watershed Nitrogen Inputs to Estuary (10^6 kg N yr^{-1})	Atmospheric Contribution to Total N Inputs to the Estuary (% of total N input)
Northeast Region				
Casco Bay	0.20	0.15	0.99	29.8
Great Bay	0.02	0.25	1.62	16.8
Merrimack River	0.01	1.43	9.94	14.4
Massachusetts Bay	0.57	0.96	10.60	13.7
Buzzards Bay	0.39	0.17	2.07	22.5
Narragansett Bay	0.30	1.13	8.69	15.9
Long Island Sound	2.70	6.12	43.17	19.2
Hudson River/Raritan Bay	0.68	6.31	71.90	9.6
Barnegat Bay	0.16	0.43	0.86	57.6
Great Bay	0.24	0.60	3.02	26.0
Mid-Atlantic Region				
Delaware Bay	1.87	5.63	45.54	15.8
Chesapeake Bay	10.53	22.41	132.22	23.1
Pamlico River	3.57	2.10	28.87	17.5
Southeast Region				
Wynah Bay	0.05	2.67	37.80	7.2
Charleston Harbor	0.04	2.05	30.47	6.9
St. Helena Sound	0.09	0.33	4.20	9.8
St. Catherines-Sapelo	0.10	0.07	0.27	46.8
Altamaha Sound	0.02	1.78	21.46	8.4
Indian River	0.36	0.53	5.93	14.1
East Gulf Coast Region				
Charlotte Harbor	0.24	0.83	13.71	7.7
Tampa Bay	0.46	0.95	12.46	10.9
Apalachee Bay	0.76	0.52	6.68	17.2
Apalachicola Bay	0.28	2.40	32.35	8.2
Mobile Bay	0.65	5.72	59.90	10.5
West Mis Sound	3.22	3.61	24.93	24.3
Barataria Bay	0.71	0.45	3.46	28.0
Terrebonne Bay	1.02	0.18	0.99	60.1
West Gulf Coast Region				
Calcasieu River	0.20	0.84	6.40	15.7
Sabine River	0.16	3.32	30.82	11.2
Galveston Bay	0.85	12.11	70.01	18.3
Matagorda Bay	0.42	13.60	44.97	30.9
Corpus Christi Bay	0.21	3.52	12.23	30.0
Upper Laguna Madre	0.34	0.51	3.44	22.6
Lower Laguna Madre	0.54	0.98	12.58	11.6

watershed was the dominant pathway for atmospheric N inputs, accounting for 75 and 68 % of the total atmospheric N loads, respectively. In contrast, direct deposition to the surface of the estuary was the dominant source of atmospheric N for the Pamlico Sound estuary, accounting for 63% of the total atmospheric N input (Figure 9).

In the Southeast, the contribution made by atmospheric deposition to the total N loads was generally lower (<14%) than in the Northeast (10 to 58%) and Mid-Atlantic regions (16 to 23%), except for the St. Catherines-Sapelo estuary. Atmospheric N deposition accounted for 47 % of the total N load to the St. Catherines-Sapelo estuary, with the majority (56%) of total atmospheric N load from direct deposition to the surface of the estuary (Figure 9).

On the East Gulf coast, the contribution made by atmospheric N deposition to total N load to the estuary ranged from 8 to 60 %. Four of the 8 estuaries (Charlotte Harbor, Tampa Bay, Apalachicola Bay and Mobile Bay) in this region had atmospheric N contributions that were less than 12% of the total N loads. The other 4 estuaries had atmospheric contributions that ranged from 17 to 60% of the total N loads and direct deposition accounted for 32 to 85% of the total atmospheric N load (Figure 9).

On the West Gulf coast, atmospheric N contributed between 11 and 31% of the total N loads to these estuaries. Direct deposition was important for the Upper and Lower Laguna Madre estuaries where it accounted ~35% % of the total atmospheric N input.

Over all 34 watershed/estuary systems, the ratio of the watershed area to the surface water area of the estuary was correlated with the contribution made by direct deposition to the total atmospheric N input to the estuary (Figure 10). Twelve watershed/estuary systems had ratios greater than 15 (Table 2). For these 12 systems, direct deposition accounted for 10 to 20% (10% on average) of the total atmospheric N inputs, while the other 80 to 90 % of atmospheric N load was derived from the watershed (Figure 10). In contrast, 22 systems had watershed to surface water area ratios less than 15 (Table 2). For these 22 systems, direct deposition of N to the surface of the estuary accounted for 20 to 86% of the total atmospheric input, with an average of 48 % (Figure 10). These results suggest that direct deposition is an important atmospheric input for systems with watershed to surface water areas less than 15.

Summary, Conclusions and Recommendations

Total anthropogenic N inputs to these 34 watersheds ranged from 5.8 kg N ha^{-1} yr^{-1} in St. Catherines-Sapelo, GA to 70.8 kg N ha^{-1} yr^{-1} in Tampa Bay, FL. Nitrogen fertilization of agricultural crops was clearly the dominant (29 to 84 % of the total N input) anthropogenic N source for 16 of the 34 watersheds. Net food import was the dominant (36 to 81% of the total watershed N inputs) anthropogenic N source for 10 of the 34 watersheds. Five watersheds (Charleston Harbor, SC, Mobile Bay, AL, West Mississippi Sound, LA, Sabine Lake, TX and Chesapeake Bay, MD) were characterized by approximately equal inputs of N from net feed import and N fertilization. Two watersheds (Terrebonne-Timbalier Bay, and St. Catherines-Sapelo, GA) were clearly dominated (34 to 62% of the total N inputs) by atmospheric nitrate deposition. These results indicate that atmospheric nitrate deposition is not a dominant anthropogenic N input to 32 of the 34-watershed/estuary systems examined in this chapter.

The two most important watershed N sources reaching the 34 estuaries were: (1) N runoff from fertilized agricultural lands and (2) point source N discharge from wastewater treatment plants. Total N loads to 14 of the 34 estuaries were dominated by point sources which accounted for 49 to 88% of the total N inputs. Nine of these estuaries are located in the Northeastern U.S.. Total N loads to 17 of the 34 estuaries were dominated (54 to 85% of total) by agricultural N sources. These results suggest a strong need for improvements in sewage treatment and for better N use efficiency in agriculture. Agricultural practices should reduce N fertilizer usage and increase recycling N in livestock waste back to crops.

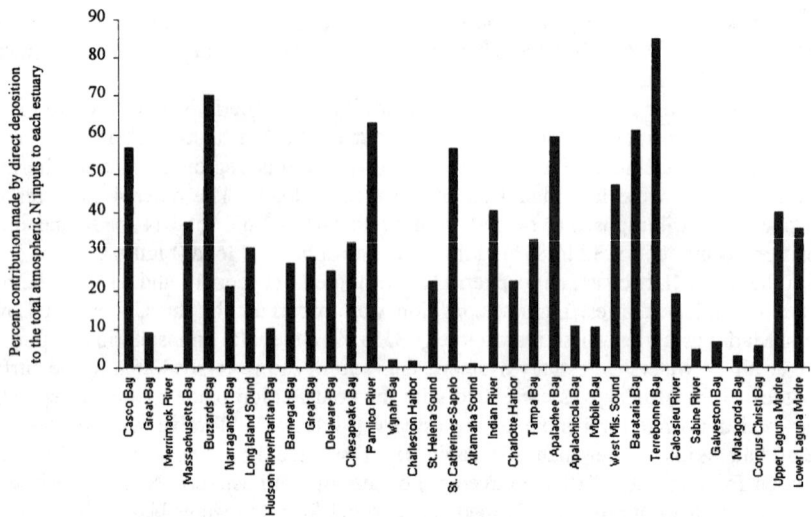

Figure 9. Percent contribution made by direct deposition of atmospheric N to the surface of the estuary to the total atmospheric N inputs to each estuary.

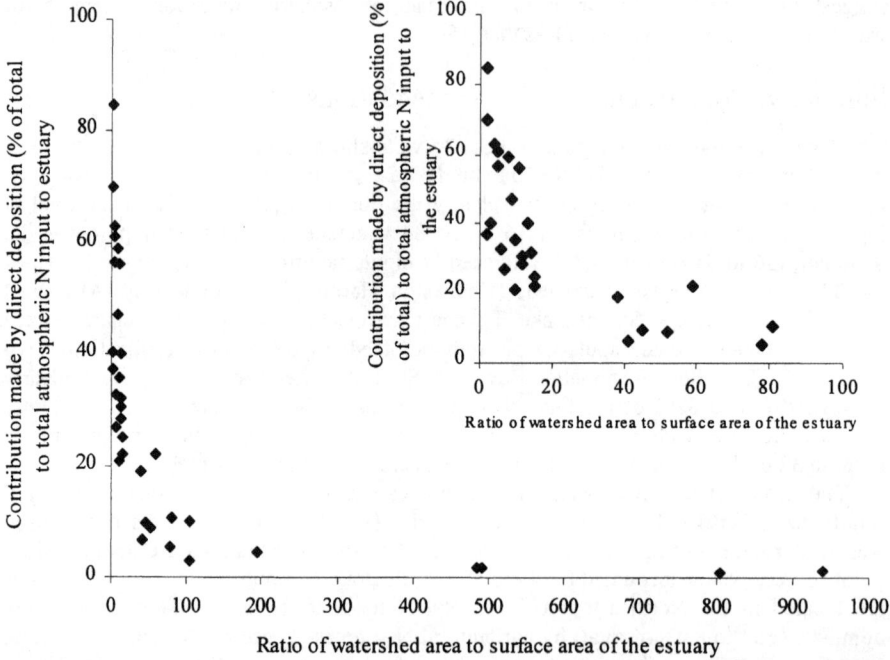

Figure 10. The relationship between the contribution made by direct deposition to the surface of the estuary to the total atmospheric N input to the estuary and the ratio of the watershed area to the surface area of the estuary. The smaller figure is an expanded version of the larger figure for ratios < 100.

Atmospheric N deposition contributed between 7 and 61% of the total N inputs to these 34 estuaries. However, only 3 watershed/estuary systems (Barnegat Bay, Terrebonne-Timbalier Bays and St. Catherines-Sapelo, GA) had atmospheric contributions that were greater than 30% of the total N inputs to the estuary. Atmospheric N accounted for 58% of the total N input to Barnegat Bay estuary because N from wastewater treatment plants in the Barnegat Bay watershed/estuary system were discharged offshore. Atmospheric deposition accounted for 47% of the total N load to Terrebonne-Timbalier Bays, LA and 61% of the total N input to St. Catherine-Sapelo, GA. The majority (85 and 56 %, respectively) of the total atmospheric N inputs to these two estuaries was from direct deposition to the surface of the estuary. This pathway was found to be an important contributor to the total atmospheric N inputs for watershed/estuary systems with ratios of watershed areas to estuary surface water areas less than 15. For the 22 watershed/estuary systems with area ratios less than 15, direct deposition accounted for 20 to 86% of the total atmospheric N input, with an average of 48 %. In contrast, direct deposition accounted from < 20% of the total atmospheric N inputs for systems with area ratios greater than 15. Thus, atmospheric N deposition is not one of the dominant watershed N sources, but it makes a significant contribution to the total N input for watershed/estuary systems with watershed to surface water areas of less than 15. Because of the importance of atmospheric N deposition for these systems, continued efforts are needed to reduce emissions of NO_x from combustion and NH_3 emissions from fertilizer and livestock waste.

Acknowledgements. We are extremely grateful for the help that we received from Percy Pacheco (NOAA-Special Projects Office, Silver Spring, MD) and R. "Srini"vasan (Texas Agriculutural Experiment Station, Temple, TX). Percy provided us with numerous data sets, without his help we would not have completed this project. Srini ran the SWAT model to estimate NPS N runoff from urban areas. We thank both Bruce Mertz and Thomas Simpson of the Maryland Department of Agriculture/University of Maryland Systems for their contributions to our agricultural budgets. We are also extremely grateful for the help of Nancy Castro, Amy Hall, and Barbara Jenkins of the Appalachian Laboratory. This book chapter is publication ES-67-99 of the Appalachian Laboratory, University of Maryland System Center for Environmental Science, and contribution no. 2255 of the Virginia Institute of Marine Science, College of William and Mary. The U.S. EPA Offices of Air and Water funded this work.

References

Alexander, R.B. and R.A. Smith. 1990. County-level estimates of nitrogen and phosphorus fertilizer use in the United States 1945-1985. Open-File Report nr 90-30. Reston (VA): US Geological Survey.

Berner, E.K. and R.A. Berner. 1987; The global water cycle. Englewood Cliffs (NJ): Prentice Hall.

Boynton, W.R., J.H. Garber, R. Summers and W.M. Kemp. 1995. Inputs, transformations, and transport of nitrogen and phosphorus in Chesapeake Bay and selected tributaries. Estuaries 18:285-314.

Bokuniewicz, H. 1980. Groundwater seepage into Great South Bay, New York. Estuarine and Coastal Marine Science 10: 437-444.

Bureau of Census. 1993. Census of agriculture. Vol. 1: Geographic area series. CD-ROM. Washington (DC): Data user services division, US Department of Commerce.

Castro, M.S. and R.P. Morgan II. 1999. Input-output budgets of major ions for a forested watershed in western Maryland, Water, Air and Soil Pollution XX:1-17.

Colbourn, P. 1992. Denitrification and N_2O production in pasture soil: the influence of nitrogen supply and moisture. Agriculture, Ecosystems and Environment 39: 267-278.

David, M.B., L.E. Gentry, D.A. Kovacic and K.M. Smith. 1997. Nitrogen balance in and export from an agricultural watershed. J. Environ. Qual. 26:1038-1048.

D'Elia, C.F., L.W. Harding, Jr., M. Leffler and G.B. Mackiernan. 1992. The role and control of nutrients in Chesapeake Bay. Wat. Sci. Tech. 26:2635-2644.

Evans, H.J. and L.E. Barber. 1977. Biological nitrogen fixation for food and fiber production. Science 197: 332-339.

Fisher, D.C. and M. Oppenheimer. 1991. Atmospheric nitrogen deposition and the Chesapeake Bay estuary. Ambio 20:102-108.

Giblin, A.E. and A.G. Gaines. 1990. Nitrogen inputs to a marine embayment: the importance of groundwater. Biogeochemistry 10: 309-328.

Groffman, P.M., G. Howard, A.J. Gold, and W.N. Nelson. 1996. Microbial nitrate processing in shallow groundwater in a riparian forest. Journal of Environmental Quality 25: 1309-1316.

Hallegraeff, G. 1993. A review of harmful algal blooms and their apparent global increase. Phycologia 32:79-99.

Harvey, A.E., M.F. Jurgensen, M.J. Larsen and R.T. Graham, 1987, Decaying organic material and soil quality in the inland northwest: a management opportunity, USDA For. Serv. Gen. Tech. Rep. INT-225.

Hendrickson, O.Q., 1990, Asymbiotic nitrogen fixation and soil metabolism in three Ontario forests, Soil Bio. Biochem. 22, 967-971.

Hinga, K.R., A.A. Keller and C.A. Oviatt. 1991. Atmospheric deposition and nitrogen inputs to coastal waters. Ambio 20:256-260.

Howarth, W.R., L.F. Elliot, J.J. Steiner, J.H. Davis, and S.M. Griffith. 1998. Denitrification in cultivated and noncultivated riparian areas of grass cropping systems. Journal of Environmental Quality 27: 225-231.

Howarth. R.W., G. Billen, D.Swaney, A. Townsend, N. Jaworski, K. Lajtha, J.A. Downing, R. Elmgren, N. Caraco, T. Jordan, F. Berendse. J. Freney, V. Kudeyarov, P. Murdoch, and Zhu, Zhao-Liang. 1996. Regional nitrogen budgets and riverine N & P fluxes for the drainages to the North Atlantic ocean: Natural and human influences. Biogeochemistry 35:75-139.

Jarvis, S.C., D. Barraclough, J. Williams, and A.J. Hook. 1991. Patterns of denitrification loss from grazed grassland: Effects of N fertilizer inputs at different sites. Plant and Soil 131: 77-88.

Jacobs, T.C. and J.W. Gilliam. 1985. Riparian losses of nitrate from agricultural drainage waters. Journal of Environmental Quality 14: 472-478.

Jaworski, N.A., R.W. Howarth and L.J. Hetling. 1997. Atmospheric deposition of nitrogen oxides into the landscape contributes to coastal eutrophication in the Northeast United States. Environmental Sci. Technol. 31:1995-2004.

Jordan, T.E., D.L. Correll, and D.E. Weller. 1993. Nutrient interception by a riparian forest receiving inputs from adjacent cropland. Journal of Environmental Quality 22: 467-473.

Jordan, T.E. and D.E. Weller. 1996. Human contributions to terrestrial nitrogen flux: Assessing the sources and fates of anthropogenic fixed nitrogen. Bioscience 46:655-664.

Jordan, T.E., D.L. Correll, and D.E. Weller. 1997a. Nonpoint Source Discharges of Nutrients from Piedmont Watersheds of Chesapeake Bay. Journal of American Water Resources Association 33(3): 631-645.

Jordan, T.E., D.L. Correll, and D.E. Weller. 1997b. Effects of Agriculture on Discharges of

Nutrients from Coastal Plain Watersheds of Chesapeake Bay. Journal of Environmental Quality 26(3): 836-848.

Keeney, D.R. 1979. A mass balance of nitrogen in Wisconsin. Wisconsin Academy of Sciences, Arts, and Letters 67:95-102.

Lapointe, B.E. and W.R. Matzie. 1996. Effects of stormwater nutrient discharges on eutrophication processes in nearshore waters of the Florida Keys. Estuaries 19(2B): 422-435.

Lowrance, R.R., R.L. Todd, and L.E. Asmussen. 1984. Nutrient cycling in an agricultural watershed: 1. Phreatic movement. Journal of Environmental Quality 13: 22-27.

Lee, V. and S. Olsen. 1985. Eutrophication and management initiatives for the control of nutrient inputs to Rhode Island coastal lagoons. Estuaries 8(2B): 191-202.

Meybeck, M., D.V. Chapman and R. Helmer. 1989. Global freshwater quality: a first assessment. World Health Organization/United Nations Environment Programme Basil Blackwell, Inc., Cambridge, MA.

Meisinger, J.J. and G.W. Randall. 1991. Estimating nitrogen budgets for soil-crop systems. In: Managing nitrogen for groundwater quality and farm profitability. R.F. Follett, D.R. Keeney, and R.M. Cruse (eds). Soil Science Society of America, Inc. Madison, Wisconsin. pp. 85-125.

Nelson, W.N., A.J. Gold, and P.M. Groffman. 1995. Spatial and temporal variation in groundwater nitrate removal in a riparian forest. Journal of Environmental Quality 24: 691-699.

Nixon, S. 1986. Nutrient dynamics and productivity of marine coastal waters. pp. 97-115. In: B. Clayton and M. Behbehani (eds.) Coastal Eutrophication, The Alden Press, Oxford.

Nixon, S. 1995. Coastal marine eutrophication: A definition, social causes and future concerns, Ophelia 41:199-220.

Pacheco, P.A. 1993. Point Source Methods Document, The National Coastal Pollutant Discharge Inventory, National Oceanic and Atmospheric Administration.

Pacheco, P.A. 1999. Coastal assessment and data synthesis (CA&DS) framework. National Coastal Assessments (NSA) Branch, Special Projects Office (SPO), National Ocean Service (NOS), National Oceanic and Atmospheric Administration (NOAA) Silver Spring, Maryland.

Paul, J.W. and B.J. Zebarth. 1997a. Denitrification during the growing season following dairy cattle slurry and fertilizer application for silage corn. Canadian Journal of Soil Science 77(2): 241-248.

Paul, J.W. and B.J. Zebarth. 1997b. Denitrification and nitrate leaching during the fall and winter following dairy cattle slurry application. Canadian Journal of Soil Science 77(2): 231-240.

Paerl, H. 1988. Nuisance phytoplankton blooms in coastal, estuarine and inland waters. Limnol. Oceanogr. 33:823-847.

Paerl, H. 1995. Coastal eutrophication in relation to atmospheric nitrogen deposition: Current perspectives. Ophelia 41:237-259.

Paerl, H. 1997. Coastal eutrophication and harmful algal blooms: Importance of atmospheric deposition and groundwater as "new" nitrogen and other nutrient sources. Limnol. Oceanogr. 42:1154-1112.

Peterjohn, W.T. and D.L. Correll. 1984. Nutrient dynamics in an agricultural watershed: Observations on the role of a riparian forest. Ecology 65: 1466-1475.

Roskoski, T.P., 1980, Nitrogen fixation in hardwood forest of the northeastern United States, Plant and Soil, 54, 33-44.

Ruz-Jerez, B.E., R.E. White, and P.R. Ball. 1994. Long-term measurement of denitrification in three contrasting pastures grazed by sheep. Soil Biology and Biochemistry 26(1): 29-39.

Ryden, J.C. 1983. Denitrification loss from a grassland soil in the field receiving different rates of nitrogen as ammonium nitrate. Journal of Soil Science 34: 355-365.

Ryther, J. and W. Dunstan, 1971. Nitrogen, phosphorus and eutrophication in the coastal marine environment. Science 171:1008-1112.

Simmons, G.M. Jr. 1989. The Chesapeake Bay's hidden tributary: submarine groundwater discharge. In: Proceedings of ground water issues and solutions in the Potomac River basin/Chesapeake Bay region. Washington DC: Cosponsored by the Association of Groundwater Scientists, et al. and George Washington University, pp. 9-28.

Schlesinger, W.L. and A.E. Hartley. 1992. A global budget for atmospheric NH_3. Biogeochemistry 15:191-211.

Schnabel, R.R. and W.L. Stout. 1994. Denitrification loss from two Pennsylvania floodplain soils. Journal of Environmental Quality 23: 344-348.

Stevenson. F.J. 1982. Origin and distribution of nitrogen in soil. In Nitrogen in agricultural soils (Editor F.J. Stevens). Madison (WI): American Society of Agronomy.

Thornewaite, C.W. and J.R. Mather. 1957. Instructions and Tables for Computing Potential Evapotranspiration and the Water Balance. Laboratory of Climatology, Centerton, NJ, Publication 10: 185-311.

U.S. Environmental Protection Agency. 1980. Design manual: onsite wastewater treatment and disposal systems. Publication Number 625/1-80-012. Office Water Program Operations and Office of Research and Development, Washington, D.C. 391pp.

U.S. EPA. 1993. Chesapeake Bay groundwater toxics loadings workshop proceedings. Basinwide toxics reduction strategy reevaluation reports CBP/TRS96/93. Annapolis, MD. U.S. EPA Chesapeake Bay Program Office.

Valiela, I. and J.E. Costa. 1988. Eutrophication of Buttermilk Bay, a Cape Cod coastal embayment: Concentrations of nutrients and watershed nutrient budgets. Environmental Management 12(4): 539-553.

Valiela, I., J.E. Costa, K. Foreman, J.M. Teal, B. Howes, and D. Aubrey. 1990. Transport of groundwater-borne nutrients from watersheds and their effects on coastal waters. Biogeochemistry 10: 177-197.

Vitousek, P.M., J.D. Aber, R.W. Howarth, G.E. Likens, P.A. Matson, D.W. Schindler, W.H. Schlesinger, and G.D. Tilman, 1997. Human alterations of the global nitrogen cycle: causes and consequences. Issues in Ecology, 1:1-15.

Valigura, R., W. Luke, R. Atrz and B. Hicks, 1996. Atmospheric nutrient inputs to coastal areas: Reducing the uncertainties. US-NOAA Coastal Ocean Program Decision Analysis Series No. 9 Washington DC.

Weiskel, P.K. and B.L. Howes. 1991. Quantifying dissolved nitrogen flux through a coastal watershed. Water Resources Research 27(11): 2929-2939.

Winchester, J.W. and J. Fu. 1992. Atmospheric deposition if nitrate and its transport to the Apalachicola Bay estuary in Florida. Water, Air, and Soil Pollution 65: 23-42.

Woodmansee, R.G. 1978. Additions and losses of nitrogen in grassland ecosystems. BioScience 28:448-453.

5

A Comparison of Independent N-loading Estimates for U.S. Estuaries

R. E. Turner, D. Stanley, D. Brock, J. Pennock, and N. N. Rabalais

Abstract

We assembled 27 recent and independently derived Total Nitrogen (TN) loading estimates for watersheds of east coast and Gulf of Mexico estuaries (watershed size range: 27 to 90,672 km^2; yield range: 44 to 2,722 kg N km^{-2} y^{-1}). These results were compared to those from two other recent studies that used national water quality data bases and computer modeling or statistical analyses to estimate loadings. The loadings from these other studies average about 50 to 60 % of the 27 independently derived TN loading estimates. In many cases the individual estimates are more likely to achieve fuller inclusion of all nitrogen sources to the estuary than these other two efforts, because of the inclusion of groundwater, sewerage, or site-specific point and non-point discharges.

The nitrogen yield from these coastal estuarine watersheds is strongly related to population density, and the per capita yield (kg N $person^{-1}$) is lower than for the large watersheds draining into the Northern Atlantic, including the Mississippi River. The variability in TN yield per capita is strongly related to the percentage of the landscape that is harvested cropland, which is generally higher in the larger watersheds. A simple statistical model of population density and % of the watershed that is harvested cropland describes 79 % of the variation in TN yield (kg N km^{-2}) for 11 basins, including the Mississippi River basin.

The amount of direct atmospheric deposition to the estuarine surface rises to about 25 % of the TN loading to the estuary, when the estuarine surface: watershed area ratio is 0.2. About 20 % of all estuarine surface area in the US, distributed in 12 estuaries, has a ratio of 0.2, or higher. The significance of direct atmospheric nitrogen loadings to the estuarine surface is thus responsive to several geomorphic and socio-economic factors that range greatly across U.S. estuaries.

Introduction

Nitrogen is an important element for many biological processes and may limit aquatic production when sparse, or stimulate eutrophic conditions when in excess. The quantity of nitrogen delivered to coastal estuaries has increased greatly this century and the form of nitrogen has changed as water use intensified, land use shifted, and population density increased [Nixon, 1992; Howarth, 1996; Jordan and Weller, 1996]. Atmospheric nitrogen deposition is a potentially significant component of the nitrogen load delivered to estuaries [e.g., Paerl, 1995, 1997, 1999], is changing with global economic development (see other Chapters in this volume), and may contribute to water quality problems in the coastal zone [Howarth, 1988; Nixon, 1992].

The results of many recent local efforts to quantify these nitrogen loadings in individual estuaries are the subject of this analysis. We gathered for as many independent estimates of nitrogen loading to estuaries as we could find, and converted those data into comparable units and categories. The resulting collection of loading estimates for individual estuaries spanned a wide range of input rates, and represent estuaries and bays ranging greatly in watershed size and other characteristics. The extent of these data was great enough that we attempted to illuminate functional relationships relevant to the questions about the significance of atmospheric inputs. We also wanted to know how these nitrogen loading estimates compared to each other and also with similar, more global, analyses conducted from a national perspective.

Methods

Data on nitrogen loading from the land to estuaries were assembled from government reports, the peer-reviewed scientific literature, and unpublished reports. These loading estimates were scrutinized for obvious errors of omission, and mathematical errors. The loadings were converted to a common basis (loading from the watershed (kg N km^{-2} y^{-1}) to the estuary). The Total Nitrogen (TN) yield is the sum of all inorganic and organic fractions of particulate and dissolved nitrogen. In practice, not all forms of nitrogen are measured through monitoring programs, which is an inadequacy of this analysis.

An annotated bibliography of these data sources is in Appendix 1, which includes a brief description of the data calculation, various contributions to the total loading (e.g., point and non-point sources, atmospheric deposition, etc.), and land use categories and/or estuarine surface area. Data on population density circa 1980 and land use circa 1987 are from the NOAA Office of Strategic Assessment, Rockville, Maryland.

These local estimates of TN loading were not computed in strictly similar ways, nor did they include the same source estimates, or calculate these sources in the same manner. Some estimates used coefficients for nitrogen sources from literature values, while others used empirically-defined loading coefficients. A few provided sparse documentation of the data derivation. Almost all local estimates used gauged stream discharge and nutrient concentrations for loading calculations, but not all included ungaged portions of the watershed. The Delaware Bay and Chesapeake Bay estimates did not include significant portions of the watershed. A watershed loading might also include a known point discharge as though it left the watershed and entered the estuary. However, a pipe might carry nitrogen seaward of the estuary (e.g., Barnegat Bay). We did not use nitrogen loading estimates when such obvious circumstances could not be corrected.

The estimate of nitrogen loading from the land to the estuary is very sensitive to the estimated area used to compute the loading. Some of the estimated areas are different from the earlier estimates by the investigator(s) and/or from state or federal agencies. In some cases the state boundaries defined a watershed, whose size was therefore underestimated. It may be difficult to define watershed boundaries where the relief is flat, e.g., south Florida. Water is diverted from one watershed to another in some instances. We used the NOAA Strategic Assessment Division estimates of watershed areas in all cases. NOAA's watershed boundaries, codes and names have changed from 1996-1998, because of several updates to NOAA's Coastal Assessment Framework (CAF) (P. Pacheco, personal communication). Population density was normalized to area by using the same population estimate given by NOAA for 1980. In this way we did not mix estimates for different areas, and minimized the chances of introducing spurious errors unknowingly. The data on the harvested cropland area for each watershed were available for the northeast U.S. and Gulf of Mexico estuaries [EPA/NOAA Team, 1987, and Anon. no date, respectively].

These local estuary estimates from the literature were compared to regional nitrogen loading data from a regional assessment of the Gulf of Mexico estuaries by NOAA [Anon, no date] and a USGS statistical modeling effort for all U.S. estuaries conducted by Smith et al. [1997]. The USGS modeling effort (the "SPARROW" model data) was updated for each estuary by R.A. Alexander (see Chapter 7). The NOAA and SPARROW models thus used national survey information on land use, fertilizer purchase, population density, water supply and water quality data, and soil types to estimate nitrogen loadings to individual estuaries. The SPARROW model is a synthetic and statistical modeling of coastal watersheds throughout the U.S. The SPARROW analysis derived output variables for individual estuaries based on the relationships between variables when analyzed as an entire data set. In other words, individual estuaries were not treated separately from one another to produce individual statistical relationships. Rather, all were included together in the production of the general statistical relationships. This general model was then used to generate estimates for individual estuaries.

The NOAA estimates for the Gulf of Mexico were derived from a consistent methodology using literature and stream gauging data for individual estuaries. The statistical or synthetic efforts, because of their global view, lacked a certain degree of intimate knowledge of the local estuary that may have been better incorporated into the independently derived local models. For example, these global models did not develop loading estimates for the following circumstances: (1) Lake Pontchartrain occasionally has river water diverted through the Bonnet Carré spillway (used to lower river stage heights above New Orleans), (2) groundwater sources may be a very important nitrogen source to parts of South Florida, and, (3) local sewer discharge diversions out of the watershed [e.g., Caloosahatchee estuary] may go unnoticed. When not included in the estimates, the significance of ungaged loadings was sometimes over-looked. In most of these examples, the result of these omissions would contribute to the under-estimation of nitrogen loading to an individual estuary.

Results

Comparisons

A comparison of various estimates of nitrogen loading is shown in Figure 1. Both the NOAA and the SPARROW estimates are much lower than the results from the regional estimates by independent efforts. A linear regression of each comparison is shown. The NOAA estimates and the results from the SPARROW model average 59 % and 48 %, respectively, of the regional estimate. The NOAA estimates are strongly correlated with the independently-derived loading estimates ($R^2 = 0.97$), but are lower. A linear regression of the results from the independent (local) models (Y) and the SPARROW model (X) has a slope of 1.02, but the $R^2 = 0.49$. Some differences between the independently derived and the NOAA and SPARROW model loading estimates may arise because loadings from groundwater, ungaged river stream elements, and downstream point sources that are not included in the NOAA and SPARROW data sets. These groundwater, ungaged or downstream sources, for example, averaged 27 % of the total regional N loading (n=10; Appendix 1).

Predicting Loading Values

The relationship between population density and the nitrogen yield from the watershed (kg N km^{-2}) is shown in Figure 2. This positive relationship is not unexpected. Others have demonstrated that these types of relationships exist for large rivers [e.g., Peierls et al., 1991; Howarth et al., 1996], and population density is sometimes used as a surrogate for non-point loading in the construction of watershed nitrogen loading estimates. What

TABLE 1. Summary of the independently developed nitrogen loading estimates of nitrogen going from the watershed (kg N km^{-2} y^{-1}) to the estuary.

Estuary (km^2)	Watershed Area	TN from Land (kg N km^{-2}y^{-1})	Notes
Casco Bay: Maquoit Bay	32	562	Horsley and Whitten, 1996
Great Bay	2,590	317	Mosher, 1995
Waquoit Bay	38	610	Valiela et al., 1997
Narragansett Bay	4,662	1,862	Nixon et al., 1995
Long Island Sound	40,943	702	Stacey, 1998
Barnegat Bay	878	415	Moser et al., 1998
Delaware Bay	34,965	1,513	Nixon et al., 1996
Delaware Inland bays	633	2,722	Horsley and Whitten, 1998; part of watershed is absent
Chesapeake Bay	56,980	2,340	Boynton et al., 1995; portions of watershed are missing
Upper Maryland bays	172	2,706	Boynton, 1993; population data unavailable
Sinepuxent Bay	27	846	Boynton, 1993
Chincoteague Bay	141	1,830	Boynton, 1993
Newport Bay	113	2,281	Boynton, 1993
Maryland coastal bays	453	2,218	Boynton, 1993
Neuse River	16,000	249	Boyer et al., 1993, 1994
Pamlico River	11,600	309	Stanley, 1997
North Inlet (Bly Creek)	4	925	Dame et al., 1991
Charleston Harbor	40,922	129	McKellar and Blood, 1997
Indian River	4,926	528	Woodward-Clyde Consultants
Charlotte Harbor	8,547	255	Coastal Environmental, Inc. 1995
Sarasota Bay	389	1,091	Heyl, 1992
Tampa Bay	5,895	585	Zarbock et al., 1996
Galveston Bay	60,597	526	Longley, 1994
Matagorda Bay	90,672	71	Longley, 1994
San Antonio Bay	26,736	301	Longley, 1994
Aransas Bay	6,424	193	Longley, 1994
Corpus Christi Bay	44,841	44	Longley, 1994
Range	4 - 90,672	44 - 2,722	

was not anticipated was the very robust coefficient of determination (R^2) of 0.78, which is excellent when one considers the uncertainty of the water quality data [Brock, this volume], and the mix of survey dates (e.g., 1980 for the population data and various years between 1980 and 1992 for the water quality data).

The Y intercept at zero population density for the small coastal watershed that is shown in Figure 2 is 78 kg N km^{-2}y^{-1}. This number may be considered an estimate of the nitrogen loading during what is sometimes called 'pristine' conditions, or the background 'natural' loading rate when Europeans first penetrated the Mississippi River valley in significant numbers. An analysis of European field notes and diaries from the 1600s suggests that the population of the Midwest in the early 1600s was no more than 106,000 [Ubelaker, 1992], or 0.1 persons km^{-2}. Diseases introduced by the Europeans

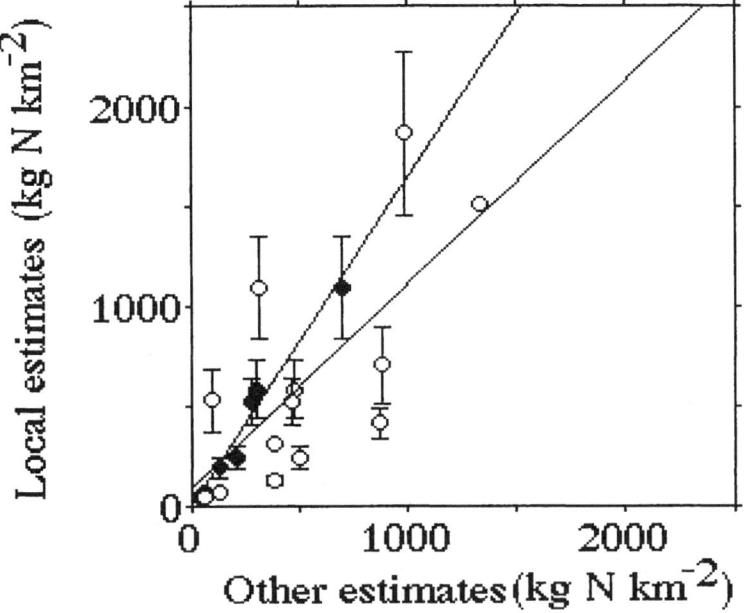

Figure 1. A comparison of the estimates of the independent or 'local' estimates of nitrogen loading from the watershed versus two different estimates of many estuaries (NOAA and SPARROW model estimates). The confidence values shown are from Chapter 7 [Brock, this volume]. Note that the local estimates are generally much higher than the estimates from general survey estimates.

reduced the native American population drastically, but it is problematic to determine how much they were affected. Even if the native American population was 100 times larger than that before the 1500s, the population density would have been less than 10 persons km^{-2}, or equal to 108 kg N km^{-2}y^{-1} suggested by the relationship shown in Figure 2. This is 'natural' loading estimate is comparable to others found in the scientific literature, which range from 76 to 110 kg N km^{-2}y^{-1} for streams in the lower 48 states [Howarth et al., 1996, Nixon 1997]. Several Texas estuaries have a TN loading below 100 kg N km^{-2}y^{-1}.

It is significant that the relationship between these two variables for the small coastal watersheds is so distinctly different for the relatively larger river watersheds draining into the Atlantic Ocean, which are discussed by Howarth et al. [1996]. The slope and intercept of a linear regression of the large river drainage basins versus population density (Figure 2) are 2 and 3 times larger, respectively, than that of the coastal systems. The larger values from large watersheds are implicitly suggested by the results of the SPARROW model, which incorporates a larger attenuation coefficient for nitrogen moving through smaller watersheds compared to larger watersheds. Land use in interior watersheds is, in general, quite different than for coastal systems. Agricultural activity, for example, tends to be a proportionally larger part of the watershed in larger watersheds, i.e., inland areas have more agriculture (Figure 3). The percentage of land that is 'harvested cropland' is a surrogate metric for a variety of factors that affect

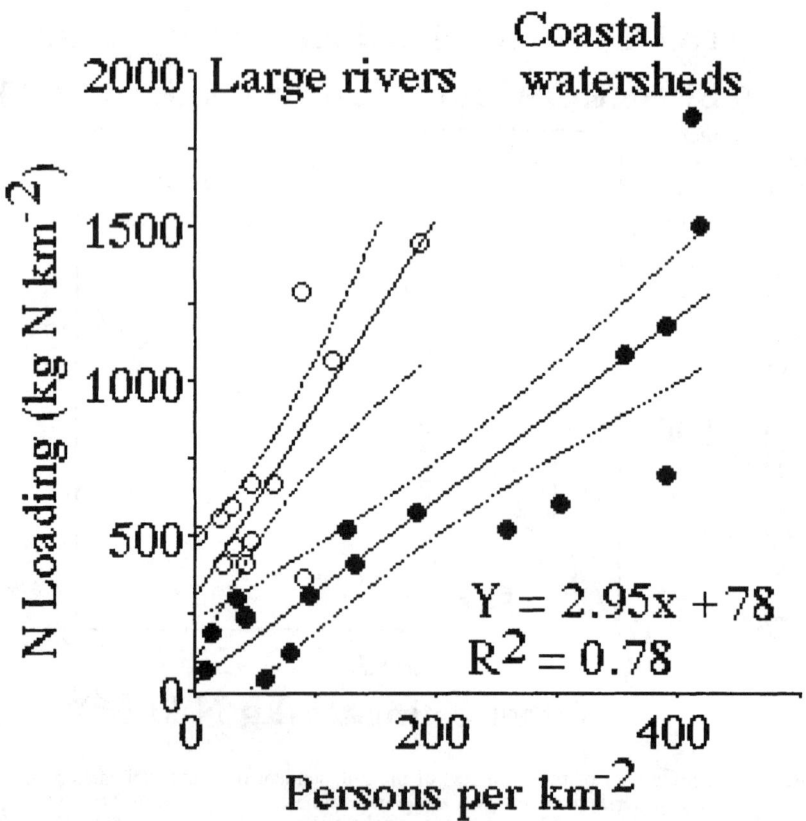

Figure 2. Annual nitrogen loading from the watershed (kg N km^{-2}) and population density (individuals km^{-2}) for these data (filled circles) and large river basins entering into the Atlantic Ocean [Howarth et al., 1996]. The coastal systems are based on the local model estimates (Table 1) and the population density circa 1980. A linear regression of the data is shown and the 95 % confidence interval. The statistics for these data are shown. Note: Mississippi River data are not included.

nitrogen loading. Obviously fertilizer use is applied to croplands and can be exported in streams [about 20 % for the Mississippi River basin, for example; Turner and Rabalais, 1991]. Other factors affected include groundwater flow and content, soil type, vegetative cover, and related activities such as livestock production [e.g., Jordan and Weller, 1996].

Much of the variation in the per capita TN loading to estuaries can be described by the variation in land use within the watershed, as shown in Figure 4. The water quality data are from regional N loading estimates referenced in the Appendix 1, except for the Mississippi River data which are from Turner and Rabalais [1991]. A multiple regression predicting the TN load per area watershed (kg N km^{-2}) as a function of (1) the dependent variables population density, and, (2) the percentage land as harvested cropland, yielded an R^2 of 0.79 (adjusted coefficient of determination; $p = 0.0009$; $n = 11$). Thus small coastal watersheds tend to have a lower proportion of their watershed in row crops and other harvested agricultural land uses, and a lower per capita nitrogen yield.

Incidentally, the relationship shown in Figure 4 should not be linear for areas with low population density and sparse agricultural activity. Under those conditions, the yield per individual is high, and dependent on the 'background' nitrogen loading, and the data

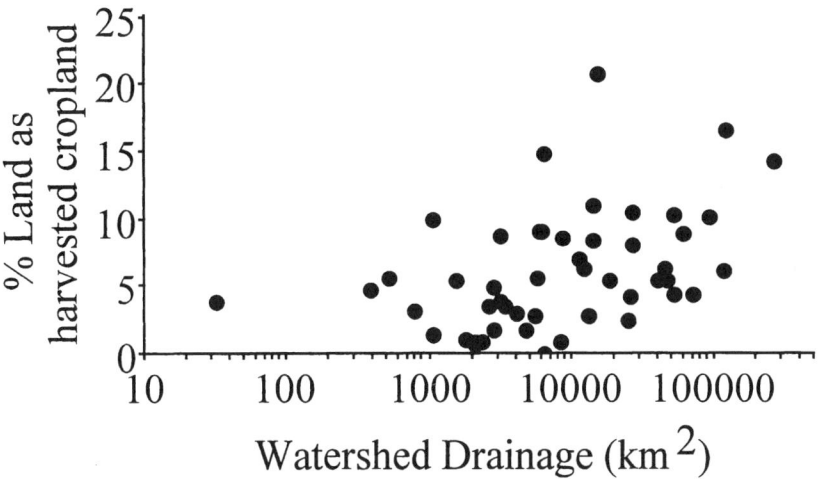

Figure 3. Agricultural land use and watershed drainage size for estuarine watersheds in the US.

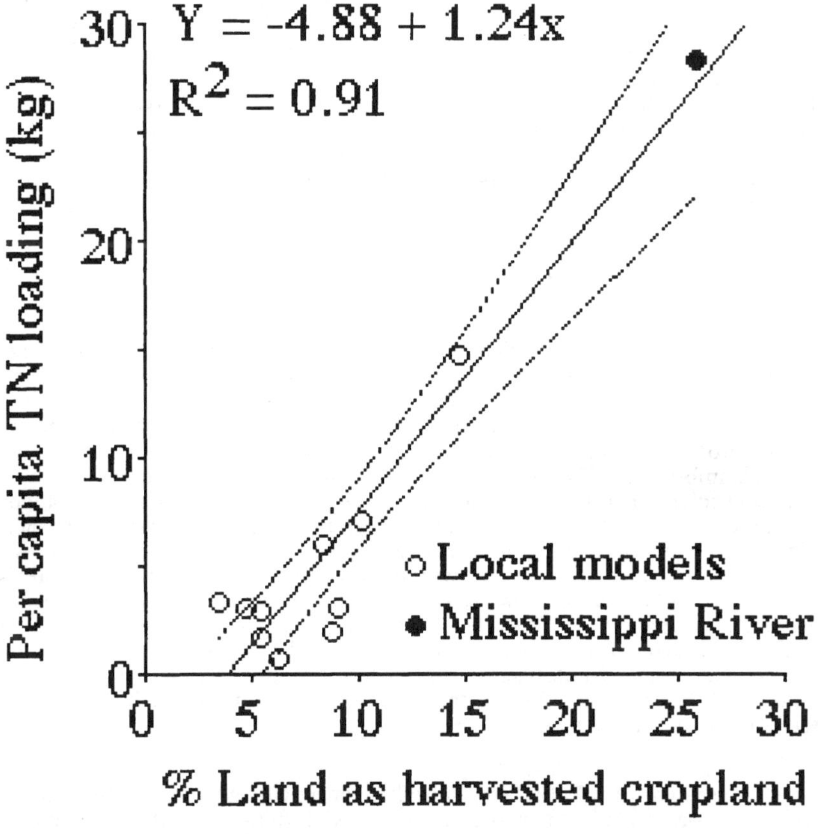

Figure 4. Annual per capita nitrogen loading from the watershed (kg N per capita) versus the percentage of the watershed area that was in harvested cropland circa 1980. A linear fit of all the data is shown, with a 95 % confidence limits for the true mean of the Y values.

points should cluster in the upper left hand quadrant of Figure 4. These data, however, are not available.

Atmospheric Sources

Paerl [1995, 1997] and others have suggested that atmospheric loading can be a very important component of N loading to some estuaries. The N loading from land was added to that from the atmosphere [Appendix 1] to derive a TN loading to the estuarine surface. Direct atmospheric N loading to the estuarine surface becomes a larger percentage of the TN loading as the size of the estuarine water surface increases relative to the watershed (Figure 5). The larger the water surface is compared to land surface, the more important the role of atmospheric deposition is to the estuarine nitrogen balance. Figure 5 shows that the contribution of atmospheric deposition is non-trivial and can be as much as 50 % for some estuaries. Figure 5 suggests that a ratio of 0.2 for the area of estuarine: watershed surface will, on the average, have 25 % of the estuarine nitrogen load (exclusive of marine sources) from atmospheric sources.

But, how much of the estuarine surface area in the U.S. presently has a significant nitrogen load from direct atmospheric loading of nitrogen? Figure 6 depicts the geographical distribution of this water: land ratio for almost all the U.S. estuaries (a few estuaries have missing values). About 10 % of all U.S. estuaries have an estuarine water surface: watershed ratio above 0.2, implying that 25 % of the TN loading to the estuarine surface is derived directly from atmospheric loading onto the estuarine surface. If we use a threshold value of 0.2 to detect the approximate ratio where 25% of the TN loading to the estuarine surface (from land and from atmosphere) is derived from atmospheric sources, then 19 % of all estuarine surface area, from 4 % of the estuarine drainage area, receives about 25 % of the estuarine nitrogen loading from atmospheric sources. These estuaries receive drainage from about 1 % of the total watershed drainage area in the U.S. and are identified in Table 2.

Conclusions

The local estimates of estuarine nitrogen loading are of sufficient quality to be useful for comparative purposes, despite the inherent differences in data sources, formulation and uncertainty in data quality, if not record length. These local, independently derived estimates are of a similar, and higher, magnitude as the estimates from more synthetic efforts, probably because of the regional investigators have good familiarity with the available information. The simple geomorphic ratio of the water surface: watershed size is a convenient first-order screening metric to scale the importance of atmospheric loading to estuaries. The quantification between nitrogen load from the watershed to the estuary and the over-riding influence of population size and land use, especially agricultural activities, is consistent with the results of other analyses. The precision of these estimates, if not accuracy, suggests that a landscape scale for management and regulation is appropriate. The role of atmospheric deposition to an estuarine nitrogen load is dependent on these land use practices, but also strongly influenced by the local geomorphic setting. Atmospheric nitrogen loading directly to the estuarine surface is estimated to account for >25% of the TN loading for approximately 19 % of the U.S. estuarine water surface area.

Acknowledgments. Dr. P. Pacheco is thanked for providing various NOAA land use data and advice on their interpretation. Support to RET from the Gulf of Mexico Program, U.S. Environmental Protection Agency and from the Lake Pontchartrain Basin Foundation is acknowledged.

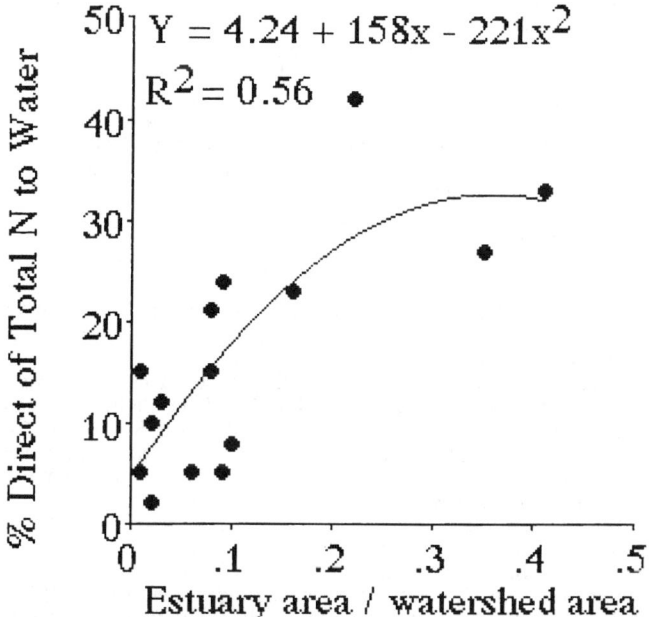

Figure 5. The relationship between the direct (atmospheric) TN loading to the estuarine water surface expressed as percentage of the total loading (the sum of nitrogen loading from the watershed and directly to the estuarine water surface) versus the ratio of estuarine : watershed area.

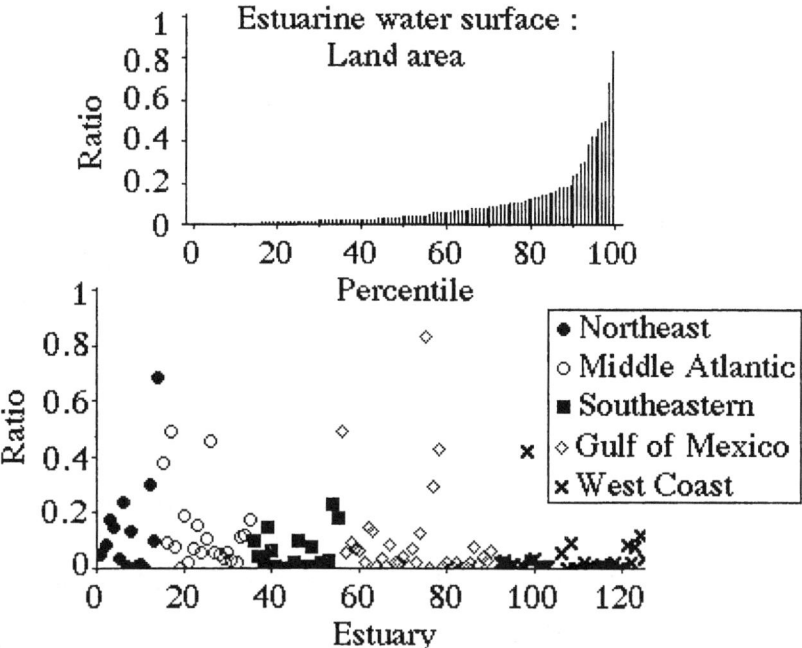

Figure 6. The ratio of estuarine water surface : watershed for U.S. estuaries (bottom) and the percentile distribution of the U.S. estuaries for this ratio (upper paper). The U.S. estuaries are arranged around the U.S. coastline from northeast to Gulf of Mexico to northwest. The data are from NOAA [Anon. no date].

TABLE 2. U.S. estuaries with a water surface area: watershed ratio of 0.2 or larger. These estuaries are likely to have 25 % of the TN load to the estuary to come from direct atmospheric deposition to the estuarine surface.

Muscongus Bay (Maine)	Chincoteague Bay (Maryland)
Massachusetts Bay (Massachusetts)	Florida Bay (Florida)
Cape Cod Bay (Massachusetts)	Breton/Chandeleur Sound (Louisiana)
Buzzards Bay (Massachusetts)	Barataria Bay (Louisiana)
Great Bay (0.19) New Jersey	Terrebonne/Timbalier Bay (Louisiana)
Gardiners Bay (New Jersey)	Santa Monica Bay (California)

References

Anonymous, *Nutrient Enrichment Potential Watershed Assessment and Comparison (NEPWAC) System Gulf of Mexico Component, Ver. 0.5*, NOAA/EPA Team on Near Coastal Waters. Strategic Assessment Branch, Ocean Assessments Division, National Ocean Service, Rockville, Maryland, No date.

Boyer, J.N., R.R. Christian, and D.W. Stanley, Patterns of phytoplankton primary productivity in the Neuse River estuary, North Carolina, USA, *Marine Ecology Progress Series*, 97, 287-297, 1993.

Boyer, J.N., D.W. Stanley, and R.R. Christian, Dynamics of NH_4 and NO_3 uptake in the water column of the Neuse River estuary, North Carolina. *Estuaries* 17, 361-371, 1994.

Boynton, W.R., J.H. Garber, R. Summers, and W.M. Kemp, Inputs, transformations and transport of nitrogen and phosphorus in Chesapeake Bay and selected tributaries, *Estuaries*, 18, 285-314, 1995.

Boynton, W.R., L. Murray, W.M. Kemp, J.D. Hagy, C. Stokes, F. Jacobs, J. Browers, S. Souza, B. Krinsky, and J. Seibel, Maryland's Coastal Bays: An Assessment of Aquatic Ecosystems, Pollutant Loadings, and Management Options. Maryland Department of the Environment, Chesapeake Bay and Special Projects Branch, Baltimore, MD, 1993.

Caraco, N.F. and J.J. Cole, Human impact on nitrate export: An analysis using major world rivers. *Ambio*, 28, 167-170, 1999.

Coastal Environmental, Inc., 1995, Estimates of Total Nitrogen, Total Phosphorus, and Total Suspended Solids Loadings to Charlotte Harbor, Florida. Prepared for Southwest Florida Water Management District, Tampa, Florida, 1999.

Dame, R.F., J.D. Spurrier, T.M. Williams, B. Kjerfve, R.G. Zingmark, T.G. Wolaver, T.H. Chrzanowski, H.N. McKellar, F.J. Vernberg, Annual material processing by a salt marsh-estuarine basin in South Carolina, USA. *Marine Ecology Progress Series* 72, 153-166, 1991

EPA/NOAA, Strategic Assessment of Near Coastal Waters. EPA/NOAA Team on Near Coastal Waters, Interim Draft, Nov. 1987, National Oceanic and Atmospheric Administration, Washington, D.C., 1987.

Heyl, M.G., Point- and Nonpoint-source Loading Assessment of Sarasota Bay, pp. 12.4-12.19, In Sarasota Bay National Estuary Program. Sarasota Bay: Framework for Action. Published by the Sarasota Bay National Estuary Program, Sarasota, FL, 1992.

Horsley & Whitten, Inc., Identification and Evaluation of Nutrient and Bacterial Loadings to Maquioit Bay, Brunswick, and Freeport, Maine, Final report submitted to Casco Bay Estuary Project, Portland, Maine, 1996.

Horsley & Witten, Inc., Assessment of Nitrogen Loading to the Delaware Inland Bays. Prepared for the Center for the Inland Bays, Nassau, DE, 1998.

Howarth, R.W. (Ed.) Nitrogen Cycling in the North Atlantic Ocean and its watersheds: Report of the International SCOPE Nitrogen Project. *Biogeochemistry*, 35, 1996.

Howarth, R.W., Nutrient limitation of net primary productivity in marine ecosystems. *Annual Rev. Ecol. and Systematics*, 19, 89-110, 1988.

Howarth, R.E., G. Billen, D. Swaney, A. Townsend, N. Jaworski, K. Lajtha, J.A. Downing, R. Elmgren, N. Caraco, T. Jordan, F. Berendse, J. Freney, V. Kudeyarov, P. Murdoch and

Z. Zhao-Liang, Regional nitrogen budgets and riverine N & P fluxes for the drainages to the North Atlantic Ocean: Natural and human influences. *Biogeochemistry*, 35, 75-139, 1996.

Jordan, T.E. and D.E. Weller, Human contributions to terrestrial nitrogen flux. *BioScience*, 46, 655-664, 1996.

Longley, W.L. (Ed.), Freshwater Inflows to Texas Bays and Estuaries: Ecological Relationships and Methods for Determination of Needs. Texas Water Development Board and Texas Parks and Wildlife Department, Austin, TX. 386 pp., 1994.

McKellar, H. and E. Blood, Nitrogen Budget for Cooper River/Charleston Harbor Estuary. Report to the Charleston Harbor Project, Office of Ocean and Coastal Resource Management, SC Department of Health and Environmental Control, Charleston, SC, 1997.

Mosher, B.W., Assessment of Atmospheric Non-point Source Nitrogen Input to the Great Bay Watershed and Estuary. Final report submitted to New Hampshire Coastal Program, Office of State Planning, Concord, NH, 1995.

Moser, F.C., S.P. Seitzinger, R.J. Murnane, and R.G. Lathrop, Local and regional nitrogen sources to a shallow coastal lagoon, Barnegat Bay, New Jersey. Unpublished manuscript submitted to *Marine Ecology Progress Series*, 1998.

Nixon, S.W., Coastal marine eutrophication: A definition, social causes, and future concerns. *Ophelia*, 41, 199-219, 1992.

Nixon, S.W., Prehistoric nutrient inputs and productivity in Narragansett Bay. *Estuaries*, 20, 253-261, 1997.

Nixon, S.W., S.L. Granger, and B.L. Nowicki, An assessment of the annual mass balance of carbon, nitrogen, and phosphorus in Narragansett Bay. Biogeochemistry, 31, 15-61, 1995.

Nixon, S.W., J.W. Ammerman, L.P. Atkinson, V.M. Berounsky, G. Billen, W.C. Boicourt, W.R. Boynton, T.M. Church, D.M. Di'toro, R. Elmgren, J.H. Garber, A.E. Giblin, R.A. Jahnke, N.J.P. Owens, M.E.Q. Pilson, and S.P. Seitzinger, The fate of nitrogen and phosphorus at the land-sea margin of the North Atlantic Ocean. *Biogeochemistry*, 35, 141-180, 1996.

Paerl, H.W., Coastal eutrophication in relation to atmospheric nitrogen deposition: Current perspectives. *Ophelia*, 41, 237-259, 1995.

Paerl, H.W., Coastal eutrophication and harmful algal blooms: Importance of atmospheric deposition and groundwater as "new" nitrogen and other sources. *Limnology and Oceanography*, 42, pt. 2, 1154-1165, 1997.

Paerl, H.W. and D.R. Whitall, Anthropogenically-derived atmospheric nitrogen deposition, marine eutrophication and harmful algal bloom expansion: is there a link?, *Ambio*, 28, 307-311, 1999.

Peierls, B., N. Caraco, M. Pace and J. Cole, Human influence on river nitrogen. *Nature*, 350, 386-387, 1991.

Smith, R.A., G.E. Schwarz and R.B. Alexander, Regional interpretation of water-quality monitoring data. *Water Resources Research*, 33, 2781-2798, 1997.

Stanley, D.W., Water Quality in the Pamlico River Estuary: 1989-1996. Institute for Coastal and Marine Resources, East Carolina University, Technical Report No. 97-02. Greenville, NC, 1997.

Stacey, P.E., (Draft) Report on Nitrogen Loads to Long Island Sound. Connecticut Department of Environmental Protection, Bureau of Water Management, Hartford, CT, 1998.

Turner, R.E. and N.N. Rabalais, Changes in the Mississippi River this century: Implications for coastal food webs. *BioScience*, 41, 140-147, 1991.

Ubelaker, D.H., The sources and methodology for Mooney's estimates of North American Indian populations, in *The Native American Populations of the Americas in 1492, edited by* W. M. Denevan, pp. 243-288, Wisconsin Press, Madison (WI), Wisconsin Press, 1992.

Valiela, I., G. Collins, J. Kremer, K. Lajtha, M. Geist, B. Seely, J. Brawley, C.H. Sham, Nitrogen loading from coastal watersheds to receiving estuaries: New method and application. *Ecological Applications*, 7, 358-380, 1997.

Woodward-Clyde Consultants, Marshall McCully & Associates, Inc., and Natural Systems Analysts, Inc., Loading Assessment of the Indian River Lagoon. Prepared for Indian River Lagoon National Estuary Program, Melbourne, FL, 1994.

Zarbock, H.W., A.J. Janicki, and S.S. Janicki, Estimates of Total Nitrogen, Total Phosphorus, and Total Suspended Solids to Tampa Bay, Florida. Technical Appendix: 1992-94 Total Nitrogen Loading to Tampa Bay, Florida. Tampa Bay National Estuary Program Technical Publication #19-96, 1996.

6

Atmospheric Nitrogen Flux From the Watersheds of Major Estuaries of the United States: An Application of the SPARROW Watershed Model

Richard B. Alexander, Richard A. Smith, Gregory E. Schwarz, Stephen D. Preston, John W. Brakebill, Raghavan Srinivasan, and Percy A. Pacheco

Abstract

To assess the atmospheric contributions of total nitrogen (TN) in riverine exports to coastal and estuarine ecosystems in the United States, we applied a nationally calibrated empirical watershed model, SPARROW (Spatially Referenced Regression on Watershed attributes), to a selected set of 40 major coastal watersheds. In contrast to conventional statistical watershed models, SPARROW uses a mechanistic structure in the correlation of observations of stream nitrogen load with spatial data on contaminant sources, landscape characteristics, and stream properties, allowing separate estimation of the quantities of nitrogen delivered to streams and the outlets of watersheds from point and diffuse sources. We calibrated the model using data from a national set of 374 watersheds. Application of the model to the 40 coastal watersheds indicates that atmospheric nitrogen contributions to riverine export range over nearly two orders of magnitude, from 4 to 326 kg km^{-2} yr^{-1}. The atmosphere is estimated to contribute from 4 to 35 percent of the TN in stream export with a median of 20 percent. The highest atmospheric contributions are observed in the northeastern and Mid-Atlantic watersheds of the United States. Uncertainties in the estimates, based on the standard error of prediction, range from 40 to 100 percent and vary inversely with watershed size. Agricultural sources typically contribute the largest share of nitrogen (more than one third in most basins), followed by the aggregate contributions of other diffuse sources.

Nitrogen Loading in Coastal Water Bodies: An Atmospheric Perspective
Coastal and Estuarine Studies, Pages 119-170
Copyright 2001 by the American Geophysical Union

Municipal and industrial point sources are similar in magnitude to atmospheric contributions in most watersheds, but represent the largest share (35-88%) of nitrogen in one half of the North Atlantic watersheds and in several watersheds of the Gulf region. Comparisons of the SPARROW model with other national and regional watershed models indicate general agreement in the predictions of TN export over a wide range of watershed sizes. Assessments of atmospheric sources to coastal watersheds are likely to benefit from a continued effort to integrate the mechanistic descriptions of deterministic models with the empirical methods of estimating watershed-scale rate processes and their uncertainties in statistical models.

1. Introduction

An increase in the flux of nitrogen to coastal marine systems during the latter half of the 20th century has caused eutrophication of many temperate estuaries, including numerous estuarine systems in the United States such as the Chesapeake Bay, Louisiana shelf, and New York bight [Diaz and Rosenberg, 1995; Vitousek et al. 1997]. Although there is ample evidence that the problem is predominantly cultural in origin [Nixon, 1995; Vitousek et al. 1997], uncertainties remain over the relative importance of the various human activities that supply nitrogen to coastal waters. Use of nitrogenous fertilizers, atmospheric emissions of nitrogenous compounds, and point-source discharges of nitrogenous wastes have all increased significantly since 1950 [CEQ, 1989; NASS, 1998; Alexander and Smith, 1990; Battaglin and Goolsby, 1994]. To date, information on the relative importance of the major anthropogenic sources of nitrogen in coastal systems has been frequently obtained by comparing the quantities of nitrogen released to the environment from those sources (e.g., Howarth, et al., 1996]. Due to denitrification, storage, and biological utilization of nitrogen in the watershed, however, only a fraction of the released nitrogen is ultimately transported to coastal waters. Moreover, the fraction transported from each local source is a function of both source-dependent and source-independent (e.g., stream channel properties) characteristics of the watershed, and has been difficult to reliably estimate [Alexander et al. 2000]. Nevertheless, such information is needed for efficient management of coastal ecosystems because the cost effectiveness of controlling individual nitrogen sources varies with the fraction of the nitrogen from each source that is transported to coastal waters.

Assessing the role of atmospheric sources of nitrogen in coastal eutrophication is an important example of both the value and difficulty of quantifying source-specific nitrogen transport in watersheds. Atmospheric emissions of nitrogenous compounds, an important source of nitrogen to coastal waters [Vitousek et al. 1997; Valigura et al. 1996; Howarth et al. 1996; Fisher and Oppenheimer, 1991], are produced in both the electric utility and transportation sectors of the economy, and are currently under environmental regulation as air pollutants. Thus, better information on the effects of these compounds on coastal water quality will provide for a more comprehensive evaluation of an existing regulatory policy. The difficulty of quantifying the movement of atmospherically deposited nitrogen through watersheds is increased by the geographic complexity of the sources, with some of the nitrogen falling directly on coastal and estuarine water surfaces and deposition occurring at varying rates throughout estuarine watersheds. Moreover, the use of different methods of assessment and the investigation of limited numbers of coastal watersheds in previous studies [Valigura et al. 1996] have prevented a consistent, comprehensive assessment of the importance of atmospheric sources to the nitrogen budgets of major coastal and estuarine ecosystems in the United States.

The general problem of tracing nitrogen flux through watersheds is complicated by the difficulty of establishing a spatially continuous mass balance between the in-stream flux of nitrogen, the rate of nitrogen supply from terrestrial and atmospheric sources, and the rate of removal due to denitrification and storage on the landscape and in stream channels. High quality stream monitoring data are frequently available for multiple sites within coastal watersheds, but these measure the integrated effects of nitrogen supply and loss processes operating continuously over the landscape and in stream channels. In this analysis, we use a watershed modeling technique [SPARROW—SPAtially-Referenced Regression On Watershed attributes; Smith et al. 1997; Alexander et al. 2000] that combines observations of stream water quality with spatial data on contaminant sources and watershed characteristics to separately estimate the quantities of nitrogen delivered to streams and the outlets of watersheds from point and diffuse sources. To provide a spatially consistent assessment of nitrogen flux from atmospheric as well as terrestrial sources to coastal waters of the conterminous United States, we calibrated the model using data from a national set of monitored watersheds. This model was previously used to quantify nitrogen deliveries to coastal waters from atmospheric and other sources in the Mississippi River Basin [Alexander et al. 2000]. The model expands on a previous national application of the SPARROW method [Smith et al. 1997; see section 2 for details]. We applied the model to 40 of the 42 major coastal watersheds of the conterminous United States (see fig. 1 in Chapter 1) selected for analysis in this book (two of the 42 watersheds lacked sufficient data), based on the use of local data on nitrogen sources and watershed attributes. We also compared the results of the national SPARROW model with those of other national and regional watershed models to assist in evaluating the model predictions. The analysis is presented in six sections. Following the introduction, the methodology and data sources for calibrating the national model are described. Section three presents the estimated model parameters. The results for the 40 U.S. estuaries are presented and discussed in section four. Section five presents the results of a comparison of model predictions with those of other large-scale watershed models. Conclusions appear in the final section.

2. Model Description and Data Sources

2.1 Background

A variety of deterministic and statistical methods have been used to develop models of nitrogen transport from human and natural sources to coastal waters. The simplest deterministic approaches [Jaworski et al. 1992; Jordan and Weller, 1996; Howarth et al. 1996] provide a static accounting of nitrogen inputs (e.g., fertilizer application, atmospheric deposition) and outputs (e.g., river export, crop removal). Where sources or sinks (e.g., denitrification in soils and streams, groundwater storage) cannot be measured, estimates are often determined as a difference between the measured inputs and outputs. These simple mass balance models assume that loss processes operate equally on all sources and that the relative contributions of sources to coastal waters are proportional to nitrogen inputs to the watersheds. More complex deterministic models of nitrogen flux [e.g., Bicknell et al. 1997; Srinivasan et al. 1993; Whitehead et al. 1998] simulate nitrogen availability, transport, and attenuation processes according to mechanistic functions and describe both spatial and temporal variations in sources and sinks. A third approach [export coefficient method; e.g. Fisher and Oppenheimer, 1991; Delwiche and Haith, 1983] has been to apply the reported yields (flux per unit area) from small, homogeneous watersheds to the variety of land types contained within larger heterogeneous basins.

There are important limitations to these approaches. First, the reported yields for various land types are highly variable [Beaulac and Reckhow, 1982; Frink, 1991; Johnson, 1992], reflecting variations in climatic conditions, nutrient supplies, and terrestrial and stream loss processes as well as methodological differences related to sampling, measurement, and statistical estimation. Thus, the extrapolation of land-use yields to unmonitored watersheds can produced imprecise and potentially biased estimates of export. A more refined version of the export coefficient method, which accounts for spatial variations in source inputs and landscape and climatic conditions, has been successfully applied to catchments in the U.K. [e.g., Johnes, 1996] although this approach typically requires considerable monitoring to calibrate and verify the model [Johnes and Heathwaite, 1997]. Second, there are potential inaccuracies in "scaling up" the results of catchment models and field-scale measurements to larger watersheds [Rastetter et al. 1992; Beaulac and Reckhow, 1982], which in addition may exclude the effects of changes in nitrogen loss rates with stream properties (see section 3). Knowledge of in-stream losses may be especially important in large watersheds to account for the quantities of nitrogen removed during the lengthy in-channel movement of water from upstream locations to coastal ecosystems. However, reported estimates of in-stream nitrogen loss show large variations, ranging from less than five percent to as much as 80 percent of the external inputs of nitrogen to streams. Although studies suggest the importance of many chemical and physical properties of streams on nitrogen loss [e.g., Seitzinger and Kroeze, 1998; Seitzinger, 1988; Howarth et al. 1996; Kelly et al. 1987; Behrendt, 1996; Rutherford et al. 1987], there is poor knowledge of how in-stream nitrogen loss varies over a range of river sizes.

Statistical approaches to modeling nitrogen flux in coastal basins have their origins in simple correlations of stream nitrogen measurements with watershed sources and landscape properties. Recent examples [Mueller et al. 1997; Jaworski et al. 1997; Peierls et al. 1991; Howarth et al. 1996] include regressions of coastal nitrogen flux on population density, atmospheric deposition, and agricultural sources. Simple correlative models consider sources and sinks to be homogeneously distributed in space, do not separate terrestrial from in-stream loss processes, and rarely account for the interactions between sources and watershed processes. In contrast to their deterministic analogs, which often have intensive data and calibration requirements, the simplest empirical models have the advantage of being more easily applied at large spatial scales. An additional noteworthy advantage of the statistical approach is the ability to quantify errors in model parameters and predictions.

2.2 The SPARROW Model

The SPARROW model used in this application is a hybrid method for empirically estimating the quantities of nitrogen delivered from point and diffuse sources to streams and watershed outlets. A spatially referenced regression technique is used to estimate in-stream flux as an exponential function of the landscape and hydraulic characteristics of watersheds. Surface water flow paths are defined according to a digital network of rivers to which stream monitoring data, nutrients sources, and watershed characteristics are spatially referenced. In contrast to conventional regression-based watershed methods [e.g., Jaworski et al. 1997; Mueller et al. 1997], this approach uses a mechanistic structure to track nitrogen transport through watersheds. Estimation is accomplished by establishing a mass balance in streams and rivers between the in-stream flux of nitrogen, the rate of nitrogen supply from atmospheric and terrestrial sources, and the rate of removal due to denitrification and storage on the landscape and in aquatic systems (i.e., channels and reservoirs). By regressing in-stream nutrient flux on watershed attributes, these rates are simultaneously estimated such that an optimal mass balance is attained between the observed and predicted flux at multiple stream monitoring locations. The method treats monitored flux as an in-stream nitrogen source in nested (i.e., overlapping) watersheds, thereby providing accurate stream data and numerous intervening river segments and drainage areas to assist in estimating the rates of nitrogen supply and removal. The contributions of various types of nitrogen sources to streams (e.g., fertilizer use, livestock wastes, municipal point sources) are quantified in the procedure from data on the magnitude and location of the source inputs. Large spatial variability in the explanatory variables improves the ability of the technique to separate "true" spatial variations in sources and processes from random variations related to measurement error and unexplained environmental factors. The empirical method also provides estimates of the uncertainty (e.g., 90% confidence intervals) in model coefficients and predictions of flux.

The model of in-stream nitrogen flux (F_i) is developed for a set of watersheds containing a defined set of stream reaches to which stream monitoring data and data on nutrient inputs and watershed characteristics are spatially referenced (see diagram in figure 1). The in-stream flux at the downstream end of a given reach i is expressed as the sum of all monitored and unmonitored sources of nitrogen in the set of upstream reaches denoted by $J(i)$. The defined set of upstream reaches for a given reach i accounts for nested watersheds in the monitoring network such that the set excludes reaches that are either located above or include monitoring stations upstream of reach i. An estimable version of the expression is written as

$$F_i = \left\{ \sum_{n=1}^{N} \sum_{j \in J(i)} S_{n,j} \beta_n \exp(-\alpha' Z_j) \exp(-k' T_{i,j}) \right\} \varepsilon_i \quad (1)$$

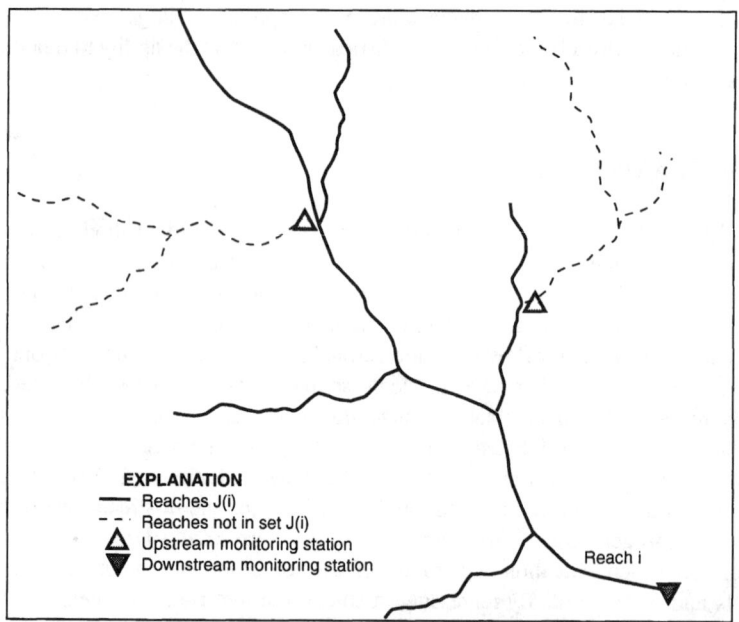

Figure 1. Schematic of hydrologic network and monitoring stations modeled by SPARROW. Modified from Smith et al. (1997).

where $S_{n,j}$ is a measure of nitrogen mass from source n applied to the drainage of reach j, β_n is a source-specific coefficient, $\exp(-\alpha' Z_j)$ is an exponential function describing the proportion of available nitrogen mass delivered to reach j as a function of land-to-water delivery coefficients (defined by vector α) and their associated terrestrial characteristics, Z_j, in the drainage to reach j, $\exp(-k' T_{i,j})$ is the proportion of nitrogen mass present in reach j that is transported to downstream reach i as a function of a first-order rate of N loss (k' defined according to a vector of four discrete classes of channel size) per unit water travel time ($T_{i,j}$), and ε is a multiplicative error term assumed to be independent and identically distributed across independent sub-basins defined by the intervening drainage located between stream monitoring sites. The product of the land-to-water delivery function (and its associated coefficients) and the nonpoint-source coefficients quantifies the fraction of the source inputs that are delivered to rivers and streams. The delivery of nitrogen to streams is a function of several landscape characteristics of watersheds (Z_j), including soil permeability, stream density, and air temperature. The reciprocol of the land-to-water delivery (Z_j^{-1}) was applied where a positive relation to in-stream flux is expected (e.g., stream density). The land-to-water delivery function is equal to one for point-source inputs. In estimating the source coefficients, upstream monitored inputs are treated as in-stream sources with their land-to-water delivery fraction, $\beta_n \exp(-\alpha' Z_j)$, constrained to unity. We assume that the in-stream attenuation of nitrogen is identical for all sources according to the estimated rates of in-stream loss.

The functional form of equation (1) dictates the use of nonlinear regression estimation methods. Coefficient estimation was performed on the log transforms of the summed quantities and error term in equation (1) using non-linear least-squares estimation in the SAS procedure PROC MODEL [SAS, 1993]. Model residuals were examined for normality, constant variance, and nonlinear patterns to determine if regression assumptions were satisfied. Robust estimates of uncertainty of model parameters and predictions (standard errors and confidence intervals) were obtained in bootstrap analyses [Efron, 1982]. Bootstrap estimates of model parameter uncertainty were made by resampling with replacement (200 iterations) from the spatial set of stream monitoring flux data. Additional uncertainty related to unexplained variability in the model was included in bootstrap estimates of model predictions by resampling with replacement from the model residuals. In using the model to predict stream flux at unmonitored locations (reaches and watershed outlets), bootstrap estimates of residual errors were added to the predicted flux values at approximately the spatial scale of the monitoring station watersheds.

The methods and the version of the model applied here expand on a previous national application of SPARROW [Smith et al. 1997] in several ways. First, we refined the empirical in-stream loss function to more accurately describe nitrogen attenuation in large rivers (> 283 m^3/s). Second, we detrended wet-fall measurements of atmospheric nitrogen deposition to reflect sources for the base year 1987 adjusted for long-term average precipitation, providing mean estimates of deposition consistent with the estimates of stream flux. Third, we calibrated the model using fewer stream monitoring stations (374 rather than 414), which were selected to provide contemporaneous records of nitrogen over a longer time period through 1992. Finally, we improved the estimates of uncertainty in the model predictions of source contributions to stream export by accounting for variability in the observed data that is unexplained by the model (i.e., residual errors).

2.3 In-Stream Monitoring Data

The SPARROW model was calibrated using U.S. Geological Survey (USGS) stream monitoring records of total nitrogen (TN) for the period 1978 to 1992 at 374 sites in the conterminous United States [see fig. 2; Alexander et al. 1998]. Estimates of TN flux in streams (F_i in the spatial model in equation 1) were computed from periodically collected water-column measurements of total nitrogen (sum of nitrate-nitrite and kjeldahl nitrogen—ammonia plus organic N) and daily measurements of streamflow. Field sampling, analytical procedures, and quality assurance methods are performed according to USGS stream monitoring protocols [Alexander et al. 1998]. Water samples were collected for nutrient analysis according to a monthly to quarterly schedule. We estimated the mean TN flux at each monitoring station by applying conventional flux-estimation techniques [Cohn et al. 1989] to measurements of total nitrogen and daily streamflow, based on a log-linear model relating stream flux to streamflow, decimal time, and season of the year [Smith et al. 1997]. This method uses the more complete daily record of streamflow in estimating flux, and provides statistically unbiased estimates with greater precision than can be obtained from methods that rely on a simple averaging of the observed concentration and streamflow data. The number of samples at monitoring

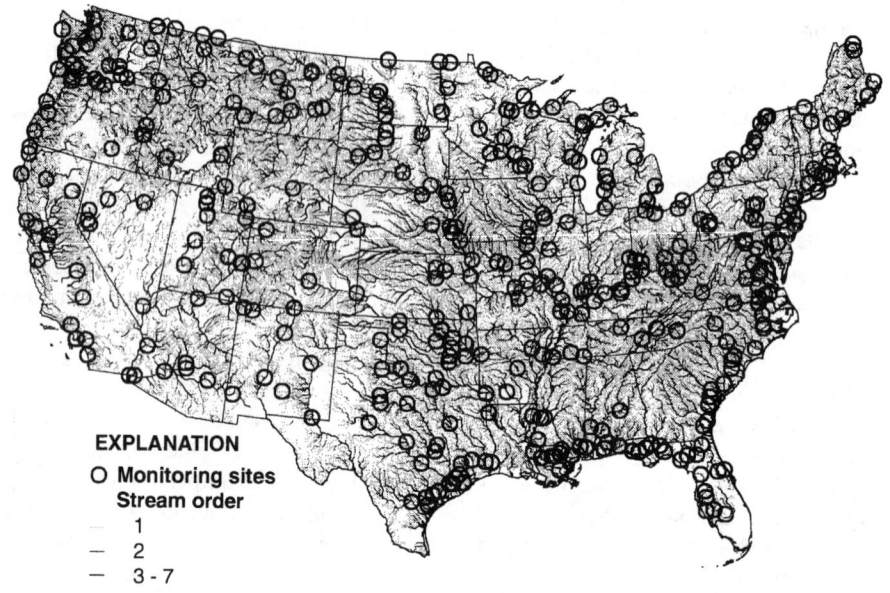

Figure 2. USGS stream monitoring locations and river reaches in the conterminous United States.

sites was typically about 90 for the period of record (interquartile range from 75 to 105). These periodic samples provided relatively good coverage of the hydrograph; more than 75 percent of the stations have nutrient records that cover more than 95 percent of the streamflow events. Mean TN flux estimates are based on 1987 nitrogen inputs, adjusted for mean streamflow conditions for the years 1970-92. Source inputs for 1987 are representative of average inputs over at least the past two decades [Alexander et al. 2000]. Estimates of uncertainty in the mean flux (i.e., standard error of estimate) are determined according to methods in Gilroy et al. [1990]. Additional details of the flux-estimation method used in this analysis are provided in Smith et al. [1997]. Watersheds for the stations range in size from 80 to 2.9 million square kilometers (median=11,700; interquartile range=3,000 to 34,000) with mean streamflow ranging from one to 18,500 cubic meters per second (median=63; interquartile range=20 to 217).

The estimates of mean TN flux at the 374 monitoring stations serve as the dependent variable in the SPARROW model. These estimates span approximately four orders of magnitude from 10^2 to slightly more than 10^6 kg day^{-1}. Yields at the stations range from 1.4 to 3,000 kg km^{-2} yr^{-1} (median=295 kg km^{-2} yr^{-1}; interquartile range from 90 to 594 kg km^{-2} yr^{-1}). The highest TN yields occur in rivers of the midwestern and northeastern portions of the United States (see fig. 3) where the largest agricultural, atmospheric, and point source inputs to watersheds are typically found. The lowest yields are found in the western rivers where both nitrogen sources and runoff tend to be low relative to other areas of the United States. Uncertainties in the estimates of mean TN flux, based on the standard error of estimate expressed as a percentage of the mean (i.e., one standard deviation of the mean), range from about 2 to 19 percent (median=6.2%; interquartile

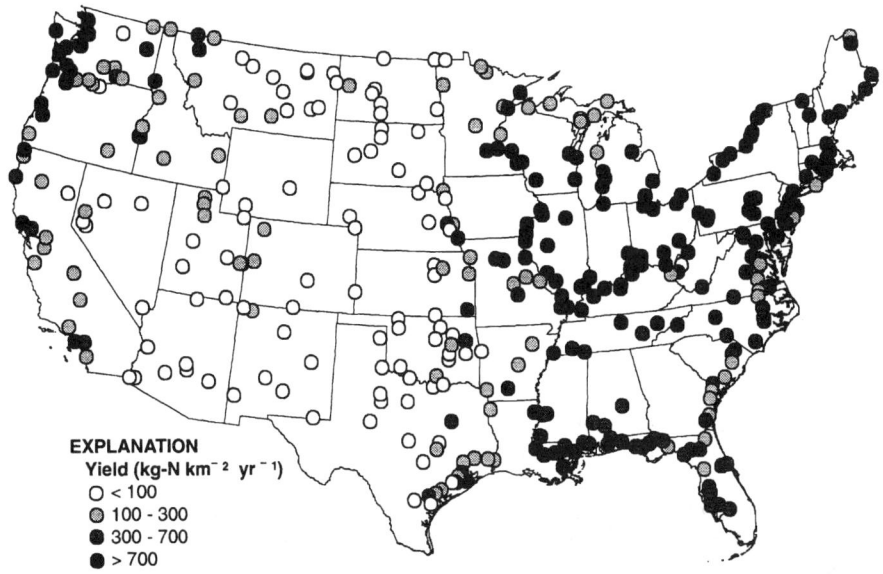

Figure 3. Mean total nitrogen yield at 374 USGS stream monitoring stations, 1978-1992. Estimates are adjusted to reflect 1987 sources and mean streamflow.

range=4.5 to 8.5%). Stations with standard errors greater than 20% of the station mean flux estimate (about 20 sites) were excluded from the spatial calibration of the SPARROW model to reduce the effects of measurement error. In general, prediction errors are lowest at those stations with the largest number of water-quality observations and in larger watersheds in the eastern portions of the United States, where less variable streamflow conditions occur.

2.4 Watershed Data

The spatial watershed data on nitrogen source inputs, physical characteristics of the landscape, and attributes of the digital stream network used in the SPARROW model have been previously described [Smith *et al.* 1997]. However, we modified the spatial estimates of wet deposition of nitrate by detrending the data according to the methods described in the subsequent section. For the base year 1987, we quantified nitrogen inputs to watersheds (the variable $S_{n,j}$ in equation 1) for five major classes of sources including fertilizer use, municipal and industrial point sources, livestock wastes, runoff from nonagricultural land, and atmospheric deposition.

Data on the source inputs and terrestrial characteristics, available for nearly 20,000 land-surface polygons, were referenced to approximately 60,000 stream reaches in a digital stream network using conventional spatial disaggregation methods in a geographic information system [see Smith *et al.* 1997]. The surface water flow paths, defined according to a 1:500,000 scale digital network of rivers for the conterminous United States, cover nearly one million kilometers of channel, and are obtained from the USGS version of the U.S. Environmental Protection Agency River Reach File 1 [ERF1; Alexander *et al.* 1999; see fig. 2]. The river reach network provides the spatial framework in the model for relating in-stream measurements of flux at monitoring stations to landscape and stream channel properties in the watersheds above these stations. The median watershed size of the reaches is 82 km^2 with an interquartile range from 40 to 150 km^2. Stream attributes of the digital network include estimates of mean streamflow and velocity from which water time of travel is computed as the quotient of stream length and mean water velocity [Alexander *et al.* 1999].

2.4.1 Nitrate wet-deposition

Data from the National Atmospheric Deposition Program [NADP, 1993] were used to estimate the long-term mean annual wet deposition within RF1 reach watersheds in the conterminous United States. We used the approximately weekly measurements of nitrate at 188 monitoring sites with continuous records over the period of record from the early 1980s through 1993. We estimated a detrended mean annual nitrate deposition for the base year 1987 similar to that used to estimate total nitrogen flux in streams. This estimate gives the mean annual deposition at each monitoring site for 1987 under mean precipitation conditions.

We computed the detrended mean by adjusting observations of nitrate wet deposition for linear time trend over the period of record at each NADP site, based on a log-linear regression of nitrate deposition on time, precipitation, and season of the year (expressed as trigonometric functions of decimal time). Weekly measurements of nitrate deposition (d_i; the product of concentration and precipitation) for the period of record were regressed on a set of five explanatory variables according to the form

$$\ln(d_i) = \lambda_0 + \lambda_1 n_i + \lambda_2 \sin(2\pi n_i) + \lambda_3 \cos(2\pi n_i) + \lambda_4 \ln(p_i) + \lambda_4 (\ln(p_i))^2 + \varepsilon_i \quad (2)$$

where n_i is decimal time for the *i*th weekly observation, p_i is the *i*th weekly precipitation value, $\sin(2\pi n_i)$ and $\cos(2\pi n_i)$ are trigonometric functions that jointly estimate seasonal variations in deposition, λ are regression coefficients, ε_i is the sampling and model error assumed to be independent and identically distributed, and ln is the natural logarithm. The detrended mean annual nitrate deposition for the base year 1987 at each NADP site (expressed as kg km^{-2} yr^{-1}) is estimated as

$$\overline{D} = (T)^{-1} (\sum_{i=1}^{T} d_i \exp[\lambda_1 (t - n_i) - 0.5 \sigma_{\lambda 1}^2 (t - n_i)^2]) \quad (3)$$

where t is the mid-year decimal value for 1987, $\sigma_{\lambda 1}$ is the standard error of the linear time model coefficient, and T is the number of observations. The models typically explained

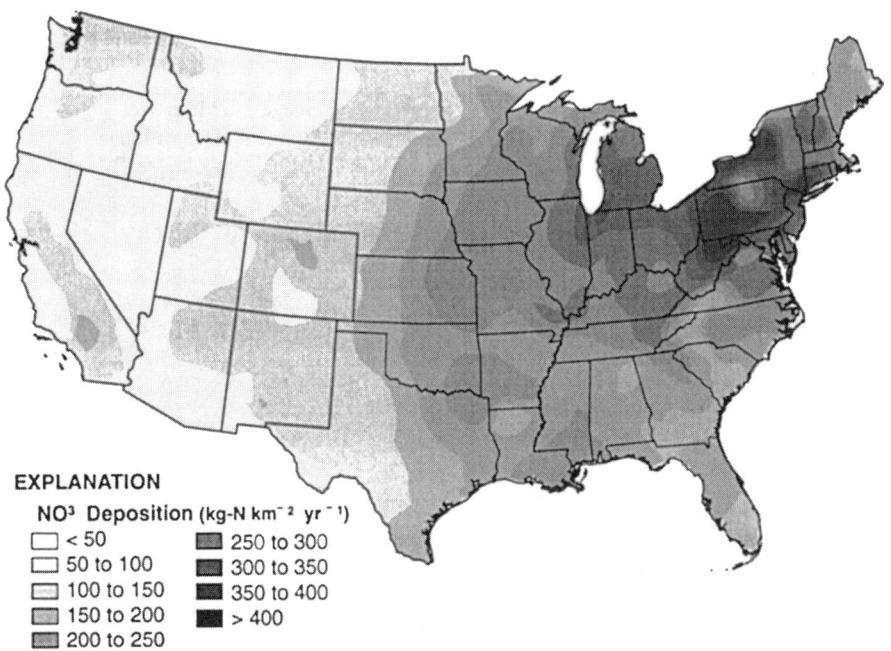

Figure 4. Mean annual nitrate wet deposition in the conterminous United States, 1980-1993. The spatially interpolated estimates at NADP sites are adjusted to reflect 1987 deposition and mean precipitation.

from 55 to 70 percent (median R-squared=64 percent) of the observed temporal variability in nitrate wet deposition. Model residuals exhibited acceptable adherence to regression assumptions.

Spatially continuous values of nitrate wet-deposition were estimated for the United States by linearly interpolating the mean annual nitrate deposition estimates for the 188 NADP monitoring locations through application of the Triangulated Irregular Network (TIN) method in Arc/Info [ESRI, 1996]. The resulting nitrate wet-deposition surface for the conterminous United States is illustrated in figure 4 for a contour interval of 50 kg km^{-2} yr^{-1}. The estimates of nitrate wet deposition span more than two orders of magnitude ranging from lows of less than 10 kilograms per square kilometer in the West to hundreds of kilograms per square kilometer in the East. A distinct pattern of high nitrate deposition occurs over the Ohio Valley and extends into the northeastern United States. The atmospheric deposition mass of contoured surface polygons (10 kg km^{-2} yr^{-1} intervals) was apportioned to the watersheds of individual RF1 reaches according to the ratio of the watershed reach length to the total reach length in deposition polygons [see Smith *et al.* 1997].

Estimates of nitrate wet deposition for the 40 estuarine drainage areas, computed as the mean of the estimated inputs to reach watersheds, range in magnitude from 0.5 to 3.4 kg ha^{-1} yr^{-1} (i.e., 50 to 340 kg km^{-2} yr^{-1}; see table 1). These estimates show a strong spatial correlation (r=0.97) with the values of nitrate wet deposition [Meyers *et al.* this volume]

TABLE 1. Nitrate wet deposition in the drainages of major estuaries of the conterminous United States.

Watershed/Estuary	Number Reaches	Drainage Area (km^2)	Mean Wet Deposition (kg ha^{-1} yr^{-1})
North Atlantic			
1. Casco Bay	30	3093	1.87
2. Great Bay	11	2378	2.02
3. Merrimack River	89	12,906	2.86
4. Massachusetts Bay	9	2,524	2.34
6. Buzzards Bay	14	3,654	2.13
7. Narragansett Bay	27	4,613	2.55
8. Gardiners Bay	1	2,192	2.79
9. Long Island Sound	525	40,289	3.12
10. Hudson River/Raritan Bay	508	41,629	3.33
11. Barnegat Bay	12	1,649	3.06
12. New Jersey Inland Bays	26	3,705	3.03
13. Delaware Bay	250	32,373	3.44
14. Delaware Inland Bays	1	726	2.81
15. Maryland Inland Bays	2	847	2.68
16. Chesapeake Bay	1293	164,156	3.10
South Atlantic			
17. Pamlico Sound	219	24,584	2.18
18. Winyah Bay	423	46,340	1.99
19. Charleston Harbor	597	40,604	1.56
20. St. Helena Sound	151	12,358	1.40
22. St. Catherines/Sapelo S.	6	2,253	1.14
23. Altamaha River	509	36,797	1.77
24. Indian River	3	1,525	1.45
Gulf of Mexico			
25. Charlotte Harbor	57	9,146	1.53
26. Sarasota Bay	2	957	1.53
27. Tampa Bay	41	6,556	1.58
28. Apalachee Bay	79	15,254	1.44
29. Apalachicola Bay	425	52,236	1.63
30. Mobile Bay	1219	115,339	1.71
31. West Mississippi Sound	302	44,448	2.17
32. Barataria Bay	20	6,151	2.26
33. Terrebonne/Timbalier Bay	6	4,095	2.25
34. Calcasieu Lake	68	11,174	1.86
35. Sabine Lake	378	54,081	1.78
36. Galveston Bay	519	63,158	1.65
37. Matagorda Bay	518	117,565	0.98
38. Corpus Christi Bay	231	44,853	0.93
39. Upper Laguna Madre	40	9,065	1.14
40. Lower Laguna Madre	20	7,179	1.15
Pacific			
41. San Francisco Bay	829	108,943	0.46
42. Puget Sound	572	31,166	0.77

TABLE 2. SPARROW total nitrogen model coefficients based on a regression of stream nitrogen flux at 374 river monitoring stations on watershed characteristics. Refer to the methods and equation 1 for an explanation of the model form and coefficients.

Model Parameters	Coefficient Units	Bootstrap Coefficient	Lower 90% CI [b]	Upper 90% CI [b]
Nitrogen source, β				
Point sources	dimensionless	0.394	0.094	0.639
Fertilizer application	dimensionless	1.37	0.605	2.34
Livestock waste production	dimensionless	0.903	0.012	1.97
Atmospheric deposition	dimensionless	4.78	1.84	8.21
Nonagricultural land	kg ha^{-1} yr^{-1}	18.6	6.18	29.3
Land-to-water loss coefficient, α				
Temperature	°F^{-1}	0.017	0.009	0.023
Soil permeability	hr. cm^{-1}	0.036	0.024	0.049
Drainage area per stream length [a]	km^{-1}	0.043	0.017	0.063
In-stream loss rate [c], k				
k_1 (Q < 28.3 m^3 s^{-1})	day^{-1}	0.455	0.344	0.579
k_2 (28.3 m^3 s^{-1} < Q < 283 m^3 s^{-1})	day^{-1}	0.118	0.063	0.176
k_3 (283 m^3 s^{-1} < Q < 850 m^3 s^{-1})	day^{-1}	0.051	0.007	0.092
k_4 (Q > 850 m^3 s^{-1})	day^{-1}	0.005	0.000	0.019
R-Squared		0.881		
Mean square error		0.435		
Number of observations		374		

[a] Variable enters the model in reciprocal form (see Smith et al. 1997).
[b] Minimum bootstrap confidence intervals (CI).
[c] In-stream loss rates fit separately for stream reaches with mean streamflow (Q) corresponding to the indicated intervals.

used in the land-use based model in this volume (also see comparisons in the final chapter). SPARROW estimates are within about 20 percent of these values in most of the watersheds (the ratio of Meyers *et al.* estimates to SPARROW estimates typically range from 1.17 to 1.30; median=1.22). Estimates of nitrate wet deposition according to Meyers *et al.* are also based on NADP data, but include an additional three years of observations (1994-96) and result from the use of different estimation methods.

3. National SPARROW Model Estimates

3.1 Calibration of the Model

Correlating stream TN flux with the spatially referenced data on nitrogen sources and watershed characteristics, we find that a model consisting of 12 explanatory variables explains approximately 88 percent of the variability in the 374 observations of mean annual TN flux. The explanatory variables and estimated coefficients for the TN model are presented in table 2. Estimates of uncertainty in the fitted coefficients in table 2 are expressed as 90 percent confidence intervals based on the bootstrap estimation procedure. Five types of nitrogen sources and three land-to-water delivery factors were statistically significant. The rates of in-stream total nitrogen removal are estimated according to four streamflow classes (table 2).

A comparison of model predictions with observed values of TN flux and yield is shown in figure 5. The differences between the observed and predicted values (i.e., model residuals) indicate acceptable adherence of the errors to the model assumptions. The model residuals are approximately normal with relatively constant variance, and no systematic linear or curvilinear patterns are visible. There is some evidence of a slight overestimation of flux in watersheds with exports less than about 1,000 kg day^{-1} (fig. 5a). Comparisons of observed and predicted yield, which adjust for the effect of basin area giving unit expressions similar to concentration, provide a more stringent evaluation of model performance. Somewhat larger variation occurs in the predicted TN yields than for flux (fig. 5b), but acceptable correlation (R-squared=0.80) is observed between the predicted and observed yield values. Moreover, relatively constant variance occurs in the residuals throughout the range of yield values suggesting the lack of any systematic biases in the estimation of TN yield (fig. 5b). Based on an examination of the distribution of the absolute percent differences between the observed and predicted yields, about half of the predictions are found to be within at least 32 percent of the observed values. A quarter of the predictions are within at least 15 percent of the observed values, whereas a quarter exceed the observed values by more than 61 percent. Only 10 percent of the model predictions exceed the observed values by more than 100 percent. Overall, the TN model performs well enough to serve as a relatively accurate predictive tool for use in estimating TN flux and yield in unmonitored watersheds. Moreover, model residuals are generally well behaved, and would be expected to provide reasonably accurate estimates of the uncertainty associated with the model predictions.

3.2 Model Parameter Estimates

The product of the nitrogen-source coefficients (table 2) and exponential land-to-water delivery function estimate the nitrogen mass that is made available from nitrogen sources and delivered to streams. The quantities of nitrogen delivered to streams may reflect the contributions from other sources not explicitly measured by the input variables, such as dry deposition, fixation by crops, crop imports, and groundwater, as well as the effects of terrestrial processes (e.g., soil denitrification, crop exports, climate, conservation tillage, and subsurface storage and transport). In addition to direct runoff of applied fertilizers, the "fertilizer" source may include fixed nitrogen in leguminous crop residues and other soil nitrogen associated with cropland. Nonagricultural nonpoint sources include nitrogen inputs quantified by the model intercept scaled for nonagricultural land, and thus include remaining nitrogen inputs unaccounted for by other sources in the model. Nonagricultural sources include nitrogen entering streams via runoff and subsurface flows from wetlands as well as from urban, range, forested, and barren lands. Nitrogen from forested and range lands may include biotic fixation (Jordan and Weller, 1996). The model coefficient for point sources (to which the land-to-water delivery function is

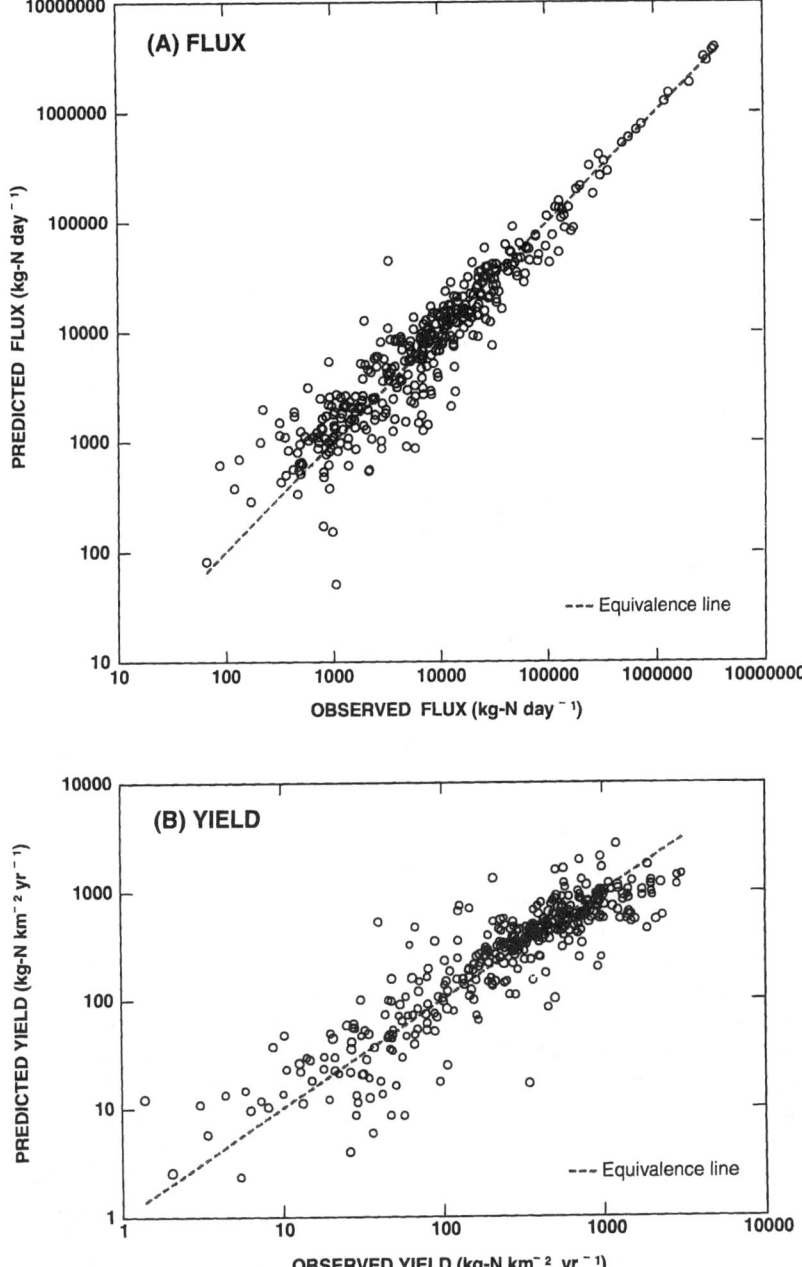

Figure 5. SPARROW model predictions at 374 stream monitoring stations in relation to the observed mean total nitrogen (A) flux and (B) yield.

not applied) is less than unity. This likely reflects model adjustments for declines in point source loads between the 1970s (the time period of the input data) and the 1987 base year used to estimate stream flux [Smith, et al. 1997]. The model estimates of municipal and industrial point-source loadings to streams expressed per capita (based on the sewered population) have a median of 3.8 kg-N person^{-1} (interquartile range of 1.7 to 7.1 kg-N person^{-1}). This compares favorably with literature estimates of per capita rates for residential wastewater effluent ranging from 2.2 to 7 kg-N person^{-1} [Thomann, 1972; US EPA, 1980].

In estimating atmospheric nitrogen contributions, we used wet-deposition measurements of nitrate nitrogen as input to the model, and excluded ammonia deposition to minimize the double accounting of agricultural sources of nitrogen in the input term [Howarth et al. 1996]. The land-to-water delivery fraction for wet nitrate deposition (product of the deposition coefficient and the exponential land-to-water delivery function) exceeds unity, and is consistent with our assumption that atmospheric sources include additional contributions from wet deposition of ammonium and organic nitrogen and dry deposition of inorganic nitrogen, which are not reflected by the input variable. The model would be expected to account for these additional sources to the extent that they are correlated with the measured inputs of nitrate. Although estimates of these other depositional forms are not widely available for the United States, available estimates for the estuarine watersheds [Meyers et al. this volume] indicate that wet nitrate deposition is highly correlated (r=0.78) with dry plus ammonium and organic wet deposition (dry is based on separate dry-fall monitoring data and model predictions). Estimates of the ratio of total (dry plus wet) deposition to nitrate wet deposition for the estuarine watersheds [Meyers et al. this volume] range from 3.2 to 4.0 with an average of 3.6 (uncertainties in estimates of total deposition may exceed a factor of two). In addition, estimates of total NOy deposition (dry plus wet oxidized forms) in the United States have been reported to range from 2 to 3 times the nitrogen in wet deposition [Fisher and Oppenheimer, 1991].

Landscape processing and transport of nitrogen to rivers are modeled as an exponential function of several physical descriptors of the watersheds, including air temperature, soil permeability, and stream density (table 2). Temperature and soil permeability are inversely correlated with stream flux; the former providing possible evidence of a large-scale temperature-related denitrification effect [Seitzinger, 1988] and the latter suggesting greater long-term storage of nitrogen or permanent loss in areas of higher soil permeability (possibly via soil denitrification or immobilization by soil microbes). Drainage density is positively correlated with in-stream nitrogen flux suggesting that watersheds with higher stream densities deliver sources more efficiently to channels; stream density may also reflect the influence of climate on nitrogen flux.

The removal of total nitrogen in rivers is estimated as a function of four first-order loss rates (expressed per unit water travel time) that vary inversely with channel size. The SPARROW loss rates span nearly two orders of magnitude from 0.455 per day of travel time in small streams to 0.005 in large rivers (table 2). The magnitude of the loss rates and their inverse relation to channel size are consistent with literature estimates of in-stream nitrogen loss based on a recent re-analysis of mass balance and denitrification studies [Alexander et al. 2000]. Benthic denitrification is expected to be the principle loss process reflected by these rates based its importance in watershed studies [Howarth et al. 1996]. The physical storage and release of particulate nitrogen, such as on flood plains and in reservoirs, may also contribute to these rates. The decline in the rate of nitrogen loss (per unit of travel time) with increasing stream size suggests that the physical and biochemical processes responsible for the removal of nitrogen in streams become progressively less effective with increases in channel depth [Alexander et al. 2000].

3.3 Model Predictions of Nitrogen Export

We applied equation 1 and the SPARROW coefficient estimates to data on nitrogen source inputs, landscape attributes, and channel time of travel to predict TN export from the five source types at the outlets of the 60,000 reach watersheds in the RF1 network. The predictions include estimates of the TN export for the entire watershed above each reach and predictions for each separate reach watershed (i.e., "local" or incremental TN export), excluding the flux from upstream basins. We estimated the fraction of the stream export contributed by each nitrogen source (i.e., source share) as the ratio of each source's TN export to the total TN export. The SPARROW predictions include estimates of the mean and standard error based on the application of the bootstrap procedure.

Predictions of TN export (expressed as yield; kg km^{-2} yr^{-1}) at the outlets of the 60,000 reach watersheds (inclusive of the entire drainage area) range over four orders of magnitude with a median of 530 kg km^{-2} yr^{-1} (interquartile range from 318 to 804 kg km^{-2} yr^{-1}). The spatial distribution of yields is similar to that observed for the monitoring sites; the highest yields occur in streams of the midwestern and northeastern states with the lowest yields predicted for streams in the western states. Model predictions are summarized for the 40 coastal watersheds in the following section.

As an initial evaluation of the accuracy of the model predictions of flux, we compared the predictions of "local" TN export (in units of yield; kg ha^{-1} yr^{-1}) for basins having a predominant land use type with yields reported in the literature for similar land uses (see table 3). Because of the large variability in the watershed characteristics, the SPARROW yields span as much as an order of magnitude or more for certain sources. Although the RF1 reach watersheds are typically larger and contain more diverse sources than those studied in the literature, the SPARROW yields for the most homogeneous watersheds lie well within the range of yields reported for various land uses and sources in North American watersheds (table 3). Watersheds dominated by urban sources and agriculture (crop and pasture land) have the largest nitrogen yields, whereas the quantities of nitrogen exported from watersheds with forest and range lands are one-tenth to one-quarter of these yields. Variability in the literature yields can be attributed to factors other than land cover, including nitrogen supply, climate, landscape characteristics, and stream properties [Ritter, 1988; Novotny and Olem, 1994; Frink, 1991; Beaulac and Reckhow, 1982].

TABLE 3. SPARROW estimates of total nitrogen (TN) export from major land types compared to literature estimates. SPARROW estimates are reported for TN exported from watersheds associated with individual stream reaches as defined by the digital river network for the conterminous United States.

Watershed Land-Cover Type	Distribution of TN Yield Exported from SPARROW Watersheds[a] (kg ha^{-1} yr^{-1})						Literature Exports[b] (kg ha^{-1} yr^{-1})
	Number	10th	25th	Median	75th	90th	Range of Values	Range of Values
Crops	203	12.1	17.4	22.2	29.3	35.5	2.2 – 42.5	0.8 – 79.6
Pasture	19	9.5	14.4	16.8	19.2	20.3	8.5 – 20.8	0.1 – 30.8
Forest	17	1.8	3.6	4.5	6.1	7.4	1.8 – 11.2	0.1 – 10.8
Range	58	1.3	2.1	2.9	4.0	5.4	0.4 – 7.4	1.5 – 6.8
Urban	22	4.6	20.0	31.6	87.0	95.2	3.6 – 175	1.6 – 38.5

[a] The land-cover types represent the following percentages of the land area in SPARROW watersheds: crops (>90%), pasture (>85%), forest (>95%), range (100%), urban (>75%).
[b] Total nitrogen export taken from ranges reported in literature reviews (Beaulac and Reckhow, 1982; Frink, 1991; Ritter, 1988). The export reported for "range" is for grasslands in Oklahoma, U.S. (Ritter, 1988).

4. Nitrogen Export and Sources in the Estuarine Watersheds

4.1 Total Nitrogen Export

The national SPARROW model estimates of mean total nitrogen flux exported from the estuarine drainage areas are presented in table 4. Included in table 4 are the mean estimates of the percentage of nitrogen export contributed by the five major types of sources. Nitrogen flux estimates are standardized by the area of the drainage basin and expressed as export or yield (kg km^{-2} yr^{-1}) to adjust for differences in area among the estuarine watersheds. Units can be converted to kg ha^{-1} yr^{-1} according to 100 kg km^{-2} = 1 kg ha^{-1}. Estimates of uncertainty in export are given in table 5.

Mean TN yield varies by about a factor of sixty among the estuarine watersheds, ranging from about 38 to 2,500 kg km^{-2} yr^{-1}. Yields more commonly vary from about 250 to 650 kg km^{-2} yr^{-1} as reflected by the interquartile range (difference between the 75th and 25th percentiles). The median TN yield is 450 kg km^{-2} yr^{-1}. Distinct regional differences exist in the nitrogen export from the estuarine watersheds among the four major geographically contiguous sections of coastline (see table 4). The highest values of export (and the largest range) are observed in the North Atlantic (NA) region (median=520 kg km^{-2} yr^{-1}), which includes coastal waters from Maine south to the Chesapeake Bay. Four watersheds display yields greater than 1,000 kg km^{-2} yr^{-1} in this region, and include the Massachusetts Bay, Narragansett Bay, Delaware Bay, and the

Hudson River. Exports of total nitrogen from the estuarine watersheds in the South Atlantic (SA) region are among the lowest and display the narrowest range of values among the regions. The median export is one half of that of the NA region and most watershed exports range from 100 to about 450 kg km^{-2} yr^{-1}. The highest export in the SA region is found in the Pamlico Sound watershed. Although the median export for watersheds in the Gulf of Mexico region (470 kg km^{-2} yr^{-1}) is similar to that of the NA, exports are generally lower than those in the NA (range of 60 to 720 kg km^{-2} yr^{-1}; interquartile range from 300 to 500 kg km^{-2} yr^{-1}). The highest yields range from 500 to about 700 kg km^{-2} yr^{-1} in the Gulf region, and occur in the watersheds of the Upper Laguna Madre, TX, Mobile Bay, West Mississippi Bay, and Calcasieu Lake. Both Pacific watersheds show moderately high yields of about 600 kg km^{-2} yr^{-1}.

4.2 Atmospheric Nitrogen

The mean quantities of atmospheric nitrogen in rivers exported from the estuarine watersheds range over nearly two orders of magnitude from 4 to 326 kg km^{-2} yr^{-1} (see table 4). Estimates typically vary from 30 to 110 kg km^{-2} yr^{-1}, based on the interquartile range. When expressed as a percentage of the total nitrogen flux, atmospheric nitrogen is estimated to represent from 4 to 35% of the total nitrogen mass exported from the estuarine watersheds (see table 6; fig. 6). Atmospheric contributions typically range from 10% to slightly more than 20% of the stream nitrogen exports.

Similar to the geographic patterns observed for total nitrogen export, the NA region shows the greatest range and highest magnitude of atmospheric export and percentage contributions to stream export (median=120 kg km^{-2} yr^{-1}; see fig. 7). In nearly one half of the estuarine watersheds in the NA region (7 watersheds), atmospheric nitrogen represents more than 20% of the total stream export, including the watersheds of the Long Island Sound (41%), the Merrimack (27%), Hudson (25%), and Chesapeake Bay (28%). Although the median contribution of atmospheric nitrogen to stream export is similar for the NA and SA regions (~17%), the atmospheric shares are somewhat lower in the SA region, typically ranging from 14% to 22%; none of the atmospheric exports are larger than about 100 kg km^{-2} yr^{-1} in this region. In the Gulf of Mexico region, atmospheric nitrogen represents a slightly smaller share of the stream export than in the SA region. The median percentage share (14%) is slightly lower than in the SA region, and atmospheric nitrogen exports are commonly less than about 80 kg km^{-2} yr^{-1}. However, there are several estuarine watersheds where atmospheric nitrogen is estimated to contribute from 20 to 27% of the total river exports, including Mobile Bay, W. Mississippi Sound, Terrebonne Bay, and Sabine Lake. The larger exports in these watersheds correspond to larger inputs of atmospheric deposition in this area of the southeastern United States as reflected in the estimates of wet deposition of nitrate (fig. 4; table 1). Despite higher wet deposition of nitrate in the more populated watersheds of the western United States (fig. 4), atmospheric nitrogen represents a very small percentage of the total nitrogen exports from the two Pacific watersheds (5% in San Francisco Bay, 12% in the Puget Sound).

TABLE 4. Total nitrogen export from sources in the drainages of major estuaries of the conterminous United States. Mean annual export (i.e., yield) is in units of kg km^{-2} yr^{-1}.

Watershed/Estuary	Total	Atmosphere	Point Sources	Fertilizer	Livestock	Nonagric. Nonpoint
North Atlantic						
1. Casco Bay	386	85	51	22	20	207
2. Great Bay	382	34	89	19	16	223
3. Merrimack River	445	123	90	22	17	193
4. Massachusetts Bay	2,489	98	2,193	27	10	161
6. Buzzards Bay	135	16	85	13	3	18
7. Narragansett Bay	1,051	110	656	59	27	200
8. Gardiners Bay	38	4	4	14	0	15
9. Long Island Sound	881	304	148	79	60	289
10. Hudson River / Raritan Bay	1,277	326	516	96	71	267
11. Barnegat Bay	864	160	367	84	10	243
12. New Jersey Inland Bays	515	136	115	104	9	151
13. Delaware Bay	1,332	296	467	225	122	222
14. Delaware Inland Bays	174	16	39	33	76	10
15. Maryland Inland Bays	243	20	105	33	67	18
16. Chesapeake Bay	814	228	62	171	173	179
South Atlantic						
17. Pamlico Sound	751	109	31	353	126	132
18. Winyah Bay	428	80	18	164	57	109
19. Charleston Harbor	107	23	12	18	11	43
20. St. Helena Sound	138	24	2	37	6	68
22. St. Catherines / Sapelo S.	234	47	4	5	2	176
23. Altamaha River	457	105	16	133	67	137
24. Indian River	89	7	11	60	2	9
Gulf of Mexico						
25. Charlotte Harbor	370	48	7	212	42	62
26. Sarasota Bay	309	35	176	33	8	56
27. Tampa Bay	481	51	106	227	53	44
28. Apalachee Bay	281	40	15	93	22	111
29. Apalachicola Bay	479	70	35	185	55	134
30. Mobile Bay	515	122	13	109	82	188

TABLE 4. Continued.

31. West Mississippi Sound	508	131	47	105	80	145
32. Barataria Bay	541	49	322	46	9	115
33. Terrebonne / Timbalier Bay	229	61	21	11	1	135
34. Calcasieu Lake	616	107	134	163	36	176
35. Sabine Lake	351	71	23	72	60	124
36. Galveston Bay	468	62	183	99	45	79
37. Matagorda Bay	123	17	3	68	17	19
38. Corpus Christi Bay	56	5	11	26	4	9
39. Upper Laguna Madre	717	83	12	226	106	290
40. Lower Laguna Madre	566	42	52	381	17	73
Pacific						
41. San Francisco Bay	585	32	74	244	82	154
42. Puget Sound	677	80	96	103	86	313

4.3 Other Source Contributions

In comparison to atmospheric sources, the other sources generally contribute greater quantities of nitrogen to the estuaries. A comparison of the percentage contributions to stream export of the five source categories estimated by SPARROW is shown in figure 6.

Agricultural sources (i.e., fertilizer use, livestock wastes) represent the largest single source of nitrogen in the estuarine watersheds, accounting for more than about a third of the nitrogen in the stream export of most basins. Agricultural source contributions to river export range from 2 to 70% of the nitrogen, although the contributions more commonly range from about 13 to 50% of the stream exports in most watersheds based on the interquartile range (median=33%). Fertilizer-related sources may include leaching of mineralized soil nitrogen and N fixation by crops. Livestock wastes represent about one-third of the total agricultural nitrogen contributions to streams.

Geographic variations in agricultural contributions to river export are shown in figure 8. Agricultural contributions are highest in the SA region where a median of 42% is observed for the estuarine watersheds. In two of the watersheds, the Pamlico Sound and Indian River, more than 60% of the stream export is derived from agricultural sources. In the Gulf of Mexico region, agricultural contributions are similar in magnitude to those in the SA region (median of 38%), and represent at least 50% of the stream export in six of the watersheds (Charlotte Harbor, Tampa Bay, Apalachicola Bay, Matagorda Bay, Corpus Christi Bay, and Lower Laguna Madre; see table 6). In the Pacific region, agricultural sources represent 61% of the nitrogen export in the San Francisco watershed and 28% in the Puget Sound. By contrast to the other regions, watershed exports to estuaries in the NA region contain much less agricultural nitrogen (typically less than 15% of the river exports). The highest contributions in the NA region are found in

TABLE 5. Estimates of uncertainty in total nitrogen export from sources in the drainages of major estuaries of the conterminous United States. Estimates of uncertainty are based on the standard error, expressed as a percentage of the mean export, and reflect error in the model coefficients and unexplained variability in the observed data (i.e., model residuals).

Watershed/Estuary	Total	Atmosphere	Point Sources	Fertilizer	Livestock	Nonagric. Nonpoint
North Atlantic						
1. Casco Bay	64	46	46	33	61	21
2. Great Bay	63	56	45	34	61	20
3. Merrimack River	46	45	44	28	58	27
4. Massachusetts Bay	71	83	10	96	104	74
6. Buzzards Bay	75	68	27	49	80	49
7. Narragansett Bay	57	61	23	42	71	42
8. Gardiners Bay	111	54	113	17	64	19
9. Long Island Sound	29	41	44	29	59	30
10. Hudson River / Raritan Bay	34	48	34	35	63	37
11. Barnegat Bay	72	50	30	33	72	31
12. New Jersey Inland Bays	49	44	43	24	63	29
13. Delaware Bay	29	48	33	30	59	31
14. Delaware Inland Bays	83	102	76	66	86	121
15. Maryland Inland Bays	81	83	43	64	91	83
16. Chesapeake Bay	19	45	51	26	53	27
South Atlantic						
17. Pamlico Sound	31	49	53	17	53	21
18. Winyah Bay	26	47	53	19	53	21
19. Charleston Harbor	35	47	48	21	53	24
20. St. Helena Sound	46	50	59	18	57	19
22. St. Catherines / Sapelo S.	73	53	74	49	80	15
23. Altamaha River	30	47	57	23	53	22
24. Indian River	87	58	77	17	69	36
Gulf of Mexico						
25. Charlotte Harbor	55	48	57	16	56	26
26. Sarasota Bay	72	59	26	41	68	39
27. Tampa Bay	37	52	49	21	56	25
28. Apalachee Bay	40	49	53	19	55	21

TABLE 5. Continued.

29. Apalachicola Bay	25	49	53	19	54	20
30. Mobile Bay	20	46	57	22	52	21
31. West Mississippi Sound	28	44	50	24	53	25
32. Barataria Bay	58	59	22	38	70	38
33. Terrebonne / Timbalier Bay	64	46	55	33	64	22
34. Calcasieu Lake	40	47	43	22	57	23
35. Sabine Lake	27	47	52	22	51	23
36. Galveston Bay	32	55	36	29	57	33
37. Matagorda Bay	32	50	58	17	56	21
38. Corpus Christi Bay	48	55	54	23	58	24
39. Upper Laguna Madre	54	52	60	19	54	17
40. Lower Laguna Madre	64	56	55	14	63	32
Pacific						
41. San Francisco Bay	26	54	50	19	53	20
42. Puget Sound	30	52	53	24	52	18

watersheds of the eastern Maryland shore, where livestock wastes contribute as much as 30 to 40% of the total nitrogen in river exports.

Non-agricultural diffuse sources are estimated to contribute slightly less to watershed exports than agricultural sources. The median contribution to watershed exports is 27% with most contributions ranging from 17 to 40%, based on the interquartile range (fig. 6). The percentage contribution is larger in the estuarine watersheds of the South Atlantic region than in other regions. Contributions from this source are based on the model intercept scaled for nonagricultural land area in the watersheds, and thus represent nitrogen inputs not explicitly accounted for by the other model sources. Nonagricultural sources may include nitrogen in the runoff from urban, forested, range, wetlands, and barren lands. Nitrogen runoff from forested and range lands may include biotic fixation. In watersheds of the western Gulf region and the Pacific region, non-agricultural diffuse sources are highly associated with range lands which constitute a predominant land type. Groundwater nitrogen, which generally reflects a more constant and less variable component of the nitrogen flux in watersheds, may also be included in this source category. Groundwater may reflect contributions from older waters originating from a variety of local and regional sources.

Municipal and industrial point sources represent the largest share of the nitrogen in stream exports in about one quarter of the estuarine watersheds, including one-half of the North Atlantic watersheds and several watersheds in the Gulf region. In the North Atlantic watersheds dominated by point sources, the shares represent from 35 to 88% of the nitrogen in stream export. The highest point source shares are found in the

TABLE 6. Source contributions to total nitrogen export in percent from the drainages of major estuaries of the conterminous United States.

Watershed/Estuary	Atmosphere	Point Sources	Fertilizer	Livestock	Nonagric. Nonpoint
North Atlantic					
1. Casco Bay	22	13	6	5	54
2. Great Bay	9	23	5	4	58
3. Merrimack River	28	20	5	4	43
4. Massachusetts Bay	4	88	1	0	6
6. Buzzards Bay	12	63	9	2	14
7. Narragansett Bay	10	62	6	3	19
8. Gardiners Bay	11	10	38	1	41
9. Long Island Sound	35	17	9	7	33
10. Hudson River / Raritan Bay	26	40	8	6	21
11. Barnegat Bay	19	43	10	1	28
12. New Jersey Inland Bays	26	22	20	2	29
13. Delaware Bay	22	35	17	9	17
14. Delaware Inland Bays	9	22	19	44	6
15. Maryland Inland Bays	8	43	14	28	7
16. Chesapeake Bay	28	8	21	21	22
South Atlantic					
17. Pamlico Sound	14	4	47	17	18
18. Winyah Bay	19	4	38	13	26
19. Charleston Harbor	22	11	17	10	40
20. St. Helena Sound	18	2	27	5	49
22. St. Catherines / Sapelo S.	20	2	2	1	75
23. Altamaha River	23	3	29	15	30
24. Indian River	8	12	68	2	10
Gulf of Mexico					
25. Charlotte Harbor	13	2	57	11	17
26. Sarasota Bay	11	57	11	3	18
27. Tampa Bay	11	22	47	11	9
28. Apalachee Bay	14	5	33	8	40
29. Apalachicola Bay	15	7	39	11	28
30. Mobile Bay	24	3	21	16	37
31. West Mississippi Sound	26	9	21	16	29
32. Barataria Bay	9	60	8	2	21
33. Terrebonne / Timbalier Bay	27	9	5	1	59
34. Calcasieu Lake	17	22	26	6	29
35. Sabine Lake	20	7	21	17	35
36. Galveston Bay	13	39	21	10	17

TABLE 6. Continued.

37. Matagorda Bay	14	2	55	14	15
38. Corpus Christi Bay	10	19	47	8	16
39. Upper Laguna Madre	12	2	31	15	40
40. Lower Laguna Madre	8	9	67	3	13
Pacific					
41. San Francisco Bay	5	13	42	14	26
42. Puget Sound	12	14	15	13	46

Massachusetts Bay (88%), Buzzards Bay (63%), and Narragansett Bay (62%). The largest point-source shares in the Gulf region include Sarasota Bay (57%), Barataria Bay (60%), and Galveston Bay (39%). In most watersheds, point sources are similar to or less than the contributions from the atmosphere (fig. 6), and typically represent less than 15% of the nitrogen in stream export.

4.4 Estimates of Uncertainty

Estimates of uncertainties in the mean nitrogen exports and source shares are presented in table 5. The estimates of the standard error (one standard deviation), expressed as a percentage of the mean, reflect two sources of uncertainty: variability in the estimates of the model coefficients and unexplained variations in the data according to the model predictions (as described by the model residuals). Estimates of the portion of the residual error associated with each of the source shares is computed by assuming that each source's share of the residual error is proportional to the source's share of mean total nitrogen export.

The standard errors on total export among the estuarine watersheds range from 19% to 117% of the mean. One half of the mean exports have standard errors less than 40%. The standard errors of the individual source contributions are typically larger. For example, errors in atmospheric nitrogen export range from 41% to 102%; one half of the exports have standard errors less than 50%. The magnitude of the uncertainty in the estimates of nitrogen export is inversely related to drainage basin size (see fig. 9). This relation reflects the intrinsic effect of averaging on error reduction. Model residual errors are systematically assigned to the predictions of stream export at the outlets of hydrologic cataloging units (HCU) to account for variability in stream flux that is unexplained by the model. Estuarine watersheds with large drainage areas have a correspondingly greater number of HCUs. Thus, the larger watersheds have lower error because the cancellation of errors increases with the number of errors averaged. The standard errors in the estimates of nitrogen export range from 19% to 50% of the mean in watersheds above 10,000 square kilometers in size (60% of the estuarine watersheds are larger than this size). Standard errors range from 50% to 70% of the mean in watersheds between 2,000

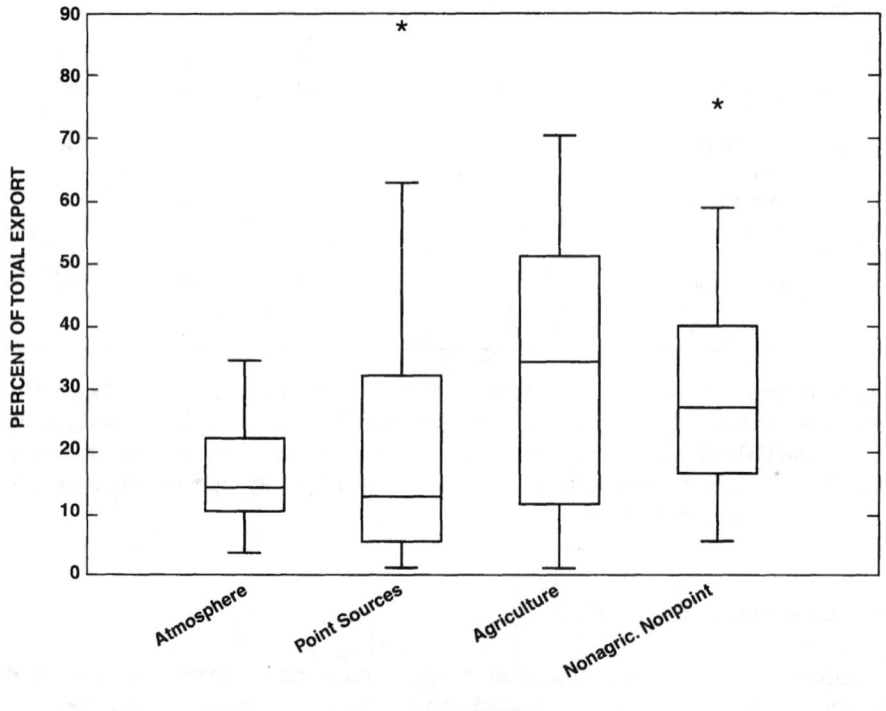

Figure 6. Contributions of sources to total nitrogen export from major estuarine drainage areas of the United States. Each box graphs the watershed quartiles, with lower and upper edges representing the 25th and 75th percentiles, respectively. The midline plots the median. The upper and lower whiskers are drawn to the watershed value within +/- 1.5 times the interquartile range. Watershed values exceeding three times the IQR appear as a "*". Agriculture is the sum of the fertilizer and livestock waste sources.

and 10,000 square kilometers in size (20% of the estuarine watersheds fall within this size range). In watersheds smaller than 2,000 square kilometers, the standard errors are typically greater than 80% of the mean. Estimates of nitrogen export and source contributions for these watersheds have the lowest reliability. Examples include the estuarine watersheds of Gardiners Bay (111%), Delaware Inland Bays (83%), Maryland Inland Bays (81%), and Indian River (87%).

4.5 Landscape and Aquatic Attenuation

Summary statistics related to nitrogen attenuation on the landscape and in streams are shown for the estuarine watersheds in table 7. The land-to-water delivery index in table 7 is an indicator of the relative proportion of a diffuse source input that is transported to streams as a function of the model's landscape properties, including soil permeability, air

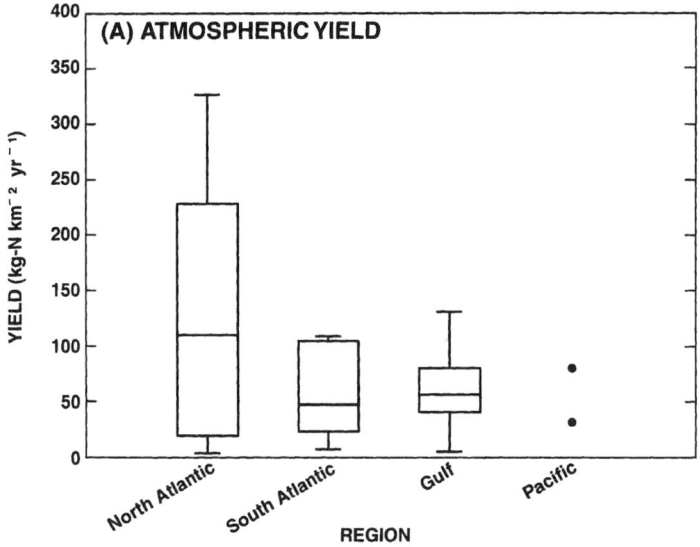

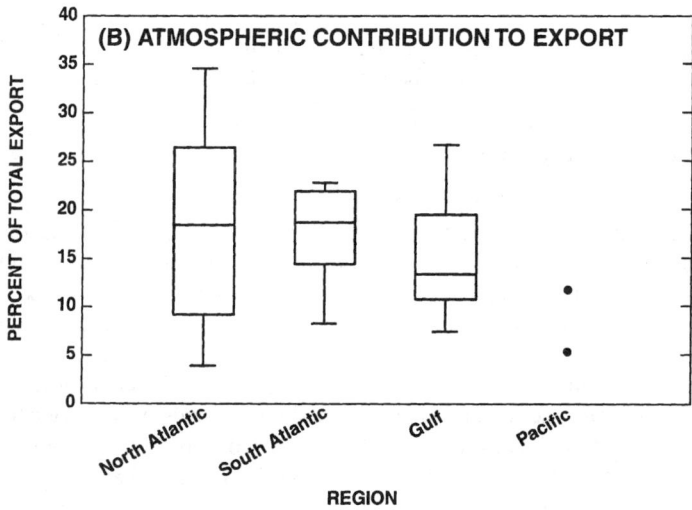

Figure 7. The atmospheric contributions to total nitrogen export from major estuarine drainage areas of the United States by region; (A) atmospheric yield, (B) percentage of total nitrogen export. See table 4 for description of estuaries and regions. Each box graphs the watershed quartiles, with lower and upper edges representing the 25th and 75th percentiles, respectively. The midline plots the median. The upper and lower whiskers are drawn to the watershed value within +/- 1.5 times the interquartile range. The dots give statistics for the two Pacific watersheds.

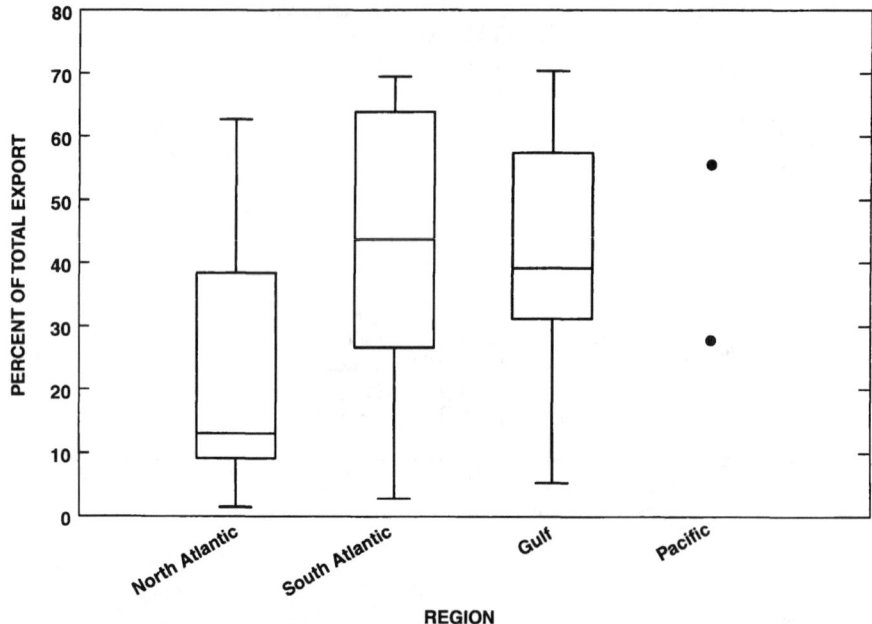

Figure 8. The contributions of agriculture to total nitrogen export from major estuarine drainage areas of the United States by region. Agriculture is the sum of fertilizer and livestock waste source contributions. Each box graphs the watershed quartiles, with lower and upper edges representing the 25th and 75th percentiles, respectively. The midline plots the median. The upper and lower whiskers are drawn to the watershed value within +/- 1.5 times the interquartile range. The dots give statistics for the two Pacific watersheds.

temperature, and drainage density. The product of the land-to-water delivery index and the model's diffuse source coefficient and input data quantify the fraction of the measured source inputs that is delivered to streams (see equation 1). Although the NA region displays the largest range in values of the land-to-water delivery index, the values are typically lower in the watersheds of this region than in other regions (median=0.13 compared with 0.20 or larger in other regions). This indicates that higher proportions of the nitrogen inputs are typically removed in this region as a function of the landscape properties of the watersheds. The higher permeability of soils in the watersheds of the NA region appears to account for an important portion of this effect. Low values of the index are also observed in several Florida watersheds where higher soil permeabilities and temperatures are generally found.

The mean quantities of atmospheric nitrogen delivered to streams in the estuarine watersheds (expressed per unit of watershed area) are predicted to range from 0.05 to 4.7 kg ha^{-1} yr^{-1} (table 7). These quantities range from a few percent to nearly 30 percent of the total (dry plus wet) atmospheric inputs estimated by Meyers *et al.* [this volume].

The quantities of nitrogen removed in streams and reservoirs of the estuarine watersheds, expressed as a percentage of the quantity of nitrogen delivered to stream

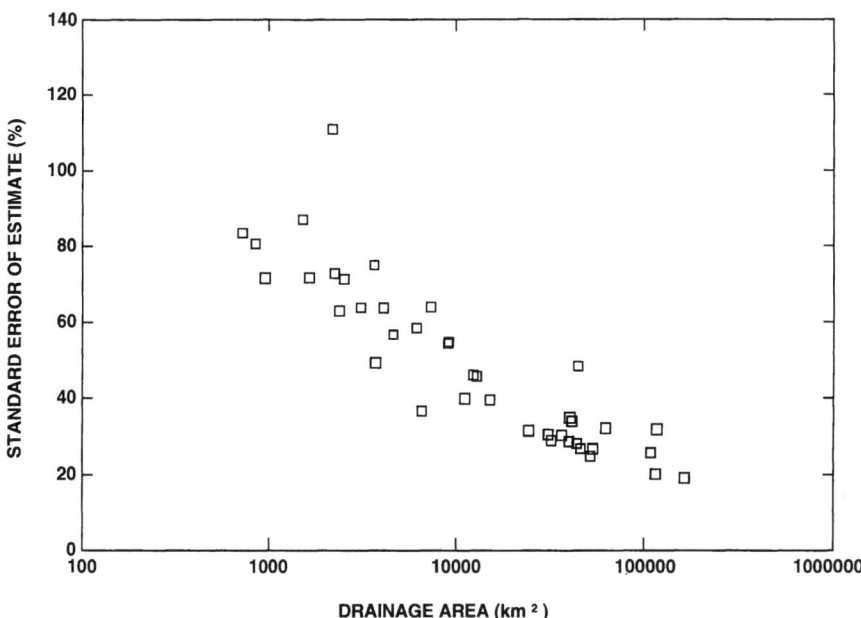

Figure 9. The standard error of estimate for mean total nitrogen yield in relation to estuarine drainage area.

channels, range from nearly zero to about 90 percent (table 7). The percentage removal more typically ranges from about 5 to 44%, based on the interquartile range (median=25%). In the smallest estuarine watersheds, especially those containing fewer than 10 reaches, in-stream losses of nitrogen are negligible. We find that in-stream losses are much lower in watersheds of the NA region, where a median of 7% of the nitrogen is removed. This reflects the effect of the shorter water travel times in this region (i.e., 0.2 to 3.5 days), which are typically one quarter to one half of the travel times estimated for the estuarine watersheds in other regions. The estuarine watersheds of the SA region show the largest range of in-stream losses spanning from negligible quantities to nearly 60% (median=44%). The loss percentages for the estuarine watersheds of the Gulf region (median=42%) are similar to those in the SA region, although the percentages typically span a narrower range from about 27 to 53%. There are notable differences in the percentage of in-stream loss between the Gulf estuarine drainages in the east and those in the west that relate to both water travel time and channel size differences. Several of the western Gulf estuarine watersheds (Galveston Bay, Matagorda Bay, and Corpus Christi Bay) have large drainage areas with long water travel times (10 to 25 days) and small rivers with relatively low flow, which collectively result in large in-stream nitrogen losses (60 to 90%). By contrast, much smaller in-stream nitrogen losses (<45%) are observed in the eastern Gulf watersheds characterized by generally smaller drainage sizes and higher streamflow.

TABLE 7. Total nitrogen (TN) delivery to streams and in-stream nitrogen loss rates for the drainages of major estuaries of the conterminous United States. "N" denotes a negligible in-stream total nitrogen loss.

Watershed/Estuary	Land-to-Water Delivery Index[a]	Atmospheric TN Delivered to Streams[b] (kg ha^{-1} yr^{-1})	In-Stream TN Loss[c] (% of stream inputs)	Mean Water Time of Travel (days)
North Atlantic				
1. Casco Bay	0.19	1.07	19	1.3
2. Great Bay	0.11	0.33	N	0.4
3. Merrimack River	0.18	1.55	19	3.1
4. Massachusetts Bay	0.13	0.92	7	0.5
6. Buzzards Bay	0.03	0.15	N	0.2
7. Narragansett Bay	0.14	1.19	4	0.7
8. Gardiners Bay	0.02	0.05	N	0.2
9. Long Island Sound	0.26	4.36	27	4.0
10. Hudson River/Raritan Bay	0.28	4.67	21	4.5
11. Barnegat Bay	0.13	1.60	N	0.6
12. New Jersey Inland Bays	0.13	1.64	16	1.0
13. Delaware Bay	0.26	4.00	21	3.5
14. Delaware Inland Bays	0.05	0.14	N	0.2
15. Maryland Inland Bays	0.05	0.15	N	0.2
16. Chesapeake Bay	0.25	3.69	37	4.9
South Atlantic				
17. Pamlico Sound	0.24	2.18	44	4.5
18. Winyah Bay	0.23	2.00	60	6.3
19. Charleston Harbor	0.25	2.14	89	8.3
20. St. Helena Sound	0.18	0.48	44	4.0
22. St. Catherines/Sapelo S.	0.14	0.42	N	0.5
23. Altamaha River	0.22	2.22	53	6.6
24. Indian River	0.02	0.07	N	0.1
Gulf of Mexico				
25. Charlotte Harbor	0.08	0.67	37	1.8
26. Sarasota Bay	0.07	0.29	N	0.5
27. Tampa Bay	0.08	0.70	26	1.3
28. Apalachee Bay	0.14	0.61	44	2.6
29. Apalachicola Bay	0.20	1.23	43	5.5
30. Mobile Bay	0.24	2.00	44	7.9
31. West Mississippi Sound	0.21	2.08	42	4.2
32. Barataria Bay	0.13	0.72	35	1.7
33. Terrebonne/Timbalier Bay	0.13	0.66	8	0.8
34. Calcasieu Lake	0.21	1.69	33	2.9
35. Sabine Lake	0.21	1.49	56	6.2
36. Galveston Bay	0.23	1.77	61	9.9
37. Matagorda Bay	0.21	0.86	81	24.9
38. Corpus Christi Bay	0.20	0.95	91	17.9

TABLE 7. Continued.

39. Upper Laguna Madre	0.20	1.07	23	1.4
40. Lower Laguna Madre	0.15	0.62	45	1.6
Pacific				
41. San Francisco Bay	0.25	0.63	45	5.3
42. Puget Sound	0.26	0.87	11	1.7

[a] The mean land-to-water delivery index, computed according to $\exp(-\alpha' Z_j)$ in equation 1, is an indicator of the relative proportion of a diffuse source input that is transported to streams as a function of the specified landscape properties (the product of the land-to-water delivery index and the model's diffuse source coefficient and associated input data quantify the fraction of the source input that is delivered to streams—see equation 1). The delivery index is the product of the delivery indices for temperature, permeability, and drainage density.

[b] Atmospheric delivery to streams is computed as the mean of model predictions for watershed reaches. The predictions of atmospheric TN delivery reflect uncertainties in model coefficients, but do not include uncertainties related to the unexplained variability in the observed data (i.e., model residuals). Note that the predictions of atmospheric export in table 4 include uncertainties based on the model coefficients and residuals, and therefore, may exceed stream deliveries of atmospheric TN in cases where in-stream losses are reported to be small.

[c] The in-stream loss is the mean percentage of the total quantity of nitrogen delivered to watershed reaches from all sources that is removed in streams. Note that for some estuaries the in-stream loss of atmospheric TN differs from that for total sources because of differences in the locations of sources.

4.6 Geographic Origins of Atmospheric Nitrogen

An understanding of the origins of nitrogen entering coastal ecosystems depends not only on knowledge of the relative contributions of the sources, but also on the location of sources in the watershed. This information is useful in evaluating and designing efficient nitrogen management strategies that provide the greatest reduction in nitrogen inputs to coastal waters in response to the least amount of control effort. Such information is complementary to other information, such as the costs of control technologies, that is necessary to assess the efficiencies of alternative strategies. Spatial variations in the quantities of atmospheric nitrogen delivered to an estuary will depend significantly on the location and magnitude of sources within the contributing watershed. As an illustration of this, we describe estimates of atmospheric nitrogen delivered to the Chesapeake Bay from inland watersheds. These estimates are derived from a previous application of the SPARROW model in the Chesapeake Bay watershed [CB SPARROW; Preston and Brakebill, 1999]. CB SPARROW is similar to the national SPARROW in terms of nitrogen sources and model structure, and is described in section 5.2.

One simple approach to examine the importance of nearby versus more distant watershed sources of atmospheric nitrogen is to consider the nitrogen contributions of watersheds to the Bay in relation to their river distance upstream of the estuary. Accordingly, we arranged the atmospheric nitrogen exports that are delivered to the Bay from the incremental drainage areas of the 1,400 interior reaches in ascending order by their river distance from the estuary. We summed these delivered exports over increasing river distances from the Bay and expressed the sum as a percentage of the total atmospheric nitrogen that enters the estuary. Figure 10a displays a map of the spatial pattern of the cumulative percentage of nitrogen that is delivered to the Bay from watersheds located over increasing distances from the estuary. In the accompanying plot (fig. 10b), we find that, when watersheds are considered in relation to their distance from the Bay, the per unit area contributions of atmospheric nitrogen to the estuary are largest in watersheds that comprise from 20 to about 50% of the total Bay drainage area. For these areas of the watershed, the curve describing the cumulative percentage of atmospheric nitrogen delivered to the Bay ranges from about 25 to nearly 55 percent, and plots above the line of equivalence for cumulative drainage area. Over this range, the contributions of nitrogen flux to the Bay are approximately 10 percent higher than the percentage of contributing drainage area (the percentage is about 20 percent for all sources as shown in figure 10b). This disproportionate contribution of nitrogen in relation to drainage area generally reflects the combined effect of relatively high atmospheric deposition and the efficient transport of nitrogen from areas in proximity to the Bay, especially from areas in the vicinity of large rivers. The areas include the lower portions of the Susquehanna, Potomac, and James rivers, which are characterized by short water travel times (per unit channel length) and low rates of in-stream nitrogen loss (per unit water travel time).

The quantities of atmospheric nitrogen delivered to the Bay can also be examined in terms of the per unit area contribution or "delivered yield" of inland watersheds (see fig. 11). The delivered yield adjusts the nitrogen export from inland watersheds by their drainage area, and thus, can be used to identify inland watersheds that contribute the largest nitrogen mass per unit area to the Bay. The higher values of delivered atmospheric yield in figure 11a show the effects of high atmospheric deposition and the more efficient transport of nitrogen from watersheds in the vicinity of large rivers. Watersheds in proximity to the Susquehanna, Potomac, and James rivers and their largest tributaries transport a disproportionately larger quantity of nitrogen per unit area (by a factor of 2 to 3) than many of the watersheds draining smaller streams in proximity to the Bay. This spatial pattern is attributed to the lower rates of in-stream nitrogen loss and the shorter water travel times (per unit channel length) in large rivers as compared to small streams. The importance of channel size to the efficiency of nitrogen transport and delivery to coastal waters was previously noted in the Mississippi River Basin [Alexander et al. 2000]. We find that inland watersheds in the northern portions of the Bay watershed generally receive the highest atmospheric deposition, but have low to moderate delivered yields. The water travel times are generally much longer for these watersheds increasing the opportunities for in-stream processes to remove nitrogen from the water column.

5. Comparisons with Other Large-Scale Spatial Models

One of the objectives of this chapter is to compare the SPARROW model predictions of total nitrogen flux in streams with other model estimates available at large spatial scales. There are no other national-scale nutrient models, but the agricultural model

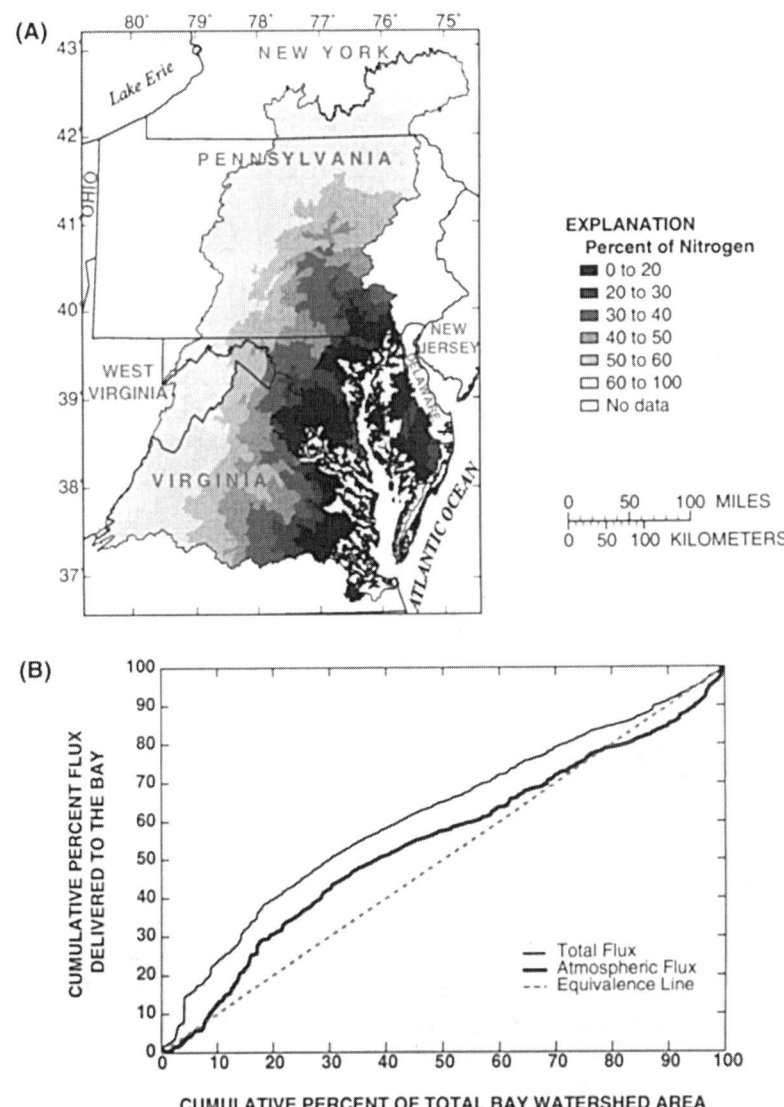

Figure 10. Cumulative percentage of the total nitrogen entering the Chesapeake Bay that originates from the outlets of interior watersheds as a function of river channel distance from the Bay: (A) map of atmospheric nitrogen by watershed; (B) total and atmospheric flux in relation to cumulative drainage area. The nitrogen export and drainage area are accumulated for incremental watershed areas defined by 1,400 reach segments.

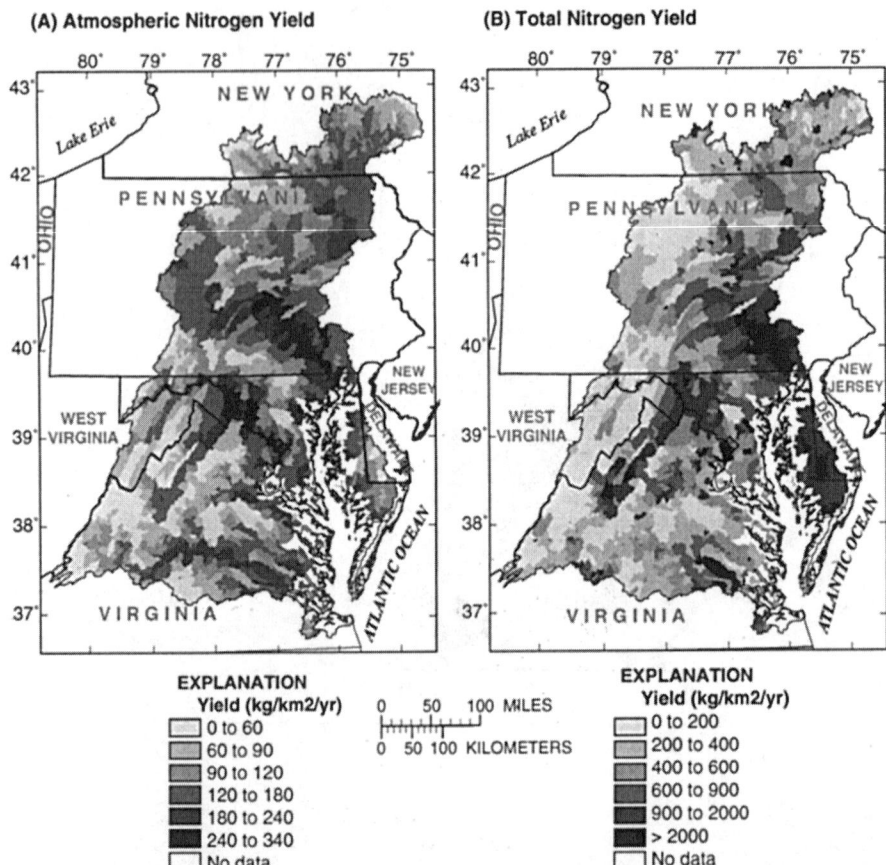

Figure 11. Mean total nitrogen yield delivered to the Chesapeake Bay from the outlets of interior basins of the watershed: (A) atmospheric nitrogen; (B) total nitrogen. Nitrogen yields are estimated for incremental watershed areas defined by 1,400 reach segments.

SWAT [Soil and Water Assessment Tool; Srinivasan et al. 1993; Arnold et al. 1990] has been recently used to simulate watershed exports of total nitrogen from 1,400 of the 2,077 hydrologic cataloging units of the central and eastern United States as part of the HUMUS (Hydrologic Unit Modeling of the United States) Project. SWAT is the only deterministic model applied at this large spatial scale. We compared SPARROW predictions to those of the SWAT model for the 1,400 hydrologic units and for 38 of the 42 estuarine watersheds selected for analysis in this book for which SWAT predictions could be obtained. We also compared results of the national SPARROW model with those available for two regional models applied in the 164,000 square kilometer Chesapeake Bay watershed: the HSPF [Hydrologic Simulation Program Fortran; Bicknell et al. 1997] hydrologic model and a recently developed SPARROW model calibrated with monitoring data from 79 locations in the Chesapeake Bay watershed.

Although comparisons of these large-scale models are possible, there are certain noteworthy limitations. First, the SWAT and HSPF models differ from SPARROW in the methods of describing nitrogen supply, terrestrial and aquatic losses, and in the spatial and temporal scales of measurement and prediction. For example, many of the nitrogen inputs in the models reflect different time periods and are based on different assumptions, such as the nitrogen inputs for fertilizer and livestock wastes in SWAT. Also, atmospheric nitrogen is not accounted for in the SWAT model, making it necessary to adjust this term to allow comparisons. Second, stream monitoring data were not available at the outlets of the hydrologic cataloging unit watersheds. Thus, SWAT and SPARROW model predictions could not be directly compared with independent observations of stream nitrogen flux (many of the monitoring stations in the SPARROW calibration data set are located near the outlets of hydrologic cataloging units [Alexander et al. 1998]; model accuracy is described in section 3.1).

5.1 Comparisons with SWAT

5.1.1 The SWAT agricultural model

The hydrologic simulation model, SWAT, is a predictive tool used as part of the Hydrologic Unit Model for the United States (HUMUS) Project [Srinivasan et al. 1993; Arnold et al. 1990], a regional and national effort to simulate the hydrologic cycle, and its related impacts on the natural resources, in the 2,077 major watersheds in the 48 conterminous. SWAT combines the features of the nonpoint-source watershed model, SWRRB [Simulator for Water Resources in Rural Basins; Williams et al. 1985; Arnold et al. 1990] and a flow routing model for channels and reservoirs, ROTO [Routing Outputs to Outlet; Arnold et al. 1995]. The SWAT model is physically based, operates on a daily time step, and simulates the effects of management changes in agricultural practices on stream water quality.

The major model components for subbasins in SWAT include hydrology, weather, sedimentation, soil temperature, crop growth, nutrients, and agricultural management. Hydrology includes surface runoff, percolation, lateral subsurface flow, groundwater flow, evapotranspiration, snow melt, channel routing, transmission losses and pond/reservoir storage. The weather variables are precipitation, air temperature, solar radiation, wind speed, and relative humidity. Estimates of the long-term mean precipitation are based on a 30 year period, 1960-89. Surface runoff volume from non-urban areas is estimated by using the Soil Conservation Service (SCS) curve number procedure and from urban areas making use of the USGS Urban Storm Runoff Loading Model [Tasker and Driver, 1988]. Sediment yield from rural watersheds is estimated for each subbasin with the Modified Universal Soil Loss Equation (MUSLE) from which the organic nitrogen content and yield are estimated. Crop use of nitrogen is based on a supply and demand approach that allows uptake to continue until the daily demand is met or nitrogen is depleted. Crop growth and nitrogen use are simulated as a function of estimates of crop conversions of energy to biomass and rates of crop harvest. The amounts of nitrate contained in runoff, lateral flow, and percolation are estimated as the products of the volume of water and the mean concentration. Soil denitrification is a function of temperature and organic carbon and soil water content. The model considers mineralization sources from the fresh active organic N pool in the crop residue of each soil layer and microbial biomass and the stable organic N pool in the soil humus. The export of total nitrogen (TN) from each land use type in a subbasin is estimated as the sum of the nitrate in surface and sub-surface runoff and organic-N in sediment. Total nitrogen export at the hydrologic unit outlet is obtained by summing the routed subbasin yields of nitrate, nitrite, organic-N and ammonium (NH4-N). Stream routing is based on the height and width of rectangular channels during two-year return flows. A neural network model [Muttiah et. al. 1997] estimates channel dimensions from modeling unit elevations and drainage areas. A storage coefficient method [Arnold et. al. 1995] routes water through streams and reservoirs, adding flows and inputting measured data and point sources. In-stream nutrient kinetics are controlled by QUAL-2E routines that use estimates of reach time of travel [Arnold et. al. 1995]. County-based municipal and industrial point sources used in the SPARROW model were accounted for in SWAT; aquatic attenuation of the point sources was applied according to SWAT estimates of channel and reservoir decay. The rate coefficients that describe watershed processes in SWAT are based on field-scale measurements; no formal model calibration is employed to adjust the rate coefficients and match model predictions with observed stream monitoring data.

Topography and channel topology are determined from 1:250,000 scale Digital Elevation Model (DEM) data. Watershed boundaries are from 1:500,000 scale State hydrologic unit maps. Land use/land cover (LULC) data are from the USGS LULC data 250 by 250 meters at 1:250,000 scale derived from aerial photography and LANDSAT images during the 1980s, updated with 1990 census population for urban areas. Cropland areas are from county-level data of the Census of Agriculture, and soils properties are from the State Soil Geographic (STATSGO) database. Model predictions of nutrient flux are made for hydrologic cataloging units, where up to 21 crop and soil polygons may exist (approximately 200 km^2 per polygon). The hydrologic response unit for estimating water balances and process interactions is defined according to these polygons. Reservoir nutrient attenuation is modeled as an aggregate component within these spatial units based on reservoir hydraulic properties.

5.1.2 Results of Comparisons with SWAT

We compared SPARROW model predictions of total nitrogen export with SWAT predictions for 1,430 hydrologic cataloging units (HCU) in the eastern and central portions of the United States covering 12 of the 18 major hydrologic regions [Seaber et al. 1987; regions 1 through 12). These estimates reflect nitrogen mass contributed by the total drainage area above each HCU, inclusive of the nitrogen contributions of all upstream HCUs. We also compared the "local" yield estimates from both SPARROW and SWAT models for the 1,430 HCUs; these reflect nitrogen mass contributed from sources within each HCU independent of inflows from upstream HCUs. Finally, we compared model predictions of total nitrogen export for 38 of the 42 estuarine drainages selected for analysis in this chapter (SWAT estimates were not computed for the two Pacific estuaries). SWAT predictions of export were computed for the estuarine watersheds by apportioning the HCU estimates of TN export to the estuarine watersheds in proportion to the HCU drainage area common to both.

Because the SWAT model does not include nitrogen inputs from atmospheric deposition, adjustments to the model predictions were necessary to allow comparisons with SPARROW; the SPARROW estimate of atmospheric nitrogen flux was added to the SWAT prediction of total nitrogen flux for each watershed. To minimize the effect of drainage area on comparisons of model predictions of TN flux, comparisons were made on TN yield (flux per unit area), adjusting the flux estimates for the drainage area of the hydrologic cataloging units used in each of the models. The model estimates of drainage area differ by less than 10 percent for more than 97 percent of the hydrologic units.

The results of the model comparisons are shown for the "total" hydrologic unit TN yield in figure 12a and for the "local" hydrologic unit yield in figure 12b. The model predictions are closely correlated for both the total (r=0.67) and local (r=0.83) yields. SPARROW predictions tend to be consistently larger than those for HUMUS over the range of yields with somewhat larger differences for the local yields. The distribution of the ratios of SPARROW to SWAT total yield has a median of 1.4 with an interquartile range from 0.9 to 2.1. Fewer than 20 percent of the ratios are smaller than 0.55 or larger than 3.0. Uncertainties in SPARROW predictions of total yield, based on the standard error of prediction, are typically 45 to 66 percent (median=59 percent), and scale inversely with watershed area (see discussion in section 4.4). The distribution of the ratios of SPARROW to SWAT local yield has a median of 2.3 with an interquartile range from 1.6 to 3.4. Fewer than 20 percent of the ratios are smaller than 1.1 or larger than 4.7. Uncertainties in SPARROW predictions of local yield typically range from 60 to 70 percent (median=65 percent). Estimates of uncertainties are not available for the SWAT model predictions.

Predictions of total nitrogen yield exported from the estuarine drainages are presented in figure 13. There is a positive correlation (r=0.55) among the model predictions over the range of yields. In contrast to the comparisons for the hydrologic units, SPARROW predictions of yield are consistently smaller (in about 2/3's of the watersheds) than those for SWAT. In addition, differences in model predictions are slightly smaller for the estuarine watersheds than for the HCU comparisons of total yield (fig. 13a). The distribution of the ratio of SPARROW to SWAT total yield has a median of 0.73 with an interquartile range from 0.5 to 1.3. Fewer than 20 percent of the ratios are smaller than 0.3 or larger than 1.5. Uncertainties in SPARROW predictions of total yield in these watersheds, based on the standard error of prediction, range from 20 to about 80 percent (see section 4.4). The largest difference between the model results (which appears as a distinct outlier in fig. 13) occurs in the Massachusetts Bay watershed, where the SWAT estimate is about twice as large as the SPARROW yield.

5.1.3 Discussion of model comparisons

There are fundamental differences between the SPARROW and SWAT models in their descriptions of nitrogen sources and sinks, methods of parameter estimation, and spatial and temporal scales of measurement and prediction, which likely account for the

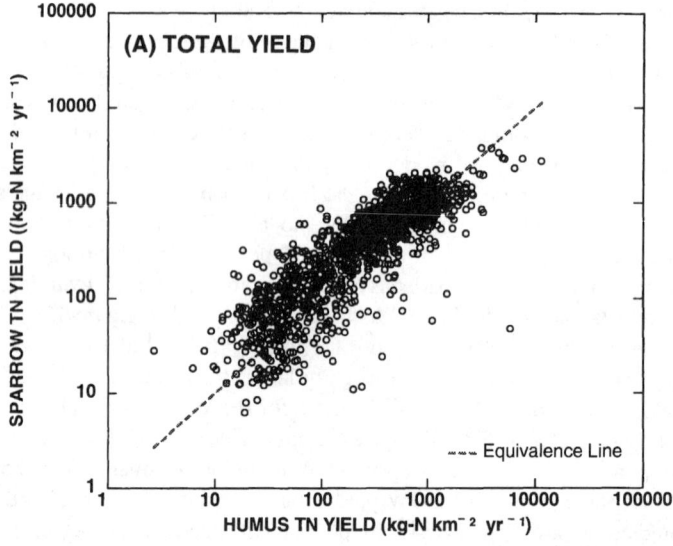

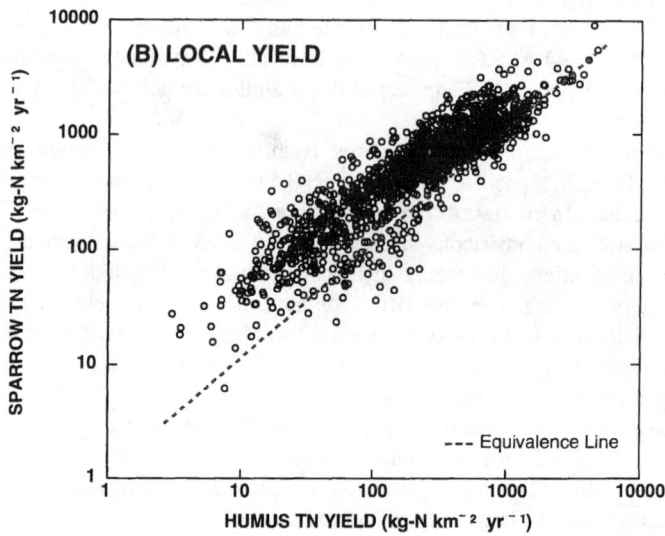

Figure 12. SPARROW and HUMUS model predictions of total nitrogen (TN) yield for 1,480 hydrologic cataloging units (HCU) in the central and eastern United States; (A) total yield for the entire drainage area above each HCU, inclusive of the nitrogen contributions of all upstream HCUs, (B) local yield reflecting the export from each HCU independent of inflows from upstream HCUs. Humus yields are adjusted to reflect atmospheric nitrogen as estimated by SPARROW.

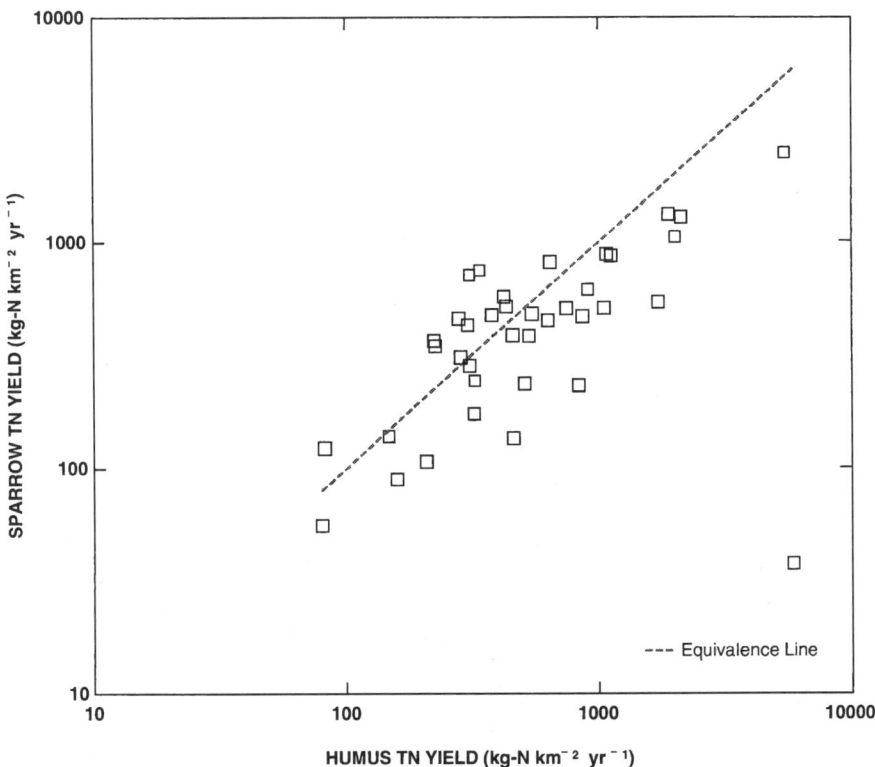

Figure 13. SPARROW and HUMUS model predictions of total nitrogen (TN) yield for major estuarine drainages of the United States. The yield reflects the total nitrogen export from the entire drainage area of each estuary.

observed differences in the model results. SWAT is a deterministic simulation model that uses field-scale experimental rates in combination with large-scale descriptions of watershed characteristics (e.g., soil, slope, crops) to quantify the supply, transformation, and delivery of nitrogen to streams and watershed outlets. By contrast, the rates of nitrogen supply and attenuation in SPARROW describe aggregate watershed-scale processes, and are quantified from a calibration of the model predictions to observed measurements of stream nitrogen flux. We highlight a few of the principle differences in the models to suggest some of the components that may contribute to the observed differences in predictions.

A prominent distinction between the models relates to descriptions of nitrogen inputs from agricultural sources. SPARROW relies on source input data (fertilizer use, livestock populations) for 1987 in combination with empirically derived rate coefficients to estimate the nitrogen supplied and delivered to streams. The model coefficients for agricultural inputs may represent the aggregate effect of multiple activities and watershed processes. By contrast, SWAT estimates agricultural inputs by applying a crop growth model that simulates nitrogen use by plants as a function of energy retention and biomass conversion rates. This approach assumes that sufficient nitrogen is available to satisfy plant requirements. Crop harvest determines the quantities of residual nitrogen available for runoff. Thus, differences in the estimates of nitrogen availability between this method and SPARROW could partially explain the observed differences in estimates. Nitrogen from livestock wastes is also handled differently in the models. SPARROW estimates nitrogen inputs from livestock populations, whereas SWAT assumes that any nitrogen consumed by livestock is part of the residual crop nitrogen available for runoff.

The use of different time periods in describing average precipitation and streamflow conditions may account for some differences in model predictions. SWAT predictions are based on a long-term average of precipitation for a 30 year period, 1960-89, whereas SPARROW describes long-term mean hydrologic conditions for a base year 1987 using streamflow data covering a shorter time period, 1970-92. In view of evidence of modest increases in precipitation [Karl and Knight, 1998] and streamflow [Lins and Slack, 1999] in the conterminous United States during the past 30 years, mean conditions described by SPARROW could be slightly larger than those characterized by SWAT.

Differences in estimates of watershed attenuation were also noted among the models in comparisons of aquatic removal rates in the hydrologic units. Some of these differences may relate to differences in the spatial scale of model descriptions of stream channels. We found that the aquatic losses according to SPARROW are commonly larger and span a wider range than those reported by SWAT. The median percentage removed by SPARROW is about four times the percentage removed by SWAT (40% vs 10% of the nitrogen delivered to streams). SPARROW estimates of in-stream removal in the HCUs typically range from 20 to 70% of the nitrogen delivered to streams, whereas SWAT estimates commonly range from 5 to 14%. Because SPARROW yield estimates for the HCUs are generally larger than those of SWAT, it would appear that SPARROW estimates of the quantities of nitrogen delivered to streams are also larger than those estimated by SWAT.

5.2 Comparison with Chesapeake Bay Watershed Models

Two models of total nitrogen flux in the Chesapeake Bay watershed are available for comparison with regional estimates from the national SPARROW model, including the deterministic simulation model, HSPF, and a regionally calibrated SPARROW model. The Chesapeake Bay watershed is 164,000 square kilometers in size, 80 percent of which is drained by three rivers, the Susquehanna, Potomac, and James. The watershed contains areas predominantly in forest (60%), crop land (20%), urban land (10%), and pasture (9%) [Donigian et al. 1994].

5.2.1 Chesapeake Bay (CB) SPARROW model

The CB SPARROW model applies a similar statistical framework as was used in the national SPARROW model. More detailed descriptions of the model, its application, and the input data sets can be found in Brakebill and Preston [1999] and Preston and Brakebill [1999]. As in the national model, the U.S. EPA River Reach File 1 (RF1) serves as the digital river network for the CB SPARROW model to which watershed attributes are spatially referenced. The model was refined by using a one-square kilometer digital elevation model (DEM) to define the watershed boundaries of each of the 1,400 reaches, thereby providing information for relating continuous spatial information on watershed characteristics to the stream network. Estimates of TN flux at 79 monitoring locations (see location map in fig. 14) were calculated from stream-discharge and TN concentrations according to methods described previously for the national SPARROW model. The drainage basin size for these monitoring stations ranges from 480 to 52,000 square kilometers. Eleven monitoring sites are common to the national SPARROW model.

Data on nutrient sources and watershed characteristics differed from those used in the national SPARROW model (but are primarily derived from those used in HSPF) with the exception of estimates of nitrate wet deposition and soil permeability. Nitrogen inputs from agricultural fertilizer and livestock wastes were quantified according to land-use data, county-level agricultural statistics, and documented nitrogen fertilizer application rates coupled with agricultural area [Gutierrez-Magness and others, 1997]. Point-source discharge information consisted of nitrogen-discharge measurements at specific locations throughout the watershed. Each point source was linked with a stream reach based on the stream network described above. Point-source loads were calculated from the average annual waste discharge for the period 1986-88. Urban area, determined from land-use data developed by EPA, NOAA and USGS, was included in the model as a possible source of nutrients.

The calibrated CB SPARROW model [Preston and Brakebill, 1999] consists of 10 explanatory variables, and explains approximately 96 percent of the variability in in-stream TN flux. The accuracy of the model predictions are similar to the national model; the distribution of the percent absolute difference between the predicted and observed values has a median of 34 percent and an interquartile range from 18 to 61 percent. The explanatory variables and estimated coefficients for the TN model are presented in table 8. Estimates of uncertainty in the fitted coefficients in table 8 are expressed as 90 percent confidence intervals based on a bootstrap estimation procedure. Five source variables were found to be significant including municipal and industrial point sources, urban land, fertilizer use, livestock wastes, and atmospheric deposition. These source categories are similar to those in the national SPARROW model with the exception of the "urban land" class. The rates of in-stream total nitrogen removal are estimated according to three streamflow classes (a smaller flow class is defined in the CB SPARROW). The in-stream loss rates are consistent with both the magnitude and the functional form (inverse relation to channel size) as those reported for the national model. The rates of in-stream nitrogen loss span approximately an order of magnitude from 0.07 day^{-1} in the largest rivers (> 28 m3/s) to 0.76 day^{-1} in the smallest streams (< 5.7 m3/s). The loss rate for the smallest stream class (< 28.3 m3/s) in the national SPARROW model (0.45 day^{-1}) lies between the CB SPARROW loss rates (0.30 to 0.76 day^{-1}) estimated for streams of this size. In contrast to the national model, the CB SPARROW model has a separate loss rate for reservoirs.

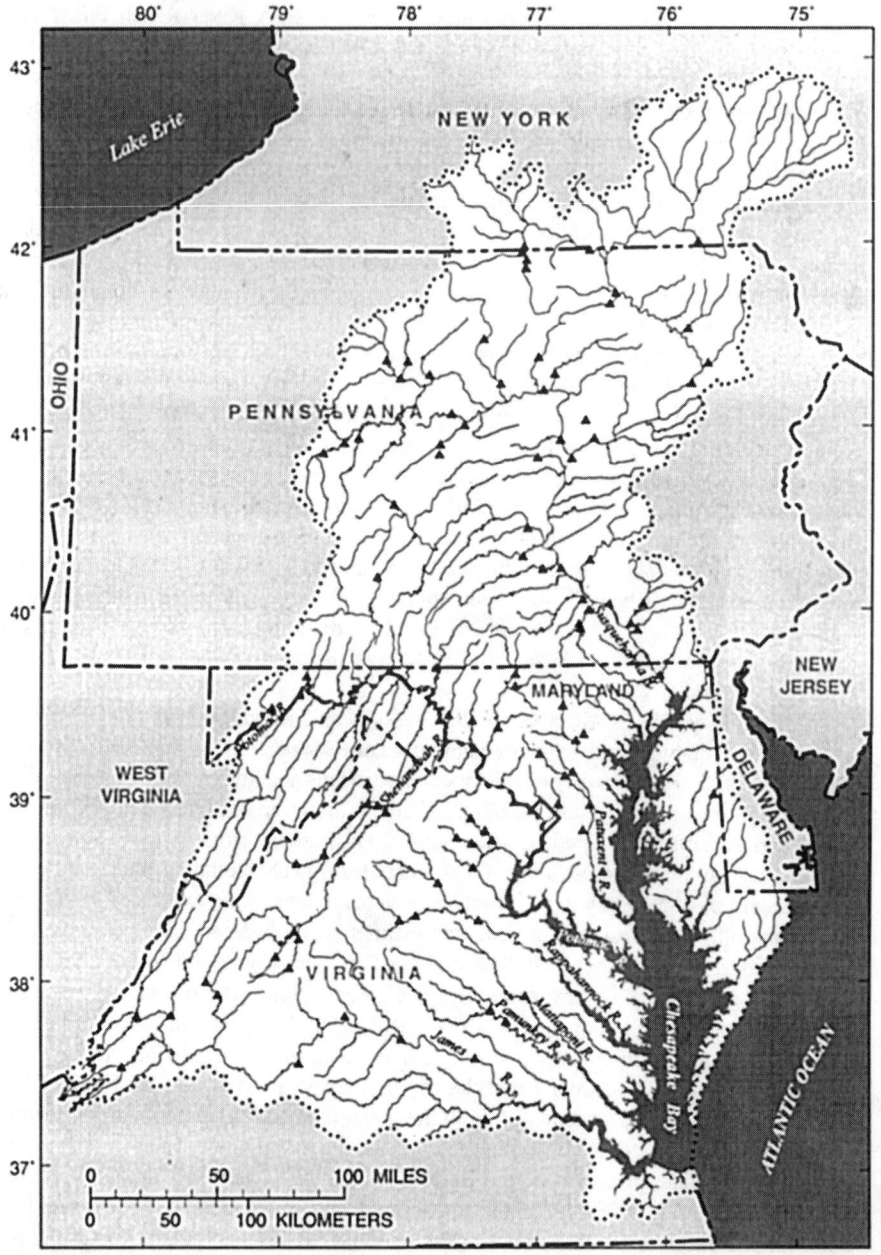

Figure 14. Locations of stream monitoring stations and RF1 reach segments in the Chesapeake Bay watershed. Station locations are denoted by a triangle.

TABLE 8. Chesapeake Bay SPARROW total nitrogen model coefficients based on a regression of stream nitrogen flux at 79 river monitoring stations on watershed characteristics.

Model Parameters	Coefficient Units	Bootstrap Coefficient	Lower 90% CI [a]	Upper 90% CI [a]
Nitrogen source, β				
Point sources	dimensionless	1.50	0.677	2.18
Fertilizer application	dimensionless	0.279	0.134	0.421
Livestock waste production	dimensionless	0.336	0.147	0.489
Atmospheric deposition	dimensionless	1.03	0.350	1.63
Urban land	kg ha^{-1} yr^{-1}	7.85	2.23	13.0
Land-to-water loss coefficient, α				
Soil permeability	hr. cm^{-1}	0.030	-0.003	0.065
In-stream loss rate[b], k				
k_1 (Q < 5.8 m^3 s^{-1})	day^{-1}	0.760	0.253	1.25
k_2 (5.8 m^3 s^{-1} < Q < 28.3 m^3 s^{-1})	day^{-1}	0.302	0.092	0.507
k_3 (> 28.3 m^3 s^{-1})	day^{-1}	0.067	0.000	0.162
Reservoir loss rate	day^{-1}	0.415	0.000	0.898
R-Squared		0.961		
Mean square error		0.167		
Number of observations		79		

[a] Minimum bootstrap confidence intervals (CI).
[b] In-stream loss rates fit separately for stream reaches with mean streamflow (Q) corresponding to the indicated intervals.

5.2.2 HSPF model

The HSPF model provides a temporal and spatial description of nutrient loads from pervious and impervious land surfaces and transport through rivers and reservoirs [Bicknell *et al.* 1997]. The simulation of nutrient load depends on a nonpoint source component which includes applications of fertilizer, manure, atmospheric deposition, crop uptake, soil adsorption, and denitrification. This component simulates the tranformations and movement of various nitrogen species in the subsurface and surface runoff. The river and reservoir components include hydraulic behavior, sediment-nutrient interactions, nitrification, denitrification, and phytoplankton growth. The Chesapeake Bay watershed is segmented into 87 watershed sections that average 1,900 sq. kilometers in size [Shenk *et al.* 1998]. Regions of similar geographic and topological characteristics were defined according to soil type, soil moisture capacity, infiltration rates, slope, and precipitation. A base year of 1985 was used for watershed characteristics, with hydrologic conditions averaged for the years 1985-94.

The hydrologic, land-use, soils, agricultural, and municipal and industrial point-source data used in the HSPF model are described in Shenk *et al.* [1998]. Atmospheric wet deposition data for nitrate and ammonia were obtained from 15 NADP monitoring sites in the region for 1984-91. Spatially continuous estimates of daily wet deposition to land and water surfaces of the watersheds were determined from a spatial regression of mean wet deposition at NADP sites on precipitation, latitude, and month of the year. Constant concentrations were assigned to each model segment according to the regression. Spatial estimates of dry deposition were obtained from an application of a metereological / chemical model, RADM [Regional Acid Deposition Model; Shenk *et al.* 1998; Gutierrez-Magness *et al.* 1997] in the eastern United States. These estimates are based on an estimated wet/dry ratio for each model cell of 400 sq. kilometers in the Chesapeake Bay region. The HSPF model was run both with and without the atmospheric deposition inputs to quantify the separate contributions of this source to total nitrogen exports in the watershed.

The CB HSPF was calibrated for 14 locations in the watershed based on the use of the observed measurements of concentration and flow. Ten of these locations were included among the 79 monitoring sites used in calibrating the CB SPARROW model. Based on the HSPF model predictions of the annual load at these 10 sites, the predicted TN yield is typically within 20 percent or less of the mean yield based on stream monitoring data (interquartile range = 14 to 30%).

5.2.3 Results and discussion of model comparisons

Figure 15 compares predictions of total nitrogen export from the national and CB SPARROW models and the HSPF model to "observations" of total nitrogen export at stream monitoring sites in the Chesapeake Bay watershed. The observed export is based on estimates of the mean annual load obtained by applying flux estimation techniques [Cohn *et al.* 1989] to stream monitoring records. SPARROW predictions are available for 79 monitoring locations; all of these sites were used in calibrating the CB SPARROW model and 11 were used in calibrating the national SPARROW model. HSPF predictions are compared to the mean annual load estimated from monitoring data at 25 of the sites; the concentration and flow data from 10 of these sites were used in calibrating the HSPF model.

The comparisons indicate that the national SPARROW predictions of total yield are within at least 39 percent of the observed yield at one half of the sites; approximately one half of the predictions are within 19 to 82 percent of the observed yield (interquartile range). The magnitude of these differences between predicted and observed values are only slightly larger than previously described for the CB SPARROW model and for the national SPARROW calibration data. Differences between the observed values and HSPF model predictions are smaller than either of the SPARROW models; HSPF predictions are within at least 18 percent of the observed yield at one half of the sites; about one half of the predictions are within 12 to 30 percent of the observed yield. In Chesapeake Bay watersheds with yields less than 500 kg ha^{-1} yr^{-1} and flux less than 2×10^5 kg yr^{-1}, there is evidence that the national SPARROW model tends to over predict export (fig. 15). In these watersheds, the national SPARROW predictions typically exceed the observed yield by a factor of 2.2 (interquartile range of 1.6 to 2.9). Uncertainties in the national predictions for watersheds of this size are typically 40 to 80 percent, based on the standard error of prediction (see section 4.4). There is less evidence of bias in the national SPARROW predictions in watersheds with higher yields and flux. All of the models tend to slightly under predict in the highest yielding watersheds, above 3,000 kg

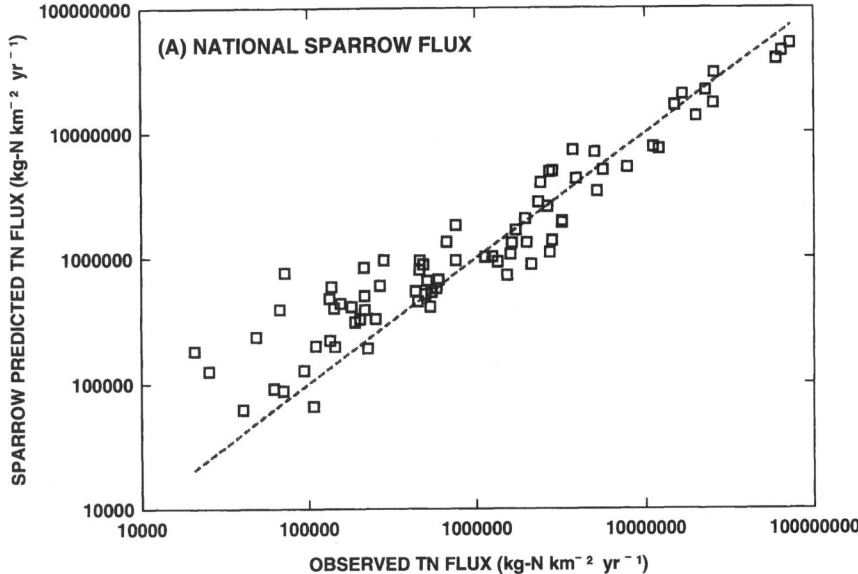

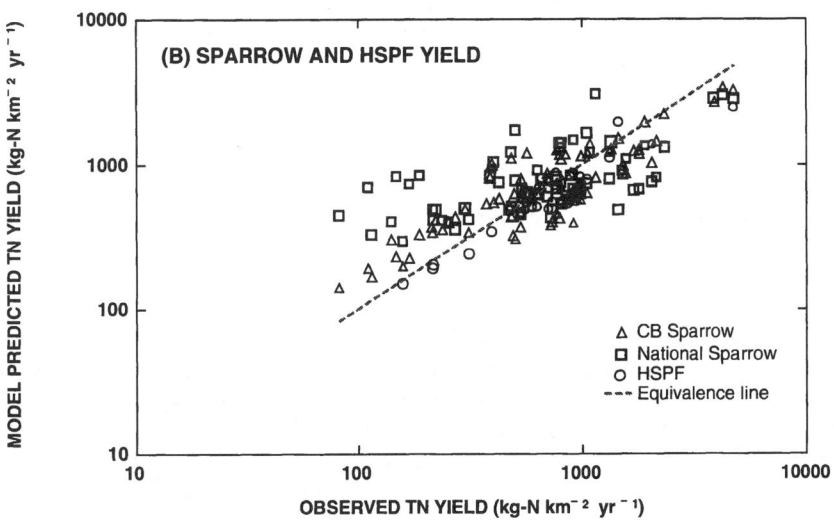

Figure 15. Model predictions of mean total nitrogen (TN) export from Chesapeake Bay (CB) watersheds in relation to the mean export observed at stream monitoring stations: (A) flux for the national SPARROW model; (B) yield for the CB and National SPARROW and HSPF models. The comparisons include 79 stream locations for the SPARROW model predictions and 26 locations for the HSPF model.

km^{-2} yr^{-1}, where comparisons are available for SPARROW at three monitoring locations and for HSPF at one monitoring location.

In figure 16, we compared the various model predictions of atmospheric TN export for 46 locations in the Chesapeake Bay watershed common to the HSPF model segments and SPARROW reaches. The model predictions of atmospheric contributions (fig. 16b) show a positive correlation over the range of percentages. National SPARROW predictions are larger than HSPF predictions at most of the outlets of the interior watersheds. The national predictions are also larger than the CB SPARROW predictions at most of the watershed outlets although the interquartile ranges of the predictions overlap considerably (table 9). The national SPARROW predictions of the atmospheric contribution to stream export (table 9) typically range from 27 to 38 percent (interquartile range) with a median of 32 percent. By comparison, CB SPARROW predictions at the 46 interior watersheds typically range from 18 to 31 percent with a median of 24 percent. Both the HSPF and CB SPARROW model predictions are within the range of uncertainty (15 to 41 percent) in the national SPARROW predictions for the Chesapeake Bay watershed, based on the standard error in the mean atmospheric contribution of 28 percent (see section 4). Literature estimates of atmospheric contributions to the Chesapeake Bay from the surrounding watershed range from 24 to 59 percent [Valigura et al. 1996; Castro et al. this volume].

The observed differences in atmospheric contributions to stream TN flux are likely explained by differences in estimates of the inputs and supply of dry and wet nitrogen deposition as well as model estimates of terrestrial and aquatic attenuation of atmospheric nitrogen. In view of the similarities in estimates of wet nitrate deposition used in the SPARROW and HSPF models, differences in deposition inputs are more likely explained by other nitrogen forms (e.g., ammonium, organic wet deposition). For example, the mean national SPARROW estimate of wet nitrate deposition in the Chesapeake Bay watershed is 3.1 kg ha^{-1} yr^{-1}. By comparison, the mean watershed estimate, based on wet nitrate deposition data used in HSPF, is 3.9 kg ha^{-1} yr^{-1}. Comparisons of the model estimates of in-stream nitrogen loss suggest that removal rates may be somewhat less in the national SPARROW model. This may partially explain the higher national SPARROW estimates of atmospheric and total nitrogen exports in the smaller watersheds. In-channel losses of TN average about 37 percent for streams in the Chesapeake Bay watershed according to the national SPARROW model (see section 4), whereas in-channel losses average from about 60 to 80 percent according to HSPF [Donigian et al. 1994] for the mean water travel time of Chesapeake Bay streams of about 5 days. Although the national SPARROW loss rate for small streams (mean flow < 28.3 m^3 s^{-1}) is consistent with the rates estimated by the CB SPARROW (see section 5.2.1), larger quantities of nitrogen are removed in the smallest streams (mean flow < 5.8 m^3 s^{-1}) according to the CB SPARROW.

6. Summary and Conclusions

An analysis of the contributions of the atmosphere to the nitrogen loads to major coastal and estuarine ecosystems in the United States was undertaken using an empirical

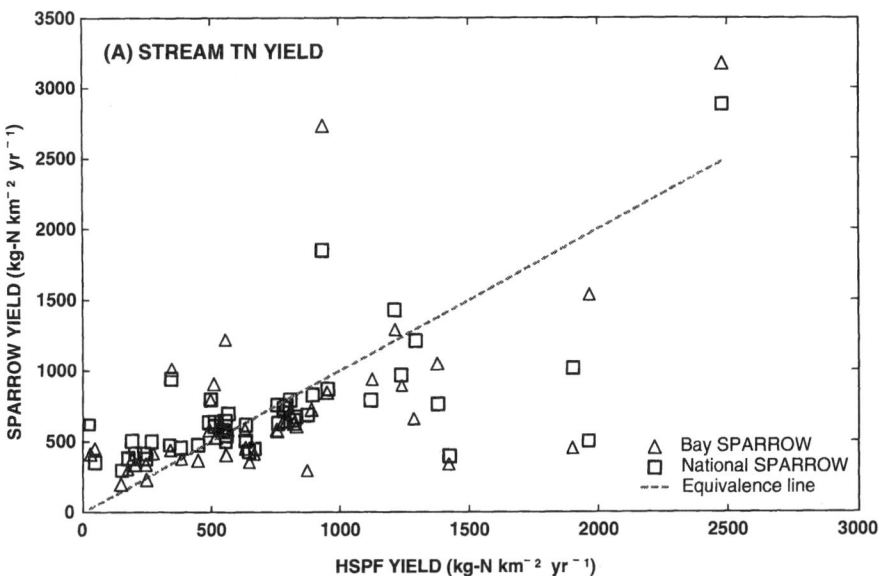

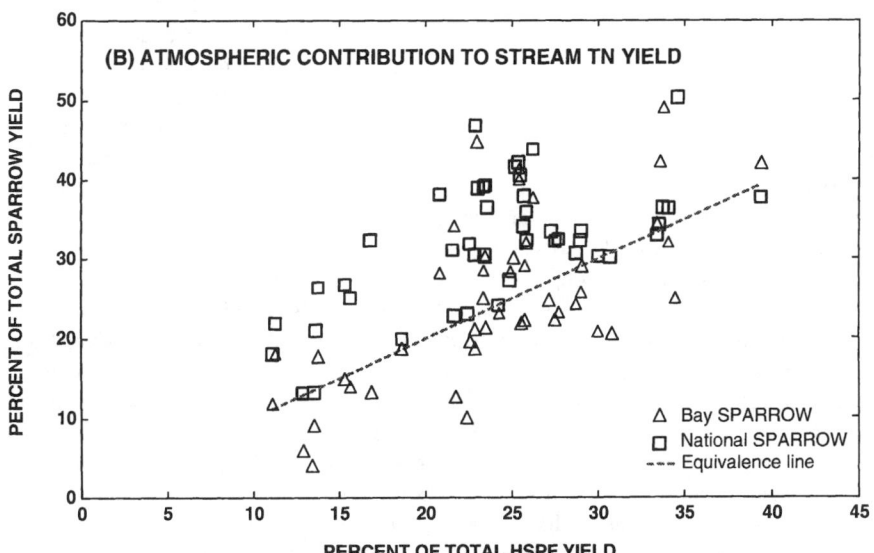

Figure 16. Model predictions for the National and Chesapeake Bay SPARROW and HSPF models for 46 interior drainage basins of the Chesapeake Bay: (A) mean total nitrogen (TN) yield; and (B) atmospheric contributions to mean total nitrogen yield.

TABLE 9. Predictions of the percentage contribution of atmospheric nitrogen sources to stream export at 46 locations in the Chesapeake Bay watershed.

Model	Atmospheric Contribution (percent)		
	25th	Median	75th
National SPARROW[a]	27	32	38
Chesapeake Bay SPARROW	18	24	31
HSPF (Hydrologic Simulation Program Fortran) Chesapeake Bay watershed model	18	20	24

[a] The estimated mean atmospheric contribution for the entire Chesapeake Bay watershed is 28 percent with a standard error of prediction ranging from 15 to 41 percent (see results in section 4).

watershed model, SPARROW (Spatially Referenced Regression on Watershed attributes). In the 40 estuarine watersheds examined, the mean total nitrogen yield ranges from 38 to 2,500 kg km^{-2} yr^{-1}. Atmospheric nitrogen contributions to riverine export range over nearly two orders of magnitude from 4 to 326 kg km^{-2} yr^{-1}. The atmosphere is estimated to contribute from 4 to 35 percent of the total nitrogen in stream export with a median contribution of about 15 percent. The highest atmospheric contributions are observed in the northeastern and Mid-Atlantic watersheds of the United States (i.e., North Atlantic region). Uncertainties in the estimates, based on the standard error of prediction, range from 40 to 100 percent and vary inversely with watershed area. Among the 40 watersheds, agricultural sources contribute the largest shares of nitrogen, accounting for more than one third of the stream export in most of the watersheds (with contributions as large as 70 percent), followed by the aggregate contributions of other diffuse sources, including runoff and subsurface discharges from urban, forested, range, and barren lands. Municipal and industrial point sources are similar in magnitude to atmospheric contributions in most watersheds, but represent the largest share (35-88%) of nitrogen in one half of the North Atlantic watersheds and in several Gulf region watersheds. Comparisons of the SPARROW model with other national and regional watershed models indicate general agreement in the predictions of TN export over a wide range of watershed sizes, but illustrate the intrinsic difficulties of comparing the flux rates of models having different temporal and spatial scales of measurement and prediction and different specifications of nitrogen supply, transformation, and transport processes.

Improvements in our knowledge of the relative contributions of nitrogen sources to coastal waters provides an important initial step in developing strategies for controlling nitrogen enrichment of coastal ecosystems. However, information that describes the watersheds and source locations responsible for coastal nitrogen inputs, as illustrated in the analysis for the Chesapeake Bay, is also needed to assist in the design of efficient management strategies. The current analysis provides an empirical framework for using stream monitoring records and data on watershed characteristics to identify the geography of nitrogen sources to coastal ecosystems and to refine model assumptions regarding the processes governing nitrogen transport through watersheds. Use of more spatially-detailed land-cover and hydrologic data and stream measurements from small watersheds are likely to enable a better separation of nitrogen sources and small-channel and terrrestrial loss processes. Moreover, the addition of model functions and data to explicitly account for the effects of subsurface transport, specific agricultural practices, and seasonal and interannual variations, may provide more accurate prediction of the origins of nitrogen delivered to coastal waters. These assessments are likely to benefit from a continued effort to integrate the mechanistic descriptions of deterministic models with the empirical methods of estimating watershed-scale rate processes and their uncertainties in statistical models.

Acknowledgments

The authors would like to thank Keith Robinson, Gary Shenk, Suzanne Bricker, and Bruce Hicks for comments on earlier versions of the chapter. Thanks to Michael Ierardi for assistance in preparing the graphics.

References

Alexander, R.B., Brakebill, J.W., Brew, R.E., and Smith, R.A., ERF1—Enhanced River Reach File 1.2, *U.S. Geological Survey Open-File Report 99-457*, Reston, Virginia (1999).

Alexander, R.B., Slack, J.R., Ludtke, A.S., Fitzgerald, K.K. & Schertz, T.L., Data from selected U.S. Geological Survey national stream water quality monitoring networks, *Wat. Resour. Res.* 34, 2401-2405 (1998).

Alexander, R.B. and Smith, R.A., County level estimates of nitrogen and phosphorus fertilizer use in the United States, 1945 to 1985, *U.S. Geological Survey Open-File Report 90-130*, Reston, Virginia (1990).

Alexander, R.B., Smith, R.A., and Schwarz, G.E., Effect of stream channel size on the delivery of nitrogen to the Gulf of Mexico, *Nature 403*, 758-761 (2000).

Arnold, J.G., R. Srinivasan, R.S. Muttiah and R.H. Griggs, Watershed Modeling and GIS with SWAT and GRASS. SWAT User's Manual. Blackland Research Center, TexasAgricultural Experiment Station, Temple, Texas (1995).

Arnold, J.G., Williams, J.R., & Maidment, D.R., A continuous-time watershed sediment routing model for large basins, *ASCE J. Hydr. Eng. 121*, 171-183 (1995).

Arnold, J.G., Williams, J.R., Nicks, A.D., & Sammons, N.B., SWRRBWQ – A basin scale simulation model for soil and water resources management, Texas A&M University Press, College Station, Texas, USA (1990).

Battaglin, W.A. & Goolsby, D.A., Spatial data in geographic information system format on agricultural chemical use, land use, and cropping practices in the United States, U.S. Geological Survey Water Resources Investigations Report 94-4176 (1994).

Beaulac, M.N. & Reckhow, K.H., An examination of land use – nutrient export relationships, *Wat. Res. Bull. 18*, 1013-1024 (1982).

Behrendt, H., Inventories of point and diffuse sources and estimated nutrient loads—a comparison for different river basins in central Europe, *Wat. Sci. Tech. 33*, 99-107 (1996).

Bicknell, B.R., Imhoff, J.C., Kittle, J.L., Jr., Donigian, A.S., Jr., & Johanson, R.C., Hydrological simulation program – fortran user's manual for release 11, U.S. Environmental Protection Agency, Environmental Research Laboratory, Athens, Georgia, USA, EPA/600/R-97/080, 755 p. (1997).

Brakebill, J.W. and Preston, S.D., Digital data used to relate nutrient inputs to water quality in the Chesapeake Bay watershed, Version 1.0: *U.S. Geological Survey Open-File Report 99-6,* Baltimore, Maryland (1999) at URL http://md.usgs.gov/publications/ofr-99-60/.

Castro, M.S., Driscoll, C., Jordan, T.E., Reay, W., Seitzinger, S., Stiles, R., & Cable, J., Assessment of the contribution made by atmospheric nitrogen deposition to the total nitrogen load to thirty-four estuaries on the Atlantic and Gulf coasts of the United States, this volume.

Cohn, T.A., DeLong, L.L., Gilroy, E.J., Hirsch, R.M. & Wells, D.K., Estimating constituent loads, *Wat. Resour. Res. 25*, 937-942 (1989).

Council on Environmental Quality (CEQ), Environmental trends, Executive Office of the President, Washington, D.C. 64-65 (1989).

Delwiche, L.L., & Haith, D.A., Loading functions for predicting nutrient losses from complex watersheds, *Water Resour. Bul. 19*, 951-959 (1983).

DeWald, T., Horn, R., Greenspun, R., Taylor, P., Manning, L., & Montalbano, A., STORET reach retrieval documentation, U.S. Environmental Protection Agency, Washington, D.C. (1985).

Diaz, R.J. & Rosenberg, R., Marine benthic hypoxia: a review of its ecological effects and the behavioural responses of benthic macrofauna, *Oceanog. Marine Biology: an Annual Review 33*, 245-303 (1995).

Donigian, A.S., Jr., Bicknell, B.R., Patwardhan A.S., Linker, L.C., & Chang, C.H., Watershed model application to calculate bay nutrient loadings: Final findings and recommendations, U.S. Environmental Protection Agency, Chesapeake Bay Program Office, Annapolis, Maryland, USA (1994).

Efron, B., The jackknife, the bootstrap and other resampling plans, in CBMS-NSF Regional Conference Series in Applied Mathematics, Soc. For Ind. And Appl. Math., Philadelphia, Pa., USA (1982).

ESRI (Environmental Systems Research Institute, Inc.), *ARC/INFO user's guide: Surface modeling with TIN,* Version 7.0, 1996.

Fisher, D.C. & Oppenheimer, M., Atmospheric nitrogen deposition and the Chesapeake Bay estuary, *Ambio. 20*, 102-108 (1991).

Frink, C.R., Estimating nutrient exports to estuaries, *J. Environ. Qual. 20*, 717-724 (1991).

Gilroy, E.J., Hirsch, R.M., & Cohn, T.A., Mean square error of regression-based constituent transport estimates, *Water Resour. Res.* 26, 2069-2077 (1990).

Gutierrez-Magness, A.L., Hannawald, J.E., & Linker, L.C., Appendix E: Phase IV watershed land uses and model linkages to the airshed and estuarine model, in Chesapeake Bay Watershed Model Application and Calculation of Nutrient and Sediment Loadings, EPA 903-R-97-019, U.S. Environmental Protection Agency, Chesapeake Bay program office, Annapolis, Maryland, USA, 142 p. (1997).

Howarth, R.W., G. Billen, D. Swaney, A. Townsend, N. Jaworski, K. Lajtha, J.A. Downing, R. Elmgren, N. Caraco, T. Jordan, F. Berendse, J. Freney, V. Kudeyarov, P. Murdoch & Zhu Zhao-liang, Regional nitrogen budgets and riverine N & P fluxes for the drainages to the North Atlantic Ocean: natural and human influences, *Biogeochem. 35*, 75-139 (1996).

Jaworski, N.A., Groffman, P.M., Keller, A.A. & Prager, J.C., A watershed nitrogen and phosphorus balance: the Upper Potomac River basin, *Estuaries 15*, 83-95 (1992).

Jaworski, N.A., Howarth, R.W. & Hetling, L.J., Atmospheric deposition of nitrogen oxides onto the landscape contributes to coastal eutrophication in the northeast United States, *Environ. Sci. Techno. 31*, 1995-2004 (1997).

Johnes, P.J. Evaluation and management of the impact of land use change on the nitrogen and phosphorus load delivered to surface waters: the export coefficient modelling approach, *J. of Hydrology 183*, 323-349 (1996).

Johnes, P.J. & Heathwaite, A.L., Modelling the impact of land use change on water quality in agricultural catchments, *Hydrological Processes 11*, 269-286 (1997).

Johnson, D.W., Nitrogen retention in forest soils, *J. Envir. Qual. 21*, 1-12 (1992).

Jordan, T.E. & Weller D.E., Human contributions to terrestrial nitrogen flux: assessing the sources and fates of anthropogenic fixed nitrogen, *Bioscience 46*, 655-664 (1996).

Karl, T.R., & Knight, R.W., Secular trends of precipitation amount, frequency, and intensity in the United States, *Bulletin of the American Meteorological Society 79*, 231-241 (1998).

Lins, H.F., & Slack, J.R., Streamflow trends in the United States, *Geophysical Research Letters 26*, 227-230 (1999).

Meyers, T., Sickles, J., Dennis, R., Russell, K., Galloway, J., & Church T., Atmospheric nitrogen deposition to coastal estuaries and their watersheds, this volume.

Mueller, D.K., Ruddy, B.C. & Battaglin, W.A., Logistic model of nitrate in streams of the upper-midwestern United States, *J. Environ. Qual. 26*, 1223-1230 (1997).

Muttiah R.S., R. Srinivasan, and P.M. Allen, Prediction of two-year peak stream discharges using neural networks. *J. Am. Water Res. Assoc. 33*, 625-630 (1997).

National Agricultural Statistics Service (NASS), Historic data on livestock, Economic Statistics System, Washington, D.C., on-line data base (1998).

National Atmospheric Deposition Program, NADP/NTN Annual data summary: Precipitation chemistry in the United States, Nat. Resour. Ecol. Lab., Colorado State Univ., Fort Collins, CO, 1983 to 1993.

Nixon, S.W., Coastal marine eutrophication: a definition, social causes, and future concerns, *Ophelia 41*, 199-219 (1995).

Novotny, V., and H. Olem, *Water quality: Prevention, identification, and management of diffuse pollution*, Van Nostrand Reinhold, New York (1994).

Peierls, B.L., Caraco, N.F., Pace, M.L. & Cole, J.J., Human influence on river nitrogen, *Nature 350*, 386-387 (1991).

Preston, S.D. and Brakebill, J.W., Application of spatially referenced regression modeling for the evaluation of total nitrogen loading in the Chesapeake Bay watershed: *U.S. Geological Survey Water Resources Investigations Report 99-4054*, Baltimore, Maryland (1999).

Rastetter, E.B., King, A.W., Cosby, B.J., Hornberger, G.M., O'Neill, R.V., & Hobbie, J.E., Aggregating fine-scale ecological knowledge to model coarser-scale attributes of ecosystems, *Ecological Apps. 2*, 55-70 (1992).

Ritter, W.F., Reducing impacts of nonpoint source pollution from agriculture: a review, *J. Environ. Sci. Health 23*, 645-667 (1988).

Rutherford, J.C., Williamson, R.B., & Cooper, A.B., *Nitrogen, phosphorus, and oxygen dynamics in rivers*, In Inland Waters of New Zealand (ed.) A.B. Viner, 139-165 (1987).

SAS Institute, Inc., *SAS/ETS user's guide*, Version 6, 2nd edition, Cary, North Carolina, USA, 509-591 (1993).

Seitzinger, S.P. & C. Kroeze, Global distribution of nitrous oxide production and N inputs in freshwater and coastal marine ecosystems, *Global Biogeochem. Cycles 12*, 93-113 (1998).

Seitzinger, S.P., Denitrification in freshwater and coastal marine ecosystems: ecological and geochemical significance, *Limnol. Oceanogr. 33*, 702-724 (1988).

Shenk, G.W., Linker, L.C., & Donigian, A.S., The Chesapeake Bay program models, Proceedings of the First Federal Interagency Hydrologic Modeling Conference, Las Vegas, Nevada, USA, U.S. Environmental Protection Agency, Chesapeake Bay Program Office, Annapolis, Maryland, USA, 8 p. (1998).

Smith, R.A., G.E. Schwarz & R.B. Alexander, Regional interpretation of water-quality monitoring data, *Wat. Resour. Res. 33*, 2781-2798 (1997).

Srinivasan, R., Arnold, J.G., Muttiah, R.S., Walker, D., & Dyke, P.T., Hydrologic unit modeling of the United States (HUMUS), In: *Advances in hydro-science and engineering*, S. Yang (ed.) Vol I, Part A., 451-456, Washington, D.C., USA (1993).

Tasker G.D., and N.E. Driver, Nationwide regression models for predicting urban runoff water quality at unmonitored sites. *Water Resources Bulletin 24*, 1091-1101 (1988).

Thomann, *Systems analysis and water quality management*, Environmental Research and Applications, Inc., New York, N.Y. (1972).

United States Environmental Protection Agency, *Design manual: Onsite wastewater treatment and disposal systems*, Office of Water Programs Operations, Washington, D.C., EPA 625/1-80-012 (1980).

Valigura, R.A., Luke, W.T., Artz, R.S., & Hicks, B.B., Atmospheric nutrient input to coastal areas: reducing the uncertainties, *NOAA Coastal Ocean Program Decision Analysis Series No. 9*, 24 p. (1996).

Vitousek, P.M., Aber, J.D., Howarth, R.W., Likens, G.E., Matson, P.A., Schindler, D.W., Schlesinger, W.H. & Tilman, D.G., Human alteration of the global nitrogen cycle: sources and consequences, *Ecological Applications 7*, 737-750 (1997).

Whitehead, P.G., Wilson, E.J., & Butterfield, D., A semi-distributed Integrated Nitrogen model for multiple source assessment in Catchments (INCA): Part I – model structure and process equations, *Sci. of the Total Environ. 210/211*, 547-558 (1998).

Williams, J.R., Nicks, A.D., & Arnold, J.G., Simulator for water resources in rural basins, *ASCE J. Hydr. Engr. 111*, 970-986 (1985).

7

Uncertainties in Individual Estuary N-loading Assessments

D. A. Brock

Abstract

The uncertainties involved in compiling nitrogen loading estimates are reviewed here. Estimates of the magnitude of uncertainty can be placed on many components. The uncertainty associated with some terms cannot be quantified, because these sources have rarely been measured, e.g., bedload transport, groundwater input, and atmospheric dry deposition. Those sources may constitute a significant portion of an estuary's nitrogen budget. Some sources of information available to estimate loading have uncertainty because the data are collected for other purposes. For example, the calculation of point source loadings from reported flows and concentrations from regulatory databases might over-estimate the actual loadings, particularly from sources with intermittent discharges. The uncertainty of tributary discharge volume and concentration typically are 10% of the mean, and may be 30% of the mean values. The inter-annual variation in loadings ranged around and above 30% for Texas estuaries spanning a range of streamflow variability. From the information compiled, surrogate confidence bounds were calculated for nitrogen loadings produced by local efforts. For many systems, the uncertainty in concentration and flow data for major tributaries is the source of greatest over-all error in loading estimates, because the major tributary inputs are large. As much as 40% of the total uncertainty is associated with the contribution by the major tributaries.

The Uncertainty in Compilations of Nitrogen Loads

Keeping in mind the confidence with which loading quantities are stated serves to keep the user's expectations of results realistic and serves to keep comparisons on a scientific basis. Comparisons of nitrogen loadings to major estuaries produced by various methods can corroborate findings. As in other scientific comparisons, judgments on the degree of discrepancies should be made in the context of the uncertainties associated with the findings. Likewise, the relative certainty associated with the data should be considered in assessing the relative magnitudes of components of the total load. Here are presented major sources of uncertainty, estimates of the relative error associated with calculated loadings, and the confidence bounds for some estimates of nitrogen inputs to several estuaries. A main focus is on the confidence in measurement data on estuarine system properties; however, the potential magnitude of year-to-year variation is also presented. Much of the information presented was developed from available information on loadings to Texas estuaries. Some of these systems may lie close to one end of the gradient of hydrologic and drainage basin types--from a national perspective--which may therefore influence the magnitude of variation in some components. But, the general considerations are common to most systems.

The elements of the loading equation discussed here include loadings from major tributaries--typically gaged streamflow--and sources below the last gage-- "below fall line". The discussion of uncertainties associated with wastewater inputs, groundwater, ungaged and other sources refer mainly to experience with below fall line loadings, the local estuary watersheds below the last gage. In equation form, total load to the estuary is the sum:

$$Load = Flux\ at\ last\ gage + Inputs\ below\ last\ gage,$$

where

$$Inputs\ below\ last\ gage = Groundwater + Local\ rainfall\ runoff + Direct\ atmospheric\ deposition + Waste\ water.$$

Sources of Uncertainty

Definition of Loading Components

Various studies of nutrient loadings have been shaped by the availability of data, by perceptions of important issues, and by the time frame over which data could be compiled. These aspects contribute to an alpha-level of certainty with which we can evaluate the results. For some estuaries there may be limited a priori information on the significance of various potential nutrient sources to guide data collection. The exercise of constructing a materials budget may be necessary to help an investigator understand if all significant nitrogen inputs have been incorporated into a loading figure. For other estuaries, the combination of gaged inflow and measured stream concentrations near the estuary head may capture most drainage basin inputs. Compiling loadings is a larger challenge for other basins, particularly those with a large shoreline to volume ratio and/or multiple, small tributaries. The following presents a few examples of how the treatment of potential nutrient sources may contribute to estimated loadings, and therefore to certainty with which some loading estimates might be viewed.

Major Tributaries, Gaged Flow

Significant loadings of nutrients may take place at times or in forms that are hard to measure. There are concerns that major amounts of nitrogenous material could be brought into the estuary on storm-water flood flows in tributary streams and rivers. Without adequate data on concentrations at "first flush", or with limited data on nutrients associated with particulates or bedload transport, the N loads from episodic inputs may be underestimated. Many loading estimates are dependent on land use and hydrography, so the absence of adequate land cover data as well as storm discharge volume and concentration data could cause a bias for some systems.

It may be possible to put limits on the potential error from an unknown flushing-flow load effect. A qualitative assessment can be made based on available data. For example, consider this test of "first flush" impacts for the Trinity River. Loads were accumulated from the first four days of high flows that followed periods of several weeks with lower flows. Applying a 90% percentile TN concentration level for this site (5.3 mg l^{-1}), these flows were found to supply 6% of the 145×10^9 kg N loading delivered by the river during the entire period, 1977 to 1987. For this major tributary to the estuary, even given a high concentration, the first-flush loadings did not bring in an extraordinarily high nitrogen load. For some studies of pollutant loadings, given the mix of various land uses that might contribute to a first-flush scenario, this look at potential flushing inputs might be part of an error bound determination.

Below Last Gage

The influence of shallow groundwater, especially from agricultural subsurface and septic tank drainage, is a potential TN source of unknown magnitude to many estuaries and coastal bays. There have been some attempts to quantify this source [e.g., Staver et al., 1996; Tampa Bay, NEP]. Variation in soils and landform surrounding each estuary make it difficult to describe groundwater loadings for any system, or to establish expectations for general loading rates applicable among a group of estuaries.

Atmospheric dry deposition to the watershed or to the estuary surface is another input unknown for many systems. For shallow, broad estuaries with limited drainage basins, dry deposition could be very significant. For the Mission-Aransas estuary on the Texas coast, dry deposition to the estuary surface amounts to greater than 20% of the average annual nitrogen input. This figure is based, however, on relationships between wet and dry deposition from other areas and estimated climatic modifiers, which are two factors with unspecified precision.

Another problem of data source definition is related to wastewater sources. Data collected by point source dischargers in compliance with regulatory requirements for monitoring may not actually be based on a design that provides all information needed for accurate loading calculations. Data from the NPDES program are available, but naive use of these data may result in a high bias in concentrations employed and resultant loading estimations [Hill, 1999]. There may be requirements for stating discharge loading in industry self-reporting data, which specify load computed per sample. For discharge points with intermittent or variable flow, extrapolating the reported load to a monthly or annual figure can greatly overestimate actual discharge. Likewise, storm sewer sampling protocol may specify data collection at times and flows when concentrations are likely to be high. This also may introduce a high bias in load calculations.

Discharge permits typically specify only data on ammonia nitrogen, if nitrogen is monitored -- at least in Texas. However, "end-of-pipe" permit data show that the total nitrogen concentration is usually much higher than for ammonia nitrogen alone. Organic N and NO_x are usually also components of the discharge. If the waste stream is chlorinated, then chloramines form quickly, thus reducing measurable ammonia.

TABLE 1. Components of loading uncertainty. Values are +/- 95% confidence bounds expressed as percentage of moderate-size estuary constituent magnitude.

Loading component	Flow Volume	TN conc.	Combined Flux
Gaged stream flow volume	5%		
Concentration data, routine monitoring		10%	
Flux via flow-conc. relationships	NA	ca. 2%	ca. 6% [a]
Ungaged watersheds flow, via rainfall-runoff model	20%		
Input from local urban ungaged areas, NURP data		10%	
Concentrations for ungaged areas		10%	
Groundwater inflow	?	?	
Point sources, Wastewater inflow, mixed types	10% - 20%	25%	
Direct precipitation to bay surface	10%	9.5%	
Direct atmospheric dry deposition	?	?	
Direct precipitation, per-area load		13%-25%	

[a] For Galveston Bay inflows from FLUX, probably conservative

Therefore, depending on data source, using permit or monitoring data from dischargers may result in either over-estimates or under-estimates of true loading.

Measurement Error

Natural systems and also anthropogenic systems are usually more variable than monitoring strategies are prepared to or able to describe. This is a major source of uncertainty when estimating their nutrient flows. The methodologies for measuring concentrations have advanced so much since many monitoring programs started that detection limits are usually not now an issue. For estuary loading studies, the problem has more to do with the expense of monitoring at the desired frequency and spatial coverage. The impact of unmeasured variation has become a critical concern.

Table 1 presents estimates of the magnitude of error for components of loading compilations developed to support a first-order uncertainty analysis for loadings to Texas bays. Uncertainty is expressed as a percentage of flow volume, concentration, or loading. (Expression of confidence bounds as a percentage may be appropriate only among estuaries similar in magnitude of loadings or other features--this may result in bounds smaller or larger than is realistic.) The data, assumptions, and rationale behind attaching confidence values to these components are discussed next.

Gaged and Ungaged Streamflow

The typical source of gaged streamflow data is from the United States Geological Survey (USGS) Water Resources Division. The records and ancillary data from gauging stations are reviewed annually and the records are classified through a performance evaluation. The typical uncertainty level published for Texas streams is 5% [e.g., US Geological Survey, 1998], although a value of 10% or more might better reflect gaging accuracy in some streams under certain conditions [Ockerman, 1999]. Winter [1981] also discusses variability of physical stream and lake measurements. Changes in bottom contours with natural shoaling and scouring processes at the gage site can cause some variation. Also, the stage to discharge-volume rating curves are likely most accurate in the mid- to high range of measured flows, with very high or low flows subject to increased error.

Ungaged flow in areas of the coastal drainage below major gages, or from small streams, are often estimated by application of rainfall runoff models, usually based on Soil Conservation Service (now National Resources Conservation Service) curve numbers [USDA, 1985]. The ability to produce reasonably accurate ungaged flows is limited by uneven distribution of local rainfall rates, as well as variation in soil properties, soil saturation level, etc. Hann et al. [1993] proposed a method for testing model uncertainty, and apply this to rainfall runoff. Srinivasan et al. [1998] found the difference between simulated (HSPF model) and observed runoff for two watersheds in Pennsylvania to be 5% and 17% for total runoff over a two-year period.(HSPF under-predicted peak flows and over-predicted low flows) Difference between observed and simulated volumes varied considerably more (8% to 167%) for individual season results. The 20% uncertainty listed in Table 1 is based on calibration and verification tests of a runoff model applied to Texas coastal watersheds, calibrated on small gaged watersheds.

Other ungaged flow contributions, such as from largely urbanized areas, or from groundwater, were assumed here to have an uncertainty equivalent to that of rainfall runoff. The effects of tides on short-term rates of shallow aquifer discharge makes it difficult to measure subsurface loadings to estuaries [Staver and Brinsfield, 1996].

Concentration Data

The uncertainty in stream nitrogen concentrations is the product of methods limitation, the vagaries of sampling, the spatial coverage and, especially, the temporal sampling frequency. Concentrations typically show some relationship with flow rate. A negative slope of concentration with flow is often observed, especially below major discharge points [Walker, 1982, 1996]. An event-mean concentration (EMC) can be calculated as the volume-weighted average concentration over a series of measurements spanning the hydrograph of an inflow episode. Data are not usually collected or reported in a way that is useful in attaching an uncertainty to EMC concentrations. In general, estimating missing concentration data using a relationship of concentrations with inflow may reduce the potential errors associated with other means of generating replacement data. The error of the regression is used to represent the confidence level for those concentrations.

A series of reports of wasteload allocation studies were reviewed to evaluate the variability in typical concentration data. A summary of TN data from several Texas streams for which a series of samples were collected at one site during one day [Kirkpatrick, 1986 1987a; Twidwell and Davis, 1987], showed that the confidence bounds were roughly 10% of mean concentrations (Table 2).

Flux

Programs such as FLUX [Walker, 1996] are available to assist calculation of loadings based on flow-concentration relationships that are derived from sample data. These methods reduce potential bias that exists in application of average concentration. In applications of FLUX in the Galveston Bay coastal watersheds, confidence bounds reported for the loading estimates were approximately 6% of the total load. Here,

TABLE 2. Variation in stream TN measurements (TN = sum NO_2, NO_3, and TKN constituents).

Stream	Mean	± 1 Std. Dev.	n	95 % C.I.	% C.I. of the Mean
San Antonio River	10.95	0.83	5	0.73	6.7
Dickinson Bayou	3.03	0.34	5	0.30	10.0
Buffalo Bayou, Sta. H	6.53	0.74	5	0.65	10.0
Buffalo Bayou, Sta. M	4.78	0.55	5	0.49	10.1

Stream	Site	Date	Reference
San Antonio River	1911.0352	June 6	Twidwell & Davis 1987
Dickinson Bayou	1104.0100	Nov. 28	Kirkpatrick 1986
Buffalo Bayou Sta. H	H	July 16	Kirkpatrick 1987a
Buffalo Bayou Sta. M	M	July 16	Kirkpatrick 1987a

available monitoring data were typically monthly or quarterly grab samples. Potential variation in concentrations with flood stage is probably not adequately represented. Since FLUX does not have an input for insertion of flow uncertainty, this degree of error reflects concentration uncertainty. Alexander et al. [1996] used log-linear regression models to compute nitrate flux in coastal rivers from periodic flow and concentration data. They report regional differences in flux variability (see below). There are probably also regional differences in typical error of flux estimation as well.

Point Source Loadings

Flows, concentrations, and loading from point sources are obtained in several ways, with different degrees of uncertainty attached to their estimation. For some basins with a limited number of dischargers, end-of-pipe determinations of flows and concentrations may make this loading category one of the more certain estimates. For estuaries like the Trinity- San Jacinto (Galveston Bay), with a larger number of dischargers composed of various municipal and industrial kinds of sources, certainty is difficult to ascertain. Pacheco et al. [1990] offer a compendium of representative concentrations from various industrial and manufacturing sources, by enterprise categories. The authors caution, however, that this sort of characterization may be based on "best management practices" older than the date of the study, and therefore are potentially no longer representative.

For Galveston Bay, local bay area end-of-pipe concentration data were available from special studies performed to determine waste load allocation [Kirkpatrick, 1986, 1987b]. A trimmed average TN from all dischargers combined was 9.84 mg l^{-1} N, the confidence interval +/- 3.5 mg l^{-1}. Municipal wastewater discharge concentration data were compiled from special studies of a number of streams [Buzan, 1990; Kirkpatrick, 1986, 1987a, 1988]. Variation was noted between samples on separate occasions, as well as among the plants. The average concentration was 9.75 mg l^{-1} TN, with a confidence interval of 2.85 mg l^{-1}, which is about 15% of the average. Industrial dischargers were more heterogeneous. The average TN in outfalls of 34 dischargers in the Houston area was 21.39, with a 95 % confidence interval of 19.6 mg l^{-1}.

TABLE 3. Multipliers applied to categories of loading to derive confidence bounds (10^3 kg y^{-1}).

Loading category	Lower Bound	Upper Bound
Gaged Streams	0.86	1.16
Gaged Streams via FLUX	0.92	1.08
Ungaged (Non-point) Inflow	0.72	1.32
Point Sources	0.68	1.34
Wet Deposition to Bay Surface	0.81	1.21
Combined Point and Stream (1:1)	0.77	1.27
Combined Point and Stream (1:5)	0.83	1.19
Combined Point and Stream (1:100)	0.85	1.16
Combined Gaged and Ungaged	0.79	1.24
Combined Gaged and Ungaged (10:1)	0.84	1.17
Combined Gaged, Point, and Ungaged	0.75	1.28

Direct Precipitation to Bay Surface

The volume of rainfall directly to the bay surface is typically a weighted average from surrounding National Weather Service rain gages. The National Atmospheric Deposition Program [NADP, 1993] provides estimates of the error associated with the reported deposition volume. For one Texas coastal station, the confidence bounds were approximately 9% of the total annual volume collected. Estimates of error are also reported with NADP deposition rates per area. For a Corpus Christi Bay area gage, this error was 13%. For a Galveston Bay area gage, the error was 25%. Chang and Flannery [1998] conducted comparisons among conventional rain gages. They found a consistent low bias to recorded rainfall as compared with a reference pit gage. The average deficiency was 11.8%. This degree of discrepancy may be related to the type of weather events producing rain in the area of the study. This potential bias deserves more investigation.

Uncertainty Incorporated in Comparisons

Information developed from sources and analyses described above was used to compute surrogate confidence bounds for use in comparing the local nitrogen loading estimates (EEG, Chapter 5) with estimates derived from the Land-use Based Group (LBG, Chapter 4) and national modeling efforts such as SPARROW (NWG, Chapter 6). Table 3 presents multipliers applied to input sources to derive upper and lower confidence bounds. The numbers are based on combinations of information listed in Table 1. Where separate loadings were reported from gaged stream-flow, waste water discharge, and direct rainfall, for example, each quantity was multiplied by a separate factor, and the results summed to a total loading bound. Loadings with confidence bounds are presented for the water bodies for which a) all three groups developed total loadings estimates and b) had enough information for the EEG to develop confidence intervals. (Table 4). Unfortunately, only 11 of the estuaries for which there were NMG and LBG loadings calculated have also had loadings compiled by independent efforts in sufficient detail to perform the error analyses presented here. However, the 11 water bodies that do have estimates span a range of location, type, and size.

TABLE 4. Estimated total annual N loads for each of the three groups, and confidence bounds developed by the NMG and EEG (10^6 kg N y^{-1}).

	NMG Load	NMG Lower Bound	NMG Upper Bound	LBG Load	EEG Load	EEG Lower Bound	EEG Upper Bound
Narragansett Bay	4.85	2.10	7.60	8.99	9.10	7.23	11.18
Long Island Sound	35.48	25.33	45.62	45.87	32.33	24.21	41.56
Barnegat Bay	1.42	0.40	2.45	1.02	0.63	0.52	0.75
Chesapeake Bay	133.63	108.08	159.18	142.75	151.57	112.66	195.75
Charleston Harbor	4.34	2.83	5.85	30.51	5.30	4.24	6.41
Indian River	0.14	0.02	0.25	6.29	2.60	1.86	3.45
Charlotte Harbor	3.38	1.53	5.23	13.95	2.72	2.10	3.41
Tampa Bay	3.15	2.00	4.30	12.92	3.45	2.64	4.37
Galveston Bay	29.58	20.10	39.07	70.86	38.35	30.38	47.24
Matagorda Bay	14.43	9.85	19.00	45.39	6.81	5.45	8.32
Corpus Christi Bay	2.50	1.29	3.71	12.44	2.29	1.77	2.88

The error bounds reported here only reflect the uncertainty about sources that were identified and compiled. It is interesting to note that while the NMG and the EEG used different techniques to develop error estimates, the magnitude of the resulting error bars are comparable (Figure 1; Chapters 6 and 7). Not included is any estimate of the certainty that all significant sources were included. Inter-annual variation also is not included. Since information about many of the loading studies was taken only from the summary presented in Appendix I, some categorizations may not be quite appropriate. However, the approach used here typically generates fairly similar relative error bounds, and probably conservative bounds. Results for Waquoit Bay were similar to the results obtained by Valiela et al. [1997], who used an exemplary approach.

There is an implicit assumption in the above estimation of confidence intervals, that variation in each parameter follows a normal distribution. For many components, at least for monthly or daily data, the distribution is better approximated as log normal, and confidence intervals should be computed as [Walker, 1996]:

$$Y_m * \exp(-2 * cv) < Y < Y_m * \exp(2 * cv),$$

where Y_m is the measurement mean, cv is the coefficient of variation, and exp denotes an exponential function. The distribution of some parameters, however, such as wastewater flow, appear normal, and the distribution of annual totals may be closer to normal, at least for the Texas data.

Figure 1--in two parts differing in scale--shows how loadings compiled by NMG and LBG efforts compare in the context of confidence bounds on local estimates. In a few cases, all estimates are similar. For the majority of cases, loadings from modeling efforts are far above the upper limits of the described confidence interval. If the procedure used

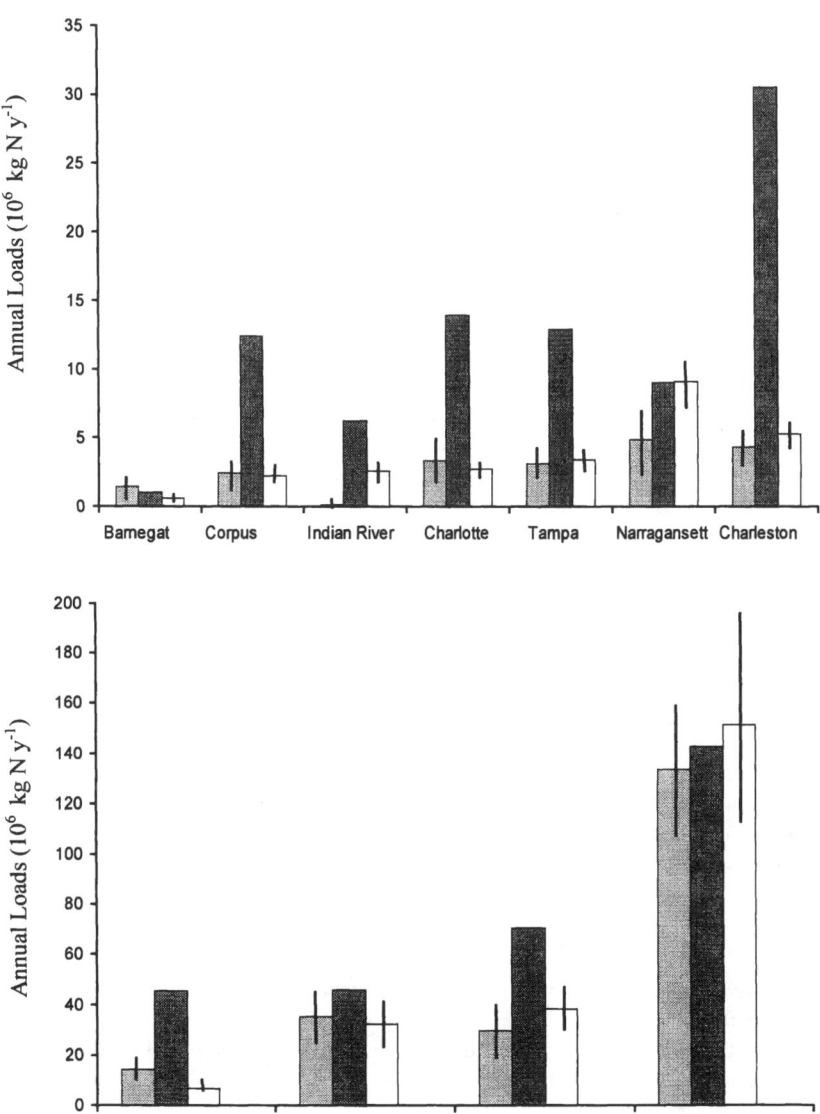

Figure 1. Comparison of annual N loading estimates developed by the NMG, the LBG, and the EEG respectively. Estimated confidence bounds are shown with solid bars for the NMG and the EEG. Data is split to accommodate large differences among estuaries in magnitude of loadings.

here to provide estimates of confidence intervals is appropriate, these results suggest that differences in loading estimates are real. As discussed elsewhere, sources of variation other than factors included in the CI estimates may be contributing to discrepancies. Different flow regimes in specific years covered by loading studies or differences in watershed areas used by the various efforts may be responsible for some differences.

Inter-annual Variation

Inter-annual variation is the inconstant backdrop for discussion of average conditions and long-term system response. The kind of inter-annual variation observed among the estuaries featured here naturally reflects the local and national climatic gradient. Contrast the variation discussed below for Texas estuaries with data for the series of years presented by Boynton et al. [1995]. With regard to comparisons among estuaries discussed here and in other chapters, the inter-annual variation may skew particular comparisons between model results and local estimates. Differences between the hydrology of the base year chosen for modeling may differ considerably from conditions during the period of local study.

There are regional differences in the degree to which any particular year may differ from the long-term mean nitrogen loading. Alexander et al. [1996] studied the variations in nitrate flux in Atlantic and Gulf of Mexico coastal basins. The year to year CV of nitrate flux typically varied from 20% to 40% of the long-term means. Western Gulf of Mexico rivers, with typically flashy hydrology, showed large variability. Rivers of the North Atlantic region showed low inter-annual nitrate flux variability. Thus, there may be regional differences in the certainty with which a particular study's results may be extended to describe average conditions.

Consideration of the annual scale of variation is appropriate to the questions of interest here, but the annual numbers may not reflect the effective variation driving productivity. The seasonal timing of inputs may enhance or diminish the effects of loading brought into a bay on flood flows, for example. Alexander et al. [1996] report large seasonal variation in nitrate flux for the Mississippi River, and for rivers of Mid Atlantic and South Atlantic regions.

In addition to the influence of changing climatic factors, real or apparent trends are manifest in many estuaries. The last few decades -- from which comes most of the data available for loading computations -- are decades during which changes in regulations and management practices have had strong water quality impacts.

Inter-annual Variation in Texas Estuaries

The series of yearly loading of TN to Texas estuaries may illustrate a few major characteristics of inter-annual variation. Loadings were compiled for each estuary using similar assumptions and data sources over the 1977-1987 period [Longley, 1994]. A summary of methods used to produce the loadings to Texas Bays is presented in the Appendix I.

Figure 2 demonstrates the series of annual TN loads during the period for estuaries at the extremes of hydrologic gradient, the Nueces estuary (Corpus Christi Bay) and the Trinity-San Jacinto Estuary (Galveston Bay). Both systems have sizable upland drainages and coastal drainages. The Trinity system shows higher loadings, in general, than the Corpus Christi system. For both systems, the series of years shows such large differences in loadings, as to weaken the utility of the average loading listed. The series also shows that for a few years, wastewater discharges are the largest source of TN to the Corpus Christi Bay Estuary. (For these compilations, wastewater (WW) includes only sources in portions of tributary basins below the lowest gage.)

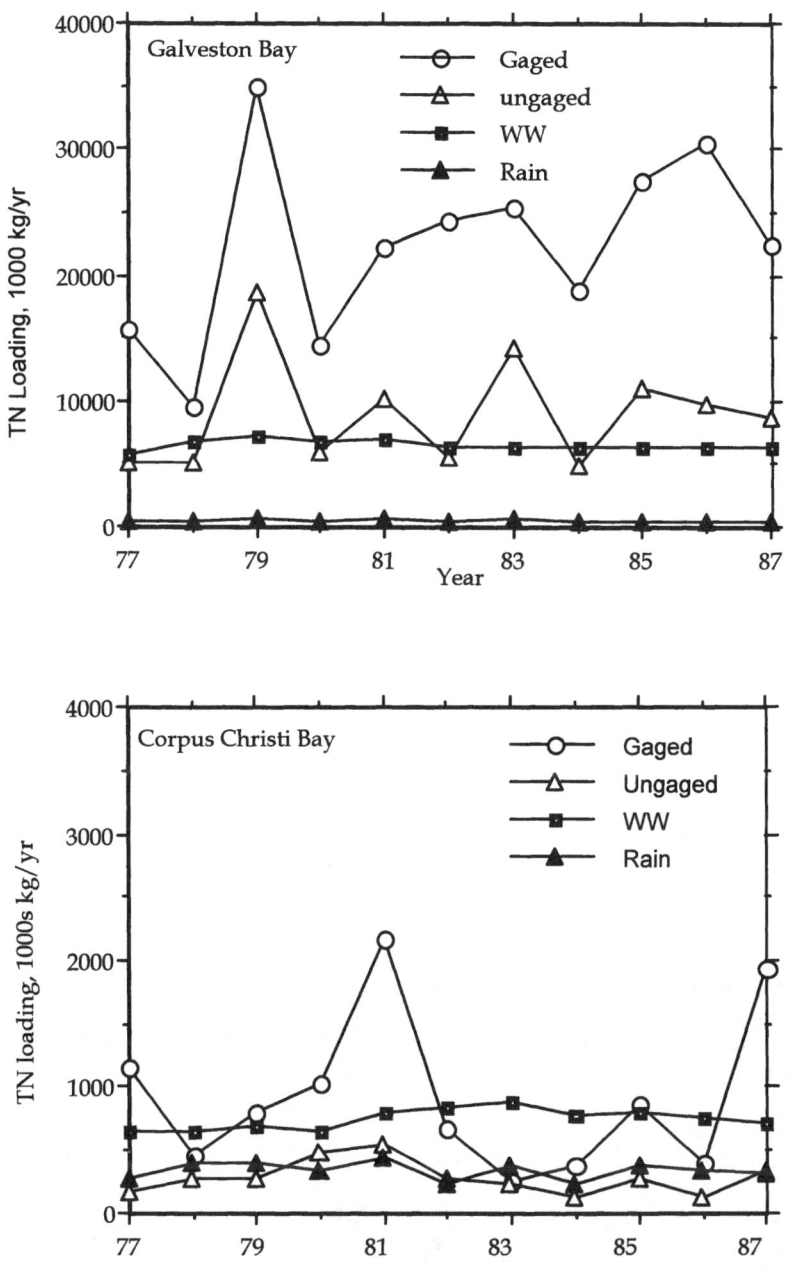

Figure 2. Inter-annual variation in TN loading sources to the Galveston Bay System (top); and to the Corpus Christi Bay System bottom).

TABLE 5. Statistics of total nitrogen annual loads (10^3 kg N y^{-1}) to Texas estuaries for 11 years for each system.

Estuary	Mean	± 1 Std. Dev.	Min.	Max.	Median
Gaged Streamflow					
Corpus Christi	920	627	257.3	2163	797
Mission-Aransas	341	242	92.9	953	269
Guadalupe	7081	3879	2489	15908	6032
Matagorda	3952	1624	2010	6904	3604
Galveston	22346	7384	9613	34873	22363
Ungaged inflow, Non-point					
Corpus Christi	286	135	121	5545	282
Mission-Aransas	8601	640	202	2061	661
Guadalupe	428	247	156	874	357
Matagorda	2125	961	1016	3845	1723
Galveston	8993	4419	4927	18651	8695
Wastewater, Point sources					
Corpus Christi	751	79.3	645	881	761
Mission-Aransas	37	5.1	26.4	43.6	38.0
Guadalupe	1173	548	562	2035	1176
Matagorda	369	65.9	265	476	365
Galveston	6523	395	5802	7296	6431
Wet Deposition direct to Bay surface					
Corpus Christi	338	71.6	228	434	330
Mission-Aransas	385	101	254	574	351
Guadalupe	233	57.2	167	356	214
Matagorda	366	79.3	264	520	337
Galveston	488	112	343	723	460
Total Load					
Corpus Christi	2295	743	1513	3945	2180
Mission-Aransas	1624	946	672	3522	1323
Guadalupe	8916	4052	3740	17024	7458
Matagorda	6813	2376	3721	10390	6005
Galveston	38350	11488	21639	61543	37803
Total Load per km^2					
Corpus Christi	.051	.017	.034	.088	.049
Mission-Aransas	.253	.147	.105	.548	.206
Guadalupe	.333	.152	.140	.637	.279
Matagorda	.075	.026	.041	.115	.066
Galveston	.633	.190	.357	1.02	.624

Table 5 presents variation in loading to five Texas estuaries, both in terms of magnitude and by area. Table 6 presents the coefficient of variation (CV) of loading by source. The analysis of variation in nitrate flux by Alexander et al. [1996] includes several Texas rivers that provide major contributions to their respective estuaries. Their

TABLE 6. Coefficients of variation (CV) of total nitrogen loading for Texas estuaries for 1977 to 1987. The total load includes wet deposition to the bay surface.

Estuary	Gaged	Ungaged	Wastewater	Rain	Total
Corpus Christi	0.68	0.47	0.11	0.22	0.32
Mission-Aransas	0.71	0.74	0.14	0.26	0.58
Guadalupe	0.55	0.58	0.47	0.25	0.45
Matagorda	0.41	0.45	0.18	0.22	0.35
Galveston	0.33	0.49	0.06	0.23	0.30

reported indices of nitrate flux variation for these rivers is high, with most CV's above 0.4, consistent with the CV's reported here for total nitrogen gaged loads. Variation in loadings of rain and return flows in this data set are limited by the use of average concentrations--the variation in these loadings for a given bay shows only the variation in flow. Given that, however, it is apparent that wastewater inputs are fairly consistent from year to year, except where there is a trend, such as occurred in the Guadalupe system. For many smaller systems, the point source loads are a new relatively high baseline, against which other variation can have a significance influence. For this series of annual loading data, it appears that a CV of at least 0.3 would be typical. Therefore, even with significant and relatively constant loadings from point sources included in total loadings, inter-annual variation is relatively high for these estuaries.

Concluding Remarks

There are other studies providing thoughtful assessments of uncertainties associated with loading analyses [e.g., Boynton et al., 1995; Valiela et al., 1997; Collins et al., ms]. Winter [1981] presents a thorough discussion of uncertainties in lake water balance work. Available also are analytical and statistical work for rigorous approaches to the problem [Beck, 1987; Scavia, 1981; Cohn et al., 1989]. The information presented here is provided to set the stage for realistic expectations when various loading estimates are presented and compared in this and other chapters. The error bounds computed here for association with others' calculated nitrogen loadings have a utility for comparison of these measurement-based loadings with loadings generated by various models. These estuaries are complex and open systems. The breadth of the confidence bounds assignable to typical loading estimates should help us see where we are in the process. As emphasis and demands of watershed management and regulatory programs shift, new concentrations, discharge, and other data are becoming available. At the same time, landuse changes and regional development may change the relative importance of various loading components, or the distribution of loadings in space and time. Tracking the certainty of our knowledge of a system should be seen as a part of the need to evaluate trends in estuary health from a well characterized baseline.

For many systems, the uncertainty in elements of standard loading measurements -- concentration and flow from major tributaries- - is the source of greatest over-all error in loading estimates, just because the major tributary inputs are so large. As much as 40%

of the total uncertainty is associated with the major tributary contributions, at least for the estuaries reviewed here. Therefore, it is important to encourage the maintenance of monitoring programs that provide the routine and mundane measurements at the core of loading computation, as we apply better techniques to more elusive components.

Some parameters not easily measured are perhaps not evaluated here appropriately. For some hard-to-measure inputs, researchers still face the possibility of unintentionally incorporating bias into loading estimates through assumptions regarding a few input categories.

Acknowledgments. This work was supported by Texas Water Development Board (TWDB). Other TWDB staff provided hydraulic modeling and hydrology data used in Texas case studies. The author thanks C. Driscoll and J. Kremer for their guidance and insights, and P. Stacey, R.E. Turner, and R. Valigura for editorial contributions.

References

Alexander, R. B., P. S. Murdoch, R. A. Smith. 1996. Streamflow-induced variations in nitrate flux in tributaries to the Atlantic Coastal Zone. Biogeochemistry 33: 149-177.

Beck, M.B. 1987. Water quality modeling; a review of the analysis of uncertainty. Water Resources Research. 23:1393-1442.

Boynton, W.R., J.H. Garber, R. Summers, W.M. Kemp. 1995. Inputs, transformations and transport of nitrogen and phosphorus in Chesapeake Bay and selected tributaries. Estuaries 18:285-314.

Buzan, D.L. 1990. Intensive Survey of the Rio Grande Segment 2304. IS 90-03. Texas Water Commission, Austin, TX. 130 pp.

Chang, M., L.A. Flannery. 1998. Evaluating the accuracy of rainfall catch by three different gages. J. Amer. Water Resources Assoc. 34(3):559-564.

Cohn, T.A., L.L. DeLong, E.J. Gilroy, R.M. Hirsch, D.K. Wells. 1989. Estimating constituent loads. Water Resources Research 25:937-942.

Collins, G. C., J. N. Kremer, I. Valiela. in press. Assessing uncertainty estimates of nitrogen loading to estuaries for research, planning, and risk management. Environmental Management.

Hann, C.T. B. Allred, D.E. Storm, G. Sabbagh, S. Prabhu. 1993. Evaluation of hydrologic/water quality models, a statistical procedure. Presented at the 1993 Winter Meeting, Paper No. 932505. American Society of Agricultural Engineers.

Hill, F.R. Mississippi Power co., Personal Communication. February 5, 1999.

Kirkpatrick, J. 1986. Intensive Survey of Dickinson Bayou Tidal Segment 1103. Report IS 86-03. Texas Water Commission, Austin, TX. 43 pp.

Kirkpatrick, J. 1987a. Intensive Survey of Segment 1013, Buffalo Bayou Tidal, and Segment 1014, Buffalo Bayou Above Tidal. Report IS 87-05. Texas Water Commission, Austin, TX. 80 pp.

Kirkpatrick, J. 1987b. Intensive Survey of the Houston Ship Channel System. IS 87-06. Texas Water Commission, Austin, TX. 130 pp.

Kirkpatrick, J. 1988. Intensive Survey of the Navasota River - Segment 1209. Report IS 88-03. Texas Water Commission, Austin, TX. 55 pp.

Longley, W.L. (ed.) 1994. Freshwater inflows to Texas bays and estuaries: ecological relationships and methods for determination of needs. Texas Water Development Board and Texas Parks and Wildlife Department, Austin, TX. 386 pp.

NADP. 1993. National Atmospheric Deposition Program (NRSP-3) National Trends Network Coordination Office, Natural Resource Ecology Laboratory, Colorado State University, Fort Collins, CO 80523.

Ockerman, D. USGS. Personal Communication, January, 1999.

Pacheco, P., D. Farrow, T. Manuelides, S. Rohmann, M. Katz, J, McLeod. 1990. Point source discharges in coastal areas of Texas: A summary by estuarine watershed for 1987. National Coastal Pollutant Discharge Inventory (NCPDI) Program, Strategic Assessment Branch, NOAA, Rockville, MD.

Scavia, D. 1981. Comparison of first order error analysis and Monte Carlo simulation in time-dependent lake eutrophication models. Water Resources. Res. 17(4):1051-1059.

Srinivasan, M.S., J.M. Hamlett, R.L. Day, J.I. Sams, G.W. Petersen. 1998. Hydrologic modeling of two glaciated watersheds in northeast Pennsylvania. J. Amer. Water. Res. Assoc. 34(4):963-978.

Staver, K.W., and R.B. Brinsfield. 1996. Seepage of groundwater nitrate from a riparian agroecosystem into the Wye River Estuary. Estuaries 19:359-370.

Twidwell, S., and J.R. Davis. 1998. Intensive Survey of Greens Bayou Segment 1016. Report AS-160/is. Texas Natural Resource Conservation Commission, Austin, TX. 60 pp.

Twidwell, S., and J.R. Davis. 1987. Intensive Survey of San Antonio River Segments 1901 and 1911. Report IS 87-04. Texas Water Commission, Austin, TX. 104 pp.

USDA, SCS. 1985. National Engineering Handbook. Section 4, Hydrology. Us Department of Agriculture, Soil Conservation Service. Washington D.C.

US Geological Survey. 1998. Water Resources data--Texas, Water Year 1997. US Geological Survey Water-Data Report TX-97. USGS, Austin, TX.

Valiela, I., G. Collins, J. Kremer, K. Lajtha, M. Geist, B. Seely, J. Brawley, C.H. Sham. 1997. Nitrogen loading from coastal watersheds to receiving estuaries: New method and application. Ecological Applications 7(2):358-380.

Walker, W.W. 1982. A sensitivity and error analysis framework for lake eutrophication modeling. Water Resources Bull. 18(1):53-60.

Walker, W.W. 1996. Simplified Procedures for eutrophication assessment and prediction: User manual. Instruction Report W-96-2. US Army Corps of Engineers, Waterways Experiment Station. 235p.

Winter, T.C. 1981. Uncertainties in estimating the water balance of lakes. Water Resources Bulletin 17 (1): 82-115.

8

Contributions of Atmospheric Nitrogen Deposition to U.S. Estuaries: Summary and Conclusions

Paul E. Stacey, Holly S. Greening, James N. Kremer, David Peterson, and David A. Tomasko

Abstract

A NOAA project was initiated in 1998, with support from the U.S. EPA, to develop state-of-the-art estimates of atmospheric N deposition to estuarine watersheds and water surfaces and its delivery to the estuaries. Work groups were formed to address N deposition rates, indirect (from the watershed) yields from atmospheric and other anthropogenic sources, and direct deposition on the estuarine waterbodies, and to evaluate the levels of uncertainty within the estimates. Watershed N yields were estimated using both a land-use based process approach and a national (SPARROW) model, compared to each other, and compared to estimates of N yield from the literature. The total N yields predicted by the national model were similar to values found in the literature and the land-use derived estimates were consistently higher. Atmospheric N yield estimates were within a similar range for the two approaches, but tended to be higher in the land-use based estimates and were not well-correlated. Median atmospheric N yields were around 15% of the total N yield for both groups, but ranged as high as 60% when both direct and indirect deposition were considered. Although not the dominant source of anthropogenic N, atmospheric N is, and will undoubtedly continue to be, an important factor in culturally eutrophied estuarine systems, warranting additional research and management attention.

Introduction

Acidification of poorly buffered lakes was intensively studied in the 1970s and 1980s and is probably the best-documented water quality problem caused by atmospheric pollution in the United States. Recently, a much broader relationship between air quality and environmental impact has been suggested. Global impacts of fossil fuel and trash burning and industrial and agricultural emissions have received increasing attention in both scientific journals and the popular press. There is a growing consensus that many terrestrial and water quality problems, including global warming, stratospheric ozone depletion, tropospheric ozone, acidification of forests and water bodies, and mercury contamination, may all be related to anthropogenic air pollution. During the last decade, concern over atmospheric deposition of nitrogen (N) and how it may be contributing to eutrophication of our nation's estuaries has become prominent. While there is an apparent commonality of cause between estuarine eutrophication and many of the other issues listed above, this effort has focused on N deposition and estuarine inputs of N.

In January 1998, the National Oceanic and Atmospheric Administration (NOAA), in conjunction with the U.S. Environmental Protection Agency (EPA), convened a workshop of national experts on atmospheric deposition, nutrient processes, and estuarine eutrophication. Their task was to report on atmospheric N deposition rates to land and water, the processes affecting N transport through watersheds, and the relative importance of N export from atmospheric sources to estuaries throughout the United States. Work groups were formed to address these topics with the goals of:

- estimating atmospheric N deposition rates to coastal estuaries and their watersheds;
- estimating N export to estuaries from major sources, including atmospheric deposition, based on an analysis of land-use based watershed processes;
- evaluating and applying a national watershed model to the estuaries considered in this analysis; and
- comparing N export estimates from independent studies with estimates generated by the national model and land-use generated estimates.

Initially, the work groups conducted their analyses and prepared their loading estimates independently. Comparative analyses followed, if appropriate and feasible, as each work group shared their results with other groups. The preceding chapters detailed the literature, rationale and approaches used to generate the atmospheric N loading estimates. The principal objectives of this summary chapter are to: 1) encapsulate the efforts of each team and, 2) where possible, compare among their atmospheric N loading estimates and identify possible reasons for observed differences.

Biogeochemical and Ecological Implications

In Chapter 2, Paerl *et al.* reviewed the biogeochemical and ecological implications of N loading to estuaries. They identified atmospheric N deposition as an important source of N that until recently may not have been adequately recognized in research and management efforts. Spatial and temporal changes in the processes that make atmospheric sources more predominant in coastal areas were identified, including development patterns, expansion of concentrated agricultural activities, and an ever-growing demand for energy derived from fossil fuels.

Unlike surface water discharges, which are not able to escape the confines of a watershed without human intervention, atmospheric sources are not so constrained. The concept of an "airshed" attempts to relate atmospheric sources to the watershed or estuary where they are predominantly deposited, but linkages between source and fate are highly variable. Nevertheless, Chapter 2 concluded that airsheds usefully define the "predomi-

TABLE 1. Continental airshed and watershed drainage areas, airshed to watershed ratios, and percent of oxidized nitrogen load deposited on the watershed explained by the airshed for selected estuaries (Chapter 2).

Estuary	Watershed Drainage Area (km^2)	Airshed Continental (over land) Area (km^2)	Airshed to Watershed Ratios	Percent of Oxidized Nitrogen Load Explained
Barnegat Bay	1,404	505,600	360	67
Narragansett Bay	4,321	595,200	140	73
Tampa Bay	5,725	256,000	45	76
Barataria-Terrebonne	7,446	409,600	55	63
Apalachee Bay	14,342	441,600	31	50
Pamlico Sound	26,829	665,600	26	63
Delaware Bay	33,381	729,600	22	75
Altamaha Sound	37,105	678,400	18	68
West Mississippi Sound	39,598	659,200	17	63
Mobile Bay	114,858	992,000	8.6	68
Chesapeake Bay	167,443	1,081,600	6.5	76

nant emissions area" relevant to a depositional area, *i.e.*, an estuarine watershed. For a subset of estuaries reviewed in this study, the ratio of airshed to watershed areas ranged from 6.5 to 360 (Table 1). A survey of recent literature referenced in Chapter 2 indicates estuaries may receive from 1 to more than 40% of their "new" (externally supplied) total N inputs from atmospheric sources. Hence, details of the transport of atmospheric N from airshed to watershed to estuary becomes key to understanding and managing N inputs to coastal waters.

Watershed processes that occur during N transport from watersheds to estuaries were outlined in Chapter 2. As N is transformed and transported through the watershed, it may reach levels above what can be assimilated by a forest. This condition of N "saturation" of forests upsets soil nutrient balances and forces NO_3^- leaching and nitrous oxide emissions. In addition, N saturation reduces forest health and productivity. In surface waters, acidification may occur from direct input of acidified forms of N and sulfur deposited with precipitation. Secondary consequences of increased nutrient loading include release of cations, which may be toxic to aquatic life (*e.g.* aluminum), or even increased production of nitrous oxide from surface waters.

In Chapter 2 it was noted that, in the estuary, N utilization dynamics are complex, and linking N inputs to trophic response "...has proven difficult and unpredictable." Nitrogen availability, form, timing, and abundance all conspire with other water quality conditions and phytoplankton speciation, presence, life cycle, and health to generate an outcome such as an algae bloom. Their conclusion was "...long-term changes in the sources and amounts of newly-delivered AD-N [atmospheric deposition of N] compounds in N-limited systems may alter both the species composition and relative size distribution of the phytoplankton community. This disturbance will have cascading effects on the trophic structure and biogeochemical cycling of impacted ecosystems."

Estuarine eutrophication, and its many symptoms, is usually the nutrient-related problem that state and federal water quality managers must address. Excess N loads to N-limited environments can stimulate the growth of phytoplankton, greatly increasing the load of organic matter delivered to the bottom of the estuary when it eventually dies or is passed through higher trophic levels. As the organic matter decays, oxygen is consumed. In those estuaries that are strongly stratified, usually during the summer, bottom waters can become hypoxic or anoxic, as oxygen rich surface waters cannot replenish bottom waters at a rate exceeding the level of respiration. According to the literature, as reported in Chapter 2 (and in Chapter 4 by Castro *et al.*), 20% or more of the dissolved inorganic N exported by rivers throughout the world is attributed to atmospheric sources.

Hypoxia is not the only ramification of anthropogenic increases in N loading to estuaries. Drawing on recent research, Chapter 2 elaborated on the consequences of altered ratios of N:phosphorus:silica, the primary nutrients that control phytoplankton growth. Increased biomass, shifts in species composition (*e.g.*, from diatoms to dinoflagellates as silica becomes limiting), harmful algae blooms, and concomitant impacts on higher trophic levels as their food resources are altered are a few of the many complex impacts that may occur.

Several case studies were presented in Chapter 2 that provided first-hand evidence of the impacts of nutrient enrichment, and atmospheric relationships. Examples included Chesapeake Bay and Albemarle-Pamlico Sound-Neuse River Estuary hypoxia and algal blooms; Waquoit Bay (Massachusetts) hypoxia, fish kills, and eelgrass losses; Barnegat Bay eelgrass losses; Tampa Bay seagrass bed losses and N control successes; and Gulf of Mexico hypoxia with its immense contributory watershed. Evidence of these effects has been reinforced in the national Estuarine Eutrophication Survey [NOAA, 1996; 1997a; 1997b; 1997c; and 1998], which documented oxygen depletion in more than half of 136 estuaries surveyed in the United States along with other typical symptoms of N over-enrichment.

An analysis of the Atlantic coastal shelf in Chapter 2 further defined the role and complexity of atmospheric deposition in N budgets. With increasing distance from the shore, terrestrial sources of N become less important than atmospheric sources because wind-borne N can be carried well offshore. Source/deposition relationships vary widely depending on local climate and wind transport patterns. A comparison of the coastal shelves (to the 200-m depth contour) of the northeast U.S. (north of Cape Hatteras) with the southeast (Cape Hatteras to Florida) and the Gulf of Mexico, identified higher atmospheric N input contributions in the northeast (40-60%) than in the other two areas (30-40% and 10%, respectively).

The Nitrogen Estimates

Nitrogen loading estimates were constructed, justified and presented in Chapters 3 through 6. The respective chapters reported on: atmospheric deposition rates to land and water compiled by the atmospheric deposition work group; N export estimates from a land-use based approach that considered watershed inputs, transformations, and export by the land-use based work group; N loading estimates from literature studies of several estuaries reviewed by the empirical estimate work group; and watershed N export predictions from a national empirical model produced by the national model work group. In Chapter 7, D. Brock presented an analysis of the uncertainties associated with the N export estimates. In this summary chapter, total N loads from all sources and the relative atmospheric contributions of N are summarized and compared among the four efforts (atmospheric deposition estimates, land-use based estimates, literature estimates, and national model estimates) to the extent practicable.

Forty-two estuaries and their watersheds were identified for analysis (Plate 1). However, every chapter did not include all 42 estuaries. Watershed drainage areas [Pacheco, 1999] ranged from less than 100 km^2 to more than 160,000 km^2 (Table 2). Estuarine surface areas (water surface) ranged from 5 km^2 to over 11,000 km^2 and watershed to water surface area ratios ranged from less than 1 to over 900 (Table 2).

Land cover data [Pacheco, 1999] reflect the wide range of urbanized, agricultural, and natural landscapes that exist throughout the U.S. (Table 2). Land cover was characteristic of predominant land uses in regions of the United States, with more urbanized land in the densely-populated northeast and portions of Florida and agricultural land cover dominating human uses in other regions (Figure 1). The Northeast region watersheds averaged 26% urban and only 9% agricultural. While the Southeast and East Gulf Coast regions averaged only 12 and 15% urban land use, respectively, the Indian River (29% urban) and Tampa Bay (24% urban) watersheds were exceptions. Agriculture land use was highest in the Mid-Atlantic region, averaging 38% for the five estuaries and ranging as high as 48% in the Maryland Inland Bays estuary. The Southeast, East Gulf Coast, and West Gulf Coast regions also had pockets of intensive agricultural land use, averaging over 20% for the three regions.

Atmospheric Nitrogen Deposition to Coastal Estuaries and Their Watersheds

The atmospheric deposition work group [Meyers *et al.* - Chapter 3] estimated atmospheric wet and dry N deposition rates (kgN/ha-yr) for NO_3^-, NH_4^+, and DON (wet only). They further separated estimates of N deposition onto terrestrial watersheds from deposition onto waterbodies, noting a difference in deposition kinetics between the two that led to differing rates. All 42 waterbodies selected for this study (Plate 1 and Table 2) were included in their analysis.

Wet deposition estimates of NH_4^+ and NO_3^- were generated using data from (National Atmospheric Deposition Program (NADP) sites. Ammonium concentrations in wet samples were multiplied by 1.15 to account for volatilization losses that may have occurred during the weekly sampling period. Wet organic N estimates were based on seven recent studies in which wet deposition of organic N ranged from 7 to 28% of the total. The atmospheric deposition group settled on a uniform organic N percentage of 25% of the total wet N deposition as an acceptable estimate. Dry deposition estimates, which did not include an organic N component because of insufficient data, were constructed from field measurements and model estimates, primarily using the EPA's Regional Acid Deposition Model (RADM) and data from EPA's Clean Air Status and Trends Network (CASTNet). Dry loading estimates of NH_4^+ were doubled to account for NH_3^+ deposition, which was not measured in CASTNet.

The atmospheric deposition group reviewed temporal changes in wet deposition of NH_4^+ and NO_3^- over the period of record (early to mid 1980s to 1994) at NADP sites to identify any bias from trends. The group judged annual loading rates to be a reasonable approximation of N deposition today because they did not identify temporal trends at most sites. Similarly, within watershed boundaries, atmospheric data showed little variability among NADP sites; hence, each estimate was believed to be representative of conditions throughout the watershed.

Nitrogen deposition to water surfaces was treated separately from the watersheds. The atmospheric deposition group estimated that the smoother surface of water compared to terrestrial roughness from vegetation and topography reduced dry deposition rates by a

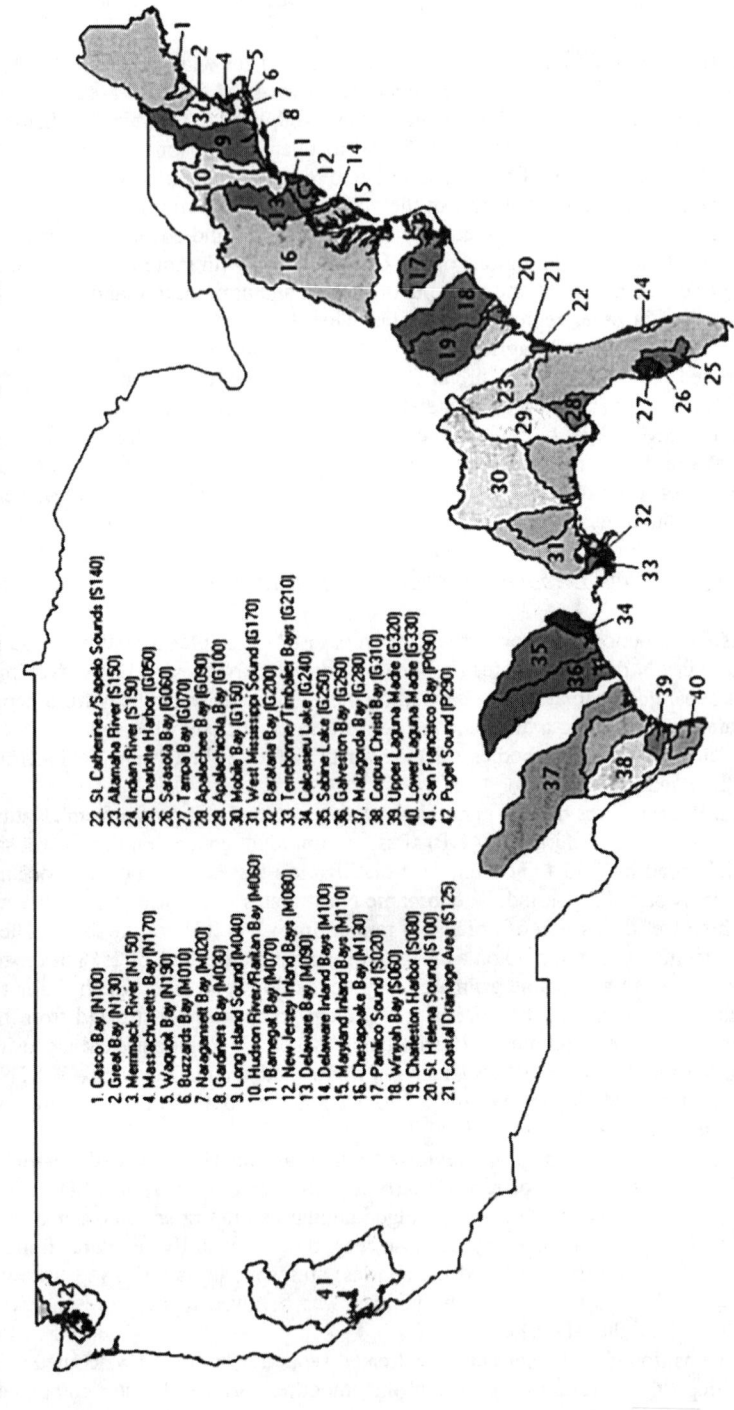

Plate 1. Watersheds chosen for the atmospheric nutrient input project [Source: Pacheco, 1999].

TABLE 2. Watershed and estuarine waterbody surface areas (km²), their ratios, and watershed land characteristics for U.S. estuaries and their watersheds [Pacheco, 1999].

Estuary	Watershed Area (km²)*	Percent Urban	Percent Agriculture	Percent Range, Forest and other	Surface Water Area (km²)	Ratio of Watershed to surface water area
Northeast Region						
1. Casco Bay	2,562	13.2	9.5	77.3	426.9	6
2. Great Bay	2,554	18.2	7.9	73.9	47.5	54
3. Merrimack River	13,002	15.8	5.2	79.0	15.5	840
4. Massachusetts Bay	2,437	72.3	1.2	26.5	953.3	3
5. Waquoit Bay	40	25.6	1.3	73.1	5.2	8
6. Buzzards Bay	1,256	17.0	6.9	76.1	639.0	2
7. Narragansett Bay	4,305	37.5	4.6	57.9	415.6	10
8. Gardiners Bay	710	29.9	22.6	47.5	512.2	1
9. Long Island Sound	41,663	15.7	11.0	73.4	3300.9	13
10. Hudson River/Raritan Bay	41,601	15.2	20.2	64.6	798.5	52
11. Barnegat Bay	1,399	33.7	5.3	61.1	182.1	8
12. New Jersey Inland Bays	3,432	15.1	8.6	76.3	278.1	12
Mid-Atlantic Region						
13. Delaware Bay	33,253	15.3	29.7	55.0	2068.9	16
14. Delaware Inland Bays	559	8.0	44.1	47.8	72.1	8
15. Maryland Inland Bays	282	8.3	48.3	43.4	54.0	5
16. Chesapeake Bay	166,799	8.1	32.1	59.8	11262.1	15
17. Pamlico Sound	26,726	6.7	34.4	59.0	5588.6	5
Southeast Region						
18. Winyah Bay	46,959	4.6	35.7	59.7	88.6	530
19. Charleston Harbor	41,119	10.6	24.1	65.3	85.1	483
20. St. Helena Sound	12,264	17.8	34.5	63.7	202.9	60
21. Ogeechee Coastal	233	12.6	0.3	87.1	666.5	<1
22. St. Catherines/Sapelo Snds	2,256	4.1	1.2	94.7	187.5	12
23. Altamaha River	36,962	5.5	25.8	68.7	39.1	946
24. Indian River	3,093	29.3	21.4	49.3	866.4	4
East Gulf Coast						
25. Charlotte Harbor	11,694	6.1	37.4	56.5	502.5	23
26. Sarasota Bay	653	41.0	20.5	38.5	123.9	5
27. Tampa Bay	5,703	24.3	30.2	45.4	902.0	6
28. Apalachee Bay	14,286	33.0	21.3	75.4	1772.9	8
29. Apalachicola Bay	52,214	6.7	27.6	65.7	593.0	88
30. Mobile Bay	114,416	4.1	23.0	72.9	1078.7	106
31. West Mississippi Sound	38,407	6.3	26.1	67.6	4203.5	9
32. Barataria Bay	4,784	7.6	14.7	77.6	852.4	6
33. Terrebonne/Timbalier Bays	2,634	6.9	10.0	83.1	1262.2	2

TABLE 2. Continued.

	West Gulf Coast					
34. Calcasieu Lake	10,865	2.9	23.3	73.8	258.2	42
35. Sabine Lake	53,675	3.4	26.2	70.4	264.7	203
36. Galveston Bay	61,826	14.0	47.7	38.3	1456.4	42
37. Matagorda Bay	121,764	3.8	29.3	66.9	1115.0	109
38. Corpus Christi Bay	44,525	1.5	17.4	81.1	570.7	78
39. Upper Laguna Madre	10,580	1.1	26.2	72.7	829.5	
40. Lower Laguna Madre	13,165	5.0	35.4	59.5	1308.4	
	West Coast					
41. San Francisco Bay	119,181	4.5	20.0	75.5	1325.1	90
42. Puget Sound	28,063	9.0	2.4	88.6	2638.1	11

*Watershed areas used by the various work groups were not always congruent or of the same size.

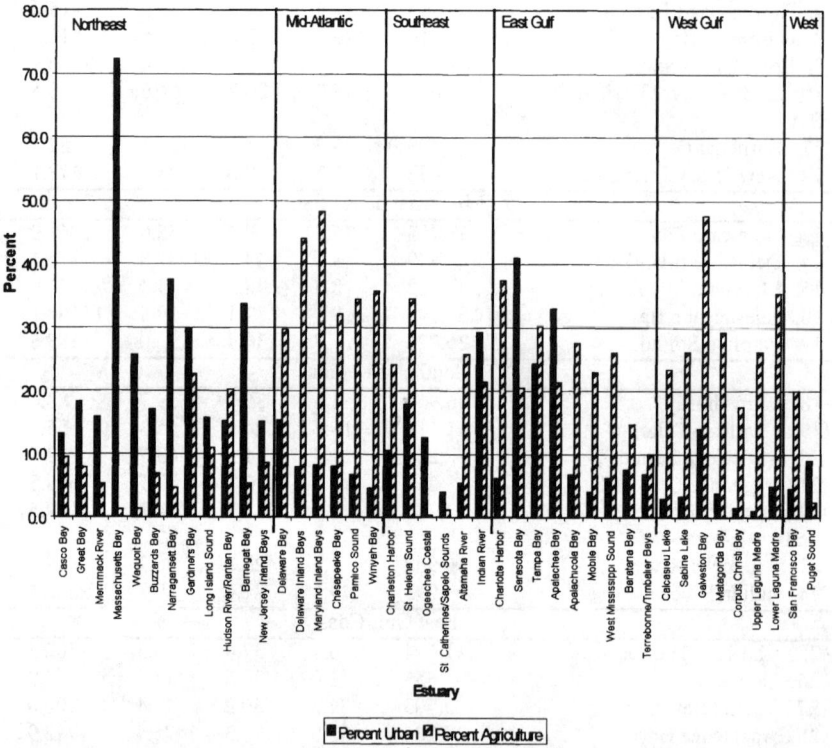

Figure 1. Percent urban and agricultural land cover in 42 estuarine watersheds. Missing bars indicate data were not available. Data provided by NOAA [Pacheco, 1999].

factor of 4. That adjustment reduced total deposition to waterbodies from 15 to 40% with an average reduction of about 30%. Wet deposition onto surface waters was assumed to equal wet deposition onto terrestrial surfaces.

Wet plus dry total N deposition rates for the 42 target watersheds ranged from less than 3 kg/ha-yr to 13.5 kg/ha-yr (Table 3) with a median value of 8.4 kgN/ha-yr and an interquartile range of 6.4-10.9 kgN/ha-yr. In general, estimated rates were higher in the Northeast and Mid-Atlantic regions (especially from Massachusetts to the Chesapeake Bay) and in the central region of the Gulf of Mexico (Figure 2). In the Northeast and Mid-Atlantic regions, average total deposition rates were 10 and 11 kgN/ha-yr, respectively. Four Gulf Coast estuaries, West Mississippi Sound, Barataria Bay, Terrebonne/Timbalier Bays and Calcasieu Lake, all had total N deposition rates in excess of 10 kg/ha-yr. The lowest total N deposition rates were estimated for the West Coast. The atmospheric deposition group placed the level of uncertainty for their estimates within a factor of two.

Based on the atmospheric deposition rate estimates, larger watersheds receive tens of millions of kilograms of N from the atmosphere each year (*e.g.*, the estimate for the Chesapeake Bay is over 215 million kgN/yr). Smaller watersheds, even in high deposition areas, may receive well under a million kilograms of N per year (*e.g.*, Maryland Inland Bays have one of the highest deposition rates, but the small watershed size results in a total N load of only about 330,000 kgN/yr) (Table 3).

Atmospheric deposition rates to the water surfaces of each estuary were lower, ranging from about 2 kgN/ha-yr up to 9.35 kgN/ha-yr. Because waterbodies are much smaller than their respective watershed areas, with watershed to water surface ratios ranging into the hundreds (Table 2), the direct atmospheric N load deposited onto the waterbodies was also much lower (Table 3). Only six estuaries (Long Island Sound, Delaware Bay, Chesapeake Bay, Pamlico Sound, West Mississippi Sound, and Terrebonne/Timbalier Bays) had direct (to the estuarine water body – atmospheric is the only direct source of N) total N loads in excess of 1 million kg/yr, all by virtue of their large waterbody size. The Chesapeake Bay estimate exceeded 10 million kgN/yr of direct atmospheric deposition.

Nitrogen Contributions from Estuarine Watersheds Using a Land-Use Based Approach

The land-use based work group [Castro *et al.* - Chapter 4] quantified dominant point and nonpoint sources of N for 34 estuaries along the Atlantic and Gulf Coasts using a land-use based or N-budget approach. The objectives of their analyses were to: 1) quantify the anthropogenic N inputs to the selected watersheds; 2) identify the dominant N sources contributing to the total N inputs to the selected estuaries; and 3) compare the contribution made by atmospheric deposition to the total N export to each estuary.

Net anthropogenic N inputs to each of the 34 watershed and estuary systems were estimated using the approach of Jordan and Weller [1996]. Anthropogenic N sources in the land-use based analysis included: 1) application of fertilizer N to crops, 2) biotic N fixation by crops and pastures, 3) atmospheric deposition of NO_3^-, 4) net import of food for humans, and 5) net import of feed for livestock. Land-use specific approaches were developed to estimate the N available for transport to the estuary from different watershed sources (agriculture, urban areas, upland forest and point sources) above and below the fall line. Atmospheric N contribution to the total runoff from the three land use catego-

TABLE 3. Total nitrogen wet and dry deposition rates (kgN/ha-yr) and total nitrogen input (million kgN/yr) to watersheds and surface water areas of U.S. estuaries estimated by the atmospheric deposition group.

Estuary	Deposition to Watershed (kgN/ha-yr)			Deposition to Waterbody (kgN/ha-yr)	Total Nitrogen Input (kgNx10⁶/yr)	
	Wet	Dry	Total		To Watershed	To Waterbody
Northeast Region						
1. Casco Bay	4.51	1.05	5.56	4.74	1.42	0.20
2. Great Bay	4.77	1.99	6.76	5.22	1.73	0.02
3. Merrimack River	6.50	2.90	9.40	7.15	12.22	0.01
4. Massachusetts Bay	5.00	4.31	9.31	5.96	2.27	0.57
5. Waquoit Bay	5.09	4.31	9.40	6.06	0.04	0.00
6. Buzzards Bay	5.09	4.31	9.40	6.06	1.18	0.39
7. Narragansett Bay	6.22	4.31	10.53	7.19	4.53	0.30
8. Gardiners Bay	8.12	4.31	12.43	9.08	0.88	0.47
9. Long Island Sound	7.38	3.52	10.90	8.18	45.41	2.70
10. Hudson River/Raritan Bay	7.55	4.40	11.95	8.55	49.71	0.68
11. Barnegat Bay	7.66	4.64	12.30	8.69	1.72	0.16
12. New Jersey Inland Bays	7.66	4.64	12.30	8.69	4.22	0.24
Mid-Atlantic Region						
13. Delaware Bay	8.01	4.64	12.65	9.04	42.07	1.87
14. Delaware Inland Bays	6.82	4.74	11.56	7.87	0.65	0.06
15. Maryland Inland Bays	6.82	4.74	11.56	7.87	0.33	0.04
16. Chesapeake Bay	8.34	4.57	12.91	9.35	215.34	10.54
17. Pamlico Sound	5.79	2.76	8.55	6.38	22.85	3.57
Southeast Region						
18. Winyah Bay	5.32	2.76	8.08	5.91	37.94	0.05
19. Charleston Harbor	3.86	2.54	6.40	4.43	26.32	0.04
20. St. Helena Sound	3.86	3.09	6.95	4.55	8.52	0.09
21. Ogeechee Coastal	4.48	3.53	8.01	5.29	0.19	0.35
22. St. Catherines/Sapelo Snds	4.48	2.63	7.11	5.08	1.60	0.10
23. Altamaha River	4.56	3.12	7.68	5.23	28.39	0.02
24. Indian River	3.63	2.17	5.80	4.13	1.79	0.36
East Gulf Coast						
25. Charlotte Harbor	4.24	2.14	6.38	4.72	7.46	0.24
26. Sarasota Bay	4.24	2.59	6.83	4.84	0.45	0.06
27. Tampa Bay	4.24	3.87	8.11	5.13	4.63	0.46
28. Apalachee Bay	3.88	1.78	5.66	4.27	8.09	0.76
29. Apalachicola Bay	4.26	2.45	6.71	4.78	35.04	0.28
30. Mobile Bay	5.40	2.79	8.19	5.99	93.71	0.65
31. West Mississippi Sound	6.91	3.54	10.45	7.66	40.14	3.22
32. Barataria Bay	6.81	6.69	13.50	8.36	6.46	0.71

TABLE 3. Continued.

33. Terrebonne/Timbalier Bays	6.81	5.83	12.64	8.12	3.33	1.02
West Gulf Coast						
34. Calcasieu Lake	6.68	3.97	10.65	7.57	11.57	0.20
35. Sabine Lake	5.22	3.46	8.68	6.00	46.59	0.16
36. Galveston Bay	4.97	3.94	8.91	5.87	55.09	0.86
37. Matagorda Bay	3.24	2.28	5.52	3.74	67.21	0.42
38. Corpus Christi Bay	3.23	1.66	4.89	3.60	21.77	0.21
39. Upper Laguna Madre	3.79	1.53	5.32	4.14	5.63	0.34
40. Lower Laguna Madre	3.79	1.23	5.02	4.07	6.61	0.53
West Coast						
41. San Francisco Bay	2.36	1.48	3.84	2.69	45.77	0.36
42. Puget Sound	1.76	1.11	2.87	2.01	8.05	0.53

ries and direct deposition to the surface of the estuary were estimated and compared to the total N input to each estuary.

Export of N from the watersheds to the estuaries was calculated using a land-use specific approach. The amount of N input available for waterborne transport was estimated for agricultural lands (crops and pastures), urban areas, and upland forests. With the exception of Barataria Bay and Terrebonne/Timbalier Bays, these three land covers accounted for 69 to 99% of the total land area in the 34 watersheds. The Barataria and Terrebonne/Timbalier watersheds were dominated (75% of the total area) by wetlands. Estimates of N export above and below the fall line were made for 20 watersheds (Table 4) and export from above the fall line was attenuated by 30% to account for N retention and losses during riverine transport to the estuary. Nitrogen export from below the fall line was not attenuated. The land-use based group made some adjustments to the attenuation of the above the fall line agricultural land N export when initial estimates exceeded measured N fluxes to the fall line of 18 watersheds. These adjustments resulted in a good agreement between predicted and measured fluxes ($r^2 = 0.67$ and $y = 1.1(x - 1.95)$). However, estimates for Texas watersheds were not greatly improved with respect to monitoring data and still overestimated measured N loads (*see* Chapter 4).

Nitrogen outputs from agricultural land included crop harvest, animal grazing, denitrification, ammonia volatilization, and runoff. Urban N flux considered effluent from sewage treatment plants, leachate from septic fields and runoff from urban lands. Because point source data were available only for watershed portions below the fall line in most areas, above the fall line estimates were derived using a per capita wastewater N load, which combines sewage treatment plant and septic field contributions. Atmospheric deposition and non-symbiotic N fixation were assumed to be the only N inputs to forested lands and export from forests was based on a correlation between wet deposition and stream export from studies in the literature.

Direct (to the surface of the estuary – atmospheric N only) and indirect (through the watershed – all sources including point sources discharging directly to the estuary) contributions of atmospheric N were estimated by applying deposition rates developed by the atmospheric deposition group.

The land-use based estimates provided a wealth of N input and loading statistics in Chapter 4 that account for major sources of N to the watershed, attenuation rates, and export to the estuary. In this summary, the focus is on atmospheric and total N (from all sources) export (total N yield and flux) to the estuary. Those results are presented both as specific yield per unit area (kgN/ha-yr) and as total N loads or export (kgN/yr), similar to the data presentation used for the atmospheric N deposition (Table 3).

Atmospheric yield estimates ranged from 0.28 to 4.6 kgN/ha-yr, with a median of 0.97 kgN/ha-yr (Table 4). Total N yield from all sources ranged from 1.4 to over 50 kgN/ha-yr with a median value of 7.7 kgN/ha-yr. Atmospheric N export ranged from about 70,000 kgN/yr to over 22 million kgN/yr. As a percent of the total N load, atmospheric N flux from estuarine watersheds ranged from 6 to 50% (Table 4) and had a median percent value of 13. The total N export from the watershed attributed to atmospheric deposition had an interquartile range of 7.9 to 16%.

Land-use based average atmospheric yield estimates were highest in the Northeast region (2 kgN/ha-yr) although, as a percentage of total N flux, atmospheric export was highest in the West Gulf Coast at 18% followed by the Northeast region at 17%. Average atmospheric contributions of N were lowest in the Southeast region as a yield (0.7 kgN/ha-yr) and in the West Gulf Coast as a percentage (10.5%).

Atmospheric Nitrogen Flux Estimates from Application of the SPARROW Watershed Model

The national model work group [Alexander *et al.* – Chapter 6] assessed the atmospheric contributions of N using the U.S. Geological Survey's SPARROW (Spatially Referenced Regression on Watershed Attributes) model. The model was applied to the watersheds of 40 of the 42 estuaries under consideration to predict estimates of N loading. SPARROW does not estimate direct atmospheric loading to the estuarine water surface. The model calculates nutrient loads from both point and nonpoint sources and transport effects through the watershed. Source components for N include point sources, fertilizer application, livestock waste production, atmospheric deposition, and export from nonagricultural lands. SPARROW estimates land-to-water delivery based on conditions such as temperature, soil permeability and stream drainage density and an in-stream loss rate during transport is also calculated to account for attenuation. In-stream loss rates are a function of water time of travel and channel size and are inversely related to channel size.

While atmospheric loading estimates in the SPARROW model are based on wet NO_3^- deposition only, the regression procedure used in SPARROW adjusts the quantities of atmospheric N delivered to stream channels to account for sources other than wet NO_3^- deposition. Wet NO_3^- deposition rates were calculated independently of those generated by the atmospheric deposition group, although both used NADP data.

The model performance was evaluated by comparing observed N yields to predicted yields. One half of the model predictions were within at least 32% of the observed values. Only 10% of the SPARROW model predictions exceeded observed values by more than 100%. SPARROW predictions were also compared with those generated by the SWAT (Soil and Water Assessment Tool), which had been applied to coastal watersheds as part of Texas A&M's HUMUS (Hydrologic Unit Modeling of the United States) Project. Local (1,430 hydrologic units) and total (38 of the 42 estuaries) watershed yields of total N were compared and correlated closely ($r = 0.67$ and 0.83, respectively). Based on their analyses (*see* Chapter 6 for details), the authors concluded there was general agree-

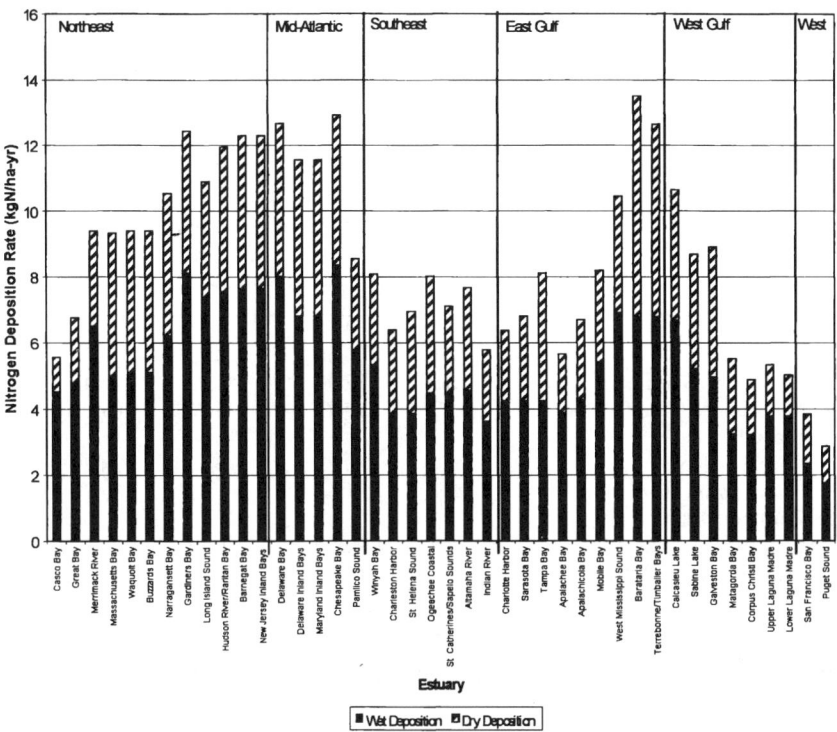

Figure 2. Wet and dry atmospheric deposition rates (kgN/ha-yr) of total N to 42 estuarine watersheds estimated by the atmospheric deposition group.

ment between the model predictions but suggested that the comparison is tenuous because of model differences.

Total N yields predicted by SPARROW for the 40 watersheds ranged from 0.38 to 24.9 kgN/ha-yr (Table 5) with a median rate of 4.63 kgN/ha-yr. There were regional differences in total N yield to the estuaries, with the highest rates generally in the Northeast and Mid-Atlantic regions where total yields averaged 7.69 and 6.63 kgN/ha-yr, respectively. Yields for the atmospheric component ranged from 0.04 to 3.26 kg/ha-yr (Table 5) with a median value of 0.66 kgN/ha-yr and an interquartile range from 0.32 to 1.1 kgN/ha-yr. Atmospheric yield exhibited a similar geographic pattern as the total yield, with the highest averages in the Northeast (1.27 kgN/ha-yr) and Mid-Atlantic (1.34 kgN/ha-yr) regions. Estimates of uncertainty in the total N yields ranged from 19% to 117% of the mean (median = 40%), based on the standard error of prediction, and varied inversely with watershed size according to the national model group. Uncertainties in the estimates of atmospheric yields were only slightly higher (41-102%; median = 50%).

The SPARROW model also predicted total N export (kgN/yr) for the 40 estuaries under study. Atmospheric N flux ranged from 10,000 kgN/yr in smaller estuaries to 37.5 million kgN/yr in the Chesapeake Bay. Atmospheric contributions to total N loads delivered to the estuary ranged from 4 to 35% (Table 5) and had a median value of 14%. The national model group did not estimate direct inputs, so these percentages reflect indirect deposition, or atmospheric N delivered from the watershed only.

TABLE 4. Atmospheric and total N yield (kgN/ha-yr) and export (million kgN/yr) from the watersheds to 34 estuaries evaluated by the land-based group and percentages of total N export derived from atmospheric sources.

Estuary	Total N Yield from Watersheds (kgN/ha-yr)		Total N Exported from the Watershed (kgN x 10^6 /yr)		Percent from Atmosphere
	Atmospheric Sources	All Sources	Atmospheric Sources	All Sources	
Northeast Region					
1. Casco Bay	0.71	4.55	0.15	0.99	16
2. Great Bay	1.01	6.50	0.25	1.62	16
3. Merrimack River*	1.15	7.98	1.43	9.94	14
4. Massachusetts Bay	4.59	50.73	0.96	10.60	9
5. Waquoit Bay					
6. Buzzards Bay	1.62	20.25	0.17	2.07	8
7. Narragansett Bay*	2.81	21.64	1.13	8.69	13
8. Gardiners Bay					
9. Long Island Sound*	1.50	10.59	6.12	43.17	15
10. Hudson River/Raritan Bay*	1.75	19.91	6.31	71.90	9
11. Barnegat Bay	3.15	6.33	0.43	0.86	50
12. New Jersey Inland Bays	1.88	9.38	0.60	3.02	20
Mid-Atlantic Region					
13. Delaware Bay*	1.83	14.79	5.63	45.54	13
14. Delaware Inland Bays					
15. Maryland Inland Bays					
16. Chesapeake Bay*	1.39	8.22	22.41	132.22	18
17. Pamlico Sound*	0.84	11.51	2.10	28.87	7
Southeast Region					
18. Winyah Bay*	0.61	8.69	2.67	37.80	7
19. Charleston Harbor*	0.50	7.41	2.05	30.47	7
20. St. Helena Sound*	0.28	3.51	0.33	4.20	8
21. Ogeechee Coastal					
22. St. Catherines/Sapelo Snds	0.38	1.35	0.07	0.27	28
23. Altamaha River*	0.48	5.84	1.78	21.46	8
24. Indian River	2.18	24.28	0.53	5.93	9
East Gulf Coast					
25. Charlotte Harbor	1.10	18.02	0.83	13.71	6
26. Sarasota Bay					
27. Tampa Bay	1.90	24.89	0.95	12.46	8
28. Apalachee Bay*	0.37	4.70	0.52	6.68	7
29. Apalachicola Bay*	0.50	6.71	2.40	32.35	7
30. Mobile Bay*	0.51	5.32	5.72	59.90	10
31. West Mississippi Sound*	0.94	6.49	3.61	24.93	14
32. Barataria Bay	1.09	8.33	0.45	3.46	13
33. Terrebonne/Timbalier Bays	0.88	4.71	0.18	0.99	19

TABLE 4. Continued.

	West Gulf Coast				
34. Calcasieu Lake*	0.86	6.51	0.84	6.40	13
35. Sabine Lake*	0.64	5.97	3.32	30.82	11
36. Galveston Bay*	2.01	11.61	12.11	70.01	17
37. Matagorda Bay*	1.18	3.91	13.60	44.97	32
38. Corpus Christi Bay*	0.80	2.76	3.52	12.23	30
39. Upper Laguna Madre	0.49	3.31	0.51	3.44	15
40. Lower Laguna Madre	0.76	9.74	0.98	12.58	8
	West Coast				
41. San Francisco Bay					
42. Puget Sound					

* Estuaries with segments above and below a fall line division defined by the land-based group.

Independent Nitrogen Loading Estimates

The empirical estimate work group [Turner et al. - Chapter 5] reviewed N export estimates generated through empirical evaluations reported in independent studies. Estimates from 27 estuaries along the Atlantic and Gulf coasts were assembled. Eighteen of those watersheds were reasonably comparable to watersheds on the list of 42 included in this study (Table 6). However, the local estimates reflected a variety of approaches and did not all categorize sources in the same way and, in some cases, watershed sizes differed considerably from those identified in this effort. Also, atmospheric load estimates were not generally available and were not reported. Consequently, total N loads were the only basis for comparison among the estimates. Some of the local models were based on export coefficients from the literature while others used an empirical approach to generate loading coefficients. Documentation was not always complete and geographic boundaries for the estimates were sometimes unclear. However, each report was carefully reviewed and estimates were adjusted if appropriate to ensure their suitability for comparison purposes.

The total N yields in the literature studies of the eighteen watersheds ranged from 0.44 to 27 kgN/ha-yr (Table 6). The median yield was 5.7 kgN/ha-yr with an interquartile range of 3.2 to 15 kgN/ha-yr. The relatively fewer estuaries compiled from the literature make these national statistics less representative than for the land-use based and SPARROW estimates and, in addition, the distribution of estuaries is balanced differently among regions.

The empirical estimate group compared their data to national model estimates of total N yield generated by the SPARROW model and a regional assessment of the Gulf of Mexico estuaries conducted by NOAA. In general, the NOAA and SPARROW estimates were much lower than were found in the independent efforts (59% and 48% lower for the NOAA and SPARROW models, respectively). However, as explained further in the detailed comparisons below, these differences calculated by the empirical estimate group may overstate the differences between the national models and the independent evaluations. This difference is caused by a disproportionate influence of a few outlying pairs on the mean concentrations for the respective sets of estimates. The correlation between the independent estimates and the NOAA predictions was high ($r^2 = 0.97$). While the slope

TABLE 5. Atmospheric and total N yield (kgN/ha-yr) and export (million kgN/yr) from the watersheds to 40 estuaries predicted by the SPARROW model and percentages of the total N export derived from atmospheric sources.

Estuary	Total N Yield from Watersheds (kgN/ha-yr)		Nitrogen Exported from the Watershed (kgN x 10^6 /yr)		Percent from Atmosphere
	Atmospheric Sources	All Sources	Atmospheric Sources	All Sources	
Northeast Region					
1. Casco Bay	0.85	3.86	0.26	1.19	22
2. Great Bay	0.34	3.82	0.08	0.91	9
3. Merrimack River	1.23	4.45	1.59	5.74	28
4. Massachusetts Bay	0.98	24.89	0.25	6.28	4
5. Waquoit Bay					
6. Buzzards Bay	0.16	1.35	0.06	0.49	12
7. Narragansett Bay	1.10	10.51	0.51	4.85	10
8. Gardiners Bay	0.04	0.38	0.01	0.08	11
9. Long Island Sound	3.04	8.81	12.27	35.48	35
10. Hudson River/Raritan Bay	3.26	12.77	13.59	53.16	26
11. Barnegat Bay	1.60	8.64	0.26	1.43	19
12. New Jersey Inland Bays	1.36	5.15	0.50	1.91	26
Mid-Atlantic Region					
13. Delaware Bay	2.96	13.32	9.59	43.11	22
14. Delaware Inland Bays	0.16	1.74	0.01	0.13	9
15. Maryland Inland Bays	0.20	2.43	0.02	0.21	8
16. Chesapeake Bay	2.28	8.14	37.49	133.63	28
17. Pamlico Sound	1.09	7.51	2.67	18.47	14
Southeast Region					
18. Winyah Bay	0.80	4.28	3.73	19.85	19
19. Charleston Harbor	0.23	1.07	0.95	4.34	22
20. St. Helena Sound	0.24	1.38	0.30	1.71	18
21. Ogeechee Coastal					
22. St. Catherines/Sapelo Snds	0.47	2.34	0.11	0.53	20
23. Altamaha River	1.05	4.57	3.85	16.82	23
24. Indian River	0.07	0.89	0.01	0.14	8
East Gulf Coast					
25. Charlotte Harbor	0.48	3.70	0.44	3.38	13
26. Sarasota Bay	0.35	3.09	0.03	0.30	11
27. Tampa Bay	0.51	4.81	0.33	3.15	11
28. Apalachee Bay	0.40	2.81	0.61	4.29	14
29. Apalachicola Bay	0.70	4.79	3.68	25.01	15
30. Mobile Bay	1.22	5.15	14.10	59.38	24
31. West Mississippi Sound	1.31	5.08	5.82	22.59	26
32. Barataria Bay	0.49	5.41	0.30	3.33	9
33. Terrebonne/Timbalier Bays	0.61	2.29	0.25	0.94	27

TABLE 5. Continued.

	West Gulf Coast				
34. Calcasieu Lake	1.07	6.16	1.19	6.89	17
35. Sabine Lake	0.71	3.51	3.86	18.98	20
36. Galveston Bay	0.62	4.68	3.91	29.58	13
37. Matagorda Bay	0.17	1.23	1.97	14.43	14
38. Corpus Christi Bay	0.05	0.56	0.24	2.50	10
39. Upper Laguna Madre	0.83	7.17	0.75	6.50	12
40. Lower Laguna Madre	0.42	5.66	0.31	4.06	8
	West Coast				
41. San Francisco Bay	0.32	5.85	3.46	63.74	5
42. Puget Sound	0.80	6.77	2.49	21.11	12

of the SPARROW/independent estimates was close to one, the r^2 was only 0.49. The work group speculated that the local estimates might have been higher because of better knowledge of local conditions by the authors, including estimates of groundwater loading and inclusion of ungaged stream segments and downstream (below the fall line) point sources that may have been omitted in the NOAA and SPARROW models.

Nitrogen Loading Estimate Comparisons

One test of the reliability of the estimates is how well they compare with one another. While this is a measure of precision rather than accuracy, a concurrence of export rates (yields) and N loads (export or flux) among the various methods used in this project provides some reassurance that the estimates are reasonable. Unfortunately, there were few opportunities for comparison between estimates. One category, total export to the estuary (direct and watershed loads combined), was only estimated by the land-use based group. Also, the direct deposition estimates made by the atmospheric deposition and land-use based groups both used the same deposition rates generated by the atmospheric deposition group, so a comparison would not be fruitful. However, some useful comparisons could be made, particularly between the land-use based and SPARROW data.

Comparisons of atmospheric and total estimates undertaken in this chapter were:

1. Wet NO_3^- deposition rates to the watersheds prepared by the atmospheric deposition group with those used in the SPARROW model.
2. Total N yield from all sources exported from the watershed (delivered to the estuary) from land-use based and SPARROW estimates and reported in the independent studies.
3. Yield of atmospheric total N from the watershed and the percent atmospheric contribution estimated by the land-use based and national model groups.

Similar comparisons of total N export or flux can also be made but would parallel yield analyses. Differences in the relationships between yield and flux estimates, when they did occur, were often a consequence of watershed size estimates, which were not identical among the groups. Those interested in N flux or other aspects of N loading not detailed in this summary are encouraged to review the earlier chapters.

TABLE 6. Total nitrogen yield (kgN/ha-yr) for 18 watersheds* evaluated in independent studies of estuarine nitrogen loading by the empirical estimate group.

Estuary	Total N Yield (kgN/ha-yr)	Estuary	Total N Yield (kgN/ha-yr)
Northeast Region		East Gulf Coast	
1. Casco Bay	5.62	25. Charlotte Harbor	2.55
2. Great Bay	3.17	26. Sarasota Bay	10.91
3. Merrimack River		27. Tampa Bay	5.85
4. Massachusetts Bay		28. Apalachee Bay	
5. Waquoit Bay	6.10	29. Apalachicola Bay	
6. Buzzards Bay		30. Mobile Bay	
7. Narragansett Bay	18.62	31. West Mississippi Sound	
8. Gardiners Bay		32. Barataria Bay	
9. Long Island Sound	7.02	33. Terrebonne/Timbalier Bays	
10. Hudson River/Raritan Bay		West Gulf Coast	
11. Barnegat Bay	4.15	34. Calcasieu Lake	
12. New Jersey Inland Bays		35. Sabine Lake	
Mid-Atlantic Region		36. Galveston Bay	5.26
13. Delaware Bay	15.13	37. Matagorda Bay	0.71
14. Delaware Inland Bays	27.22	38. Corpus Christi Bay	0.44
15. Maryland Inland Bays	27.06	39. Upper Laguna Madre	
16. Chesapeake Bay	23.40	40. Lower Laguna Madre	
17. Pamlico Sound		West Coast	
Southeast Region		41. San Francisco Bay	
18. Winyah Bay		42. Puget Sound	
19. Charleston Harbor	1.29		
20. St. Helena Sound			
21. Ogeechee Coastal			
22. St. Catherines/Sapelo Snds			
23. Altamaha River			
24. Indian River	5.28		

*The independent-study watersheds did not always appear to provide a good geographic match to land-based group and national model group watersheds based on watershed size.

Wet Nitrate Atmospheric Deposition Rate Comparison

The atmospheric deposition and national model groups both estimated wet atmospheric deposition rates for NO_3^-. Both groups relied heavily on NADP data to formulate their estimates (*see* Chapters 3 and 6 for complete discussions). Because of their reliance on the same database to generate estimates, the wet NO_3^- deposition estimates were significantly correlated ($r = 0.97$; $p < 0.01$ based on rank correlation [Snedecor and Cochran, 1967]) between the two groups (Figure 3). However, all the national model group esti-

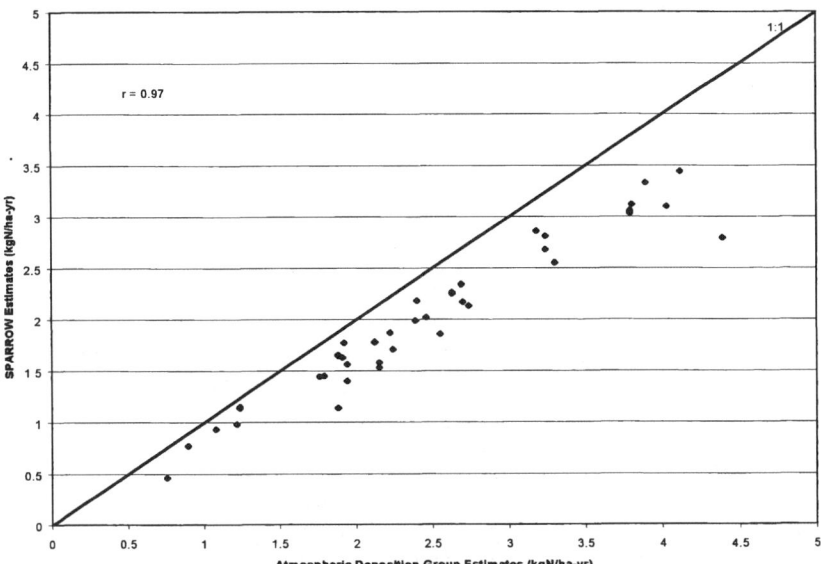

Figure 3. Comparison of SPARROW wet NO_3^- atmospheric deposition rates (kgN/ha-yr) with atmospheric deposition group estimates. SPARROW predictions were lower than those derived by the atmospheric deposition group. No pair of estimates differed by more than 40%.

mates of NO_3^- deposition were 7 to 40% lower than those of the atmospheric deposition group (Table 7).

Median wet NO_3^- deposition rates were 2.32 and 1.87 kgN/ha-yr for the atmospheric deposition group and national model group estimates, respectively. Interquartile ranges for the respective group estimates were 1.88-3.24 kgN/ha-yr and 1.45-2.68 kgN/ha-yr. Both groups' estimates reflect a pattern that appears to parallel wet deposition monitoring reported by NADP (*i.e.*, higher rates in the Mid-Atlantic, moderate rates in the northern portion of the Northeast and along the Gulf Coast, and lower rates to the west). The two sets of estimates are well within the atmospheric deposition group's factor of two level of uncertainty, but the consistently lower estimates from the national model group suggest a systematic difference.

The likely causes for the differences between the atmospheric deposition group and national model group estimates include the period of NADP data used by each group and the methods of interpretation. The atmospheric deposition group used an average of the annual means from the early to mid 1980s through 1996 to develop their estimates. The national model group used 1984-1993 data but computed a 1987 mean NO_3^- deposition rate from the data based on average precipitation and a trend analysis (data were "detrended" – *see* Chapter 6 for methodology).

The national model interpretation identified upward trends in wet NO_3^- deposition at a majority of the NADP sites that typically range from 1 to 4% per year [R. Alexander, personal communication]. By adjusting NO_3^- deposition rates to reflect 1987 conditions, three years earlier than the ca. 1990 mid point of the atmospheric deposition group data, the SPARROW methodology produced lower rates and may explain a difference of from 3 to 12%. The atmospheric group did not detect trends in deposition at most stations,

TABLE 7. Wet atmospheric nitrate deposition rates (kgN/ha-yr) as estimated by the atmospheric deposition group and the national model group.

Estuary	Atmospheric Nitrate Deposition (kgN/ha-yr)		Estuary	Atmospheric Nitrate Deposition (kgN/ha-yr)	
	Atmospheric Deposition Group	National Model Group		Atmospheric Deposition Group	National Model Group
Northeast Region			**East Gulf Coast**		
1. Casco Bay	2.22	1.87	25. Charlotte Harbor	2.15	1.53
2. Great Bay	2.46	2.02	26. Sarasota Bay	2.15	1.53
3. Merrimack River	3.18	2.86	27. Tampa Bay	2.15	1.58
4. Massachusetts Bay	2.69	2.34	28. Apalachee Bay	1.76	1.44
5. Waquoit Bay	2.74		29. Apalachicola Bay	1.91	1.63
6. Buzzards Bay	2.74	2.13	30. Mobile Bay	2.24	1.71
7. Narragansett Bay	3.30	2.55	31. West Mississippi Sound	2.70	2.17
8. Gardiners Bay	4.39	2.79	32. Barataria Bay	2.63	2.26
9. Long Island Sound	3.80	3.12	33. Terrebonne/Timbalier Bays	2.63	2.25
10. Hudson River/Raritan Bay	3.89	3.33	**West Gulf Coast**		
11. Barnegat Bay	3.79	3.06	34. Calcasieu Lake	2.55	1.86
12. New Jersey Inland Bays	3.79	3.03	35. Sabine Lake	2.12	1.78
Mid-Atlantic Region			36. Galveston Bay	1.88	1.65
13. Delaware Bay	4.12	3.44	37. Matagorda Bay	1.22	0.98
14. Delaware Inland Bays	3.24	2.81	38. Corpus Christi Bay	1.08	0.93
15. Maryland Inland Bays	3.24	2.68	39. Upper Laguna Madre	1.24	1.14
16. Chesapeake Bay	4.03	3.10	40. Lower Laguna Madre	1.24	1.15
17. Pamlico Sound	2.40	2.18	**West Coast**		
Southeast Region			41. San Francisco Bay	0.76	0.46
18. Winyah Bay	2.39	1.99	42. Puget Sound	0.90	0.77
19. Charleston Harbor	1.94	1.56			
20. St. Helena Sound	1.94	1.40			
21. Ogeechee Coastal	1.88				
22. St. Catherines/Sapelo Snds	1.88	1.14			
23. Altamaha River	1.92	1.77			
24. Indian River	1.79	1.45			

perhaps because of the difference in time frame analyzed. Also, the atmospheric deposition group assumed deposition rates for the most proximate NADP sites would reflect NO_3^- deposition throughout each watershed. The national model group's SPARROW model also interpolates between NADP sites to generate NO_3^- deposition rates within a watershed, which would cause some differences as well, but would not systematically lower the rates. The differences in trend analysis, time frame and interpolation would appear to explain the lower NO_3^- deposition rates reported by the national model group.

Comparison of Total Nitrogen Yield from the Watersheds

Total N yields (all sources, including indirect atmospheric N deposition) delivered to the estuaries from 34 common watersheds estimated by both the land-use based group and SPARROW model were compared. The general correlation of total N yield between the two estimates was good (r = 0.68), but misleading because of one high pair of values for Massachusetts Bay (Tables 4 and 5). Removal of that pair reduced the correlation coefficient to 0.30 for total N yield (Figure 4) but ranks were significantly correlated (p <0.05) using the rank correlation test [Snedecor and Cochran, 1967]. Despite this non-parametric correlation of ranks, the SPARROW predictions were considerably and consistently lower than the land-use based estimates.

The SPARROW median total N yield was 4.74 kgN/ha-yr, 38% lower than the land-use based median estimate of 7.70 kgN/ha-yr. Interquartile ranges were also disparate, ranging from 5.32 to 11.61 kgN/ha-yr for the land-use based estimates compared to 2.81-7.17 kgN/ha-yr for the SPARROW predictions. Only three SPARROW estimates exceeded the land-use based estimates: Barnegat Bay, Upper Laguna Madre and St. Catherines/Sapelo Sounds. Of those three, the Upper Laguna Madre deviation was the greatest at 54%. Land-use based estimates that exceeded SPARROW estimates by 70% or more were the Indian River, Buzzards Bay, Charleston Harbor, Tampa Bay, Charlotte Harbor and Corpus Christi Bay estuaries (Figure 4). Of the 34 estuaries compared, those five ranked 27, 34, 11, 23, 20, and 8 in watershed size, according to NOAA data [Pacheco, 1999], showing little relationship between estimate difference and watershed size.

A few factors may have caused the differences between the land-use based and SPARROW total N yields. Watershed areas reported by the two groups were not equivalent (Table 8), and four of the six estuaries with the largest (>70%) differences in total N yield had area differences greater than 20%. While watershed area datasets were both from NOAA, they were chronologically different with the land-use based set being more recent. It was not possible to update the areas used in the SPARROW model within the time frame of this assessment [R. Valigura, personal communication]. The national model group reported a watershed area for Buzzards Bay more than three times as large as the one reported by the land-use based group and Tampa Bay and Charlotte Harbor were 31% and 20% larger, respectively. The Indian River national model area estimate was 38% smaller than the land-use based estimate. These differences may have affected yield if biases in land use characteristics were created in the incongruent areas. For example, the national model (and NOAA statistics) include areas to the south of Sarasota Bay, which are almost entirely range and pasture land, that are not included in the independent study reviewed by the empirical estimate group. Consequently, the SPARROW model total N yield would be reduced because of the lower N export rates from range and pasture lands compared to the urban lands that were more predominant in the independent study model [D. Tomasko, personal communication].

Twenty-seven of the 34 common watersheds had larger areas reported by the national model group compared to the land-use based group. The (unsigned) average difference was about 14% and the median deviation was 10%. Some of this disparity is related to the exclusion of inland water surfaces in the land-use based watershed area estimates,

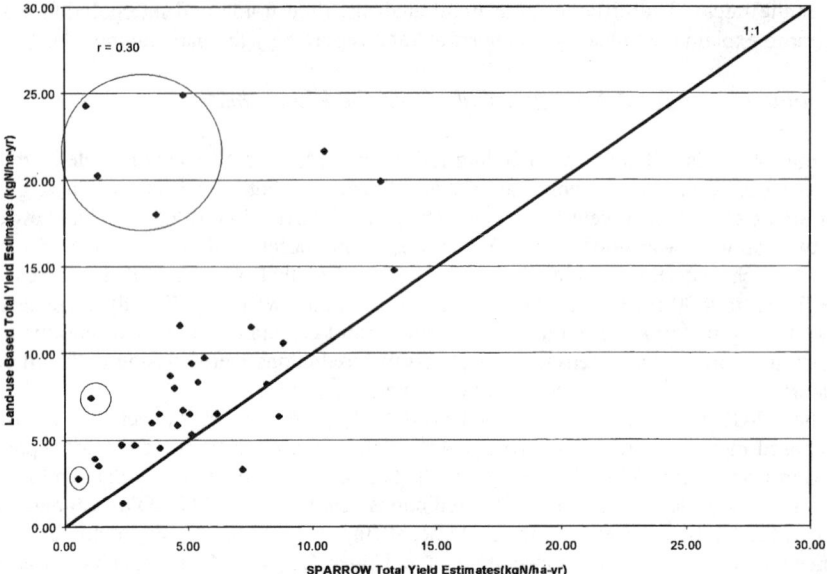

Figure 4. Comparison of SPARROW and land-use based total N (all sources) yield estimates (kgN/ha-yr) to U.S. estuaries. All but three land-use based estimates were higher than the SPARROW estimates. The Massachusetts Bay estimate was excluded for scaling purposes. Circled outliers (>70% deviation) are (from top to bottom) Tampa Bay, Indian River, Buzzards Bay, Charlotte Harbor, Charleston Harbor, and Corpus Christi Bay.

which results in an average 7% reduction in area compared to the NOAA statistics. Exclusion of water surfaces would translate to higher N yields when the smaller watershed areas were used to calculate the per unit area yield.

Other differences in the land-use based and SPARROW total N yield may be related to treatment of attenuation in the two approaches. As mentioned earlier, the land-use based approach does not riverine attenuation of N below the fall line. Since attenuation processes will occur below the fall line, the land-use based estimates may overestimate N delivery to the estuary. To test the effect of attenuation, total N yield from 20 estuarine watersheds with fall line contributions in the land-use based dataset were compared to national model group estimates for the same 20 estuaries (Tables 4 and 5; Figure 5). These 20 estuaries incorporated a 30% level of attenuation for the portions above the fall line. Although the land-use based estimates were still higher, the two estimates were strongly correlated (r = 0.84) and their ranks were significantly correlated as well (p <0.01). This suggests that systematically increasing or accounting for attenuation of land-use based estimates below the fall line might resolve some of the differences with SPARROW predictions. In addition, of the six estuaries identified above with total N yield estimate deviations >70%, only two (Charleston Harbor and Corpus Christi Bay) had fall line divisions in the land-use based analysis.

The independent studies summarized by the empirical estimate group provide another opportunity to evaluate the observed differences between the land-use based and SPAR-

TABLE 8. Estuarine watershed surface areas (km^2) reported by NOAA [Pacheco, 1999], the land-based group, the national model group, and the empirical estimate group.

Estuary	Watershed Size (km^2)			
	NOAA	Land-based Work Group	National Model Group	Independent Studies
Northeast Region				
1. Casco Bay	2,562	2,188	3,093	32
2. Great Bay	2,554	2,491	2,378	2,590
3. Merrimack River	13,002	12,458	12,906	
4. Massachusetts Bay	2,437	2,089	2,524	
5. Waquoit Bay	40			38
6. Buzzards Bay	1,256	1,021	3,654	
7. Narragansett Bay	4,305	4,018	4,613	4,662
8. Gardiners Bay	710		2,192	
9. Long Island Sound	41,663	40,774	40,289	40,943
10. Hudson River/Raritan Bay	41,601	36,114	41,629	
11. Barnegat Bay	1,399	1,365	1,649	878
12. New Jersey Inland Bays	3,432	3,215	3,705	
Mid-Atlantic Region				
13. Delaware Bay	33,253	30,792	32,373	34,965
14. Delaware Inland Bays	559		726	633
15. Maryland Inland Bays	282		847	172
16. Chesapeake Bay	166,799	160,765	164,156	56,980
17. Pamlico Sound	26,726	25,090	24,584	
Southeast Region				
18. Winyah Bay	46,959	43,492	46,340	
19. Charleston Harbor	41,119	41,143	40,604	40,922
20. St. Helena Sound	12,264	11,970	12,358	
21. Ogeechee Coastal	233			
22. St. Catherines/Sapelo Snds	2,256	1,973	2,253	
23. Altamaha River	36,962	36,711	36,797	
24. Indian River	3,093	2,441	1,525	4,926
East Gulf Coast				
25. Charlotte Harbor	11,694	7,610	9,146	8,547
26. Sarasota Bay	653		957	389
27. Tampa Bay	5,703	5,005	6,556	5,895
28. Apalachee Bay	14,286	14,215	15,254	
29. Apalachicola Bay	52,214	48,216	52,236	
30. Mobile Bay	114,416	112,665	115,339	
31. West Mississippi Sound	38,407	38,407	44,448	
32. Barataria Bay	4,784	4,156	6,151	
33. Terrebonne/Timbalier Bays	2,634	2,097	4,095	

TABLE 8. Continued.

	West Gulf Coast			
34. Calcasieu Lake	10,865	9,820	11,174	
35. Sabine Lake	53,675	51,657	54,081	
36. Galveston Bay	61,826	60,322	63,158	60,597
37. Matagorda Bay	121,764	114,981	117,565	90,672
38. Corpus Christi Bay	44,525	44,256	44,853	44,841
39. Upper Laguna Madre	10,580	10,395	9,065	
40. Lower Laguna Madre	13,165	12,916	7,179	
	West Coast			
41. San Francisco Bay	119,181		108,943	
42. Puget Sound	28,063		31,166	

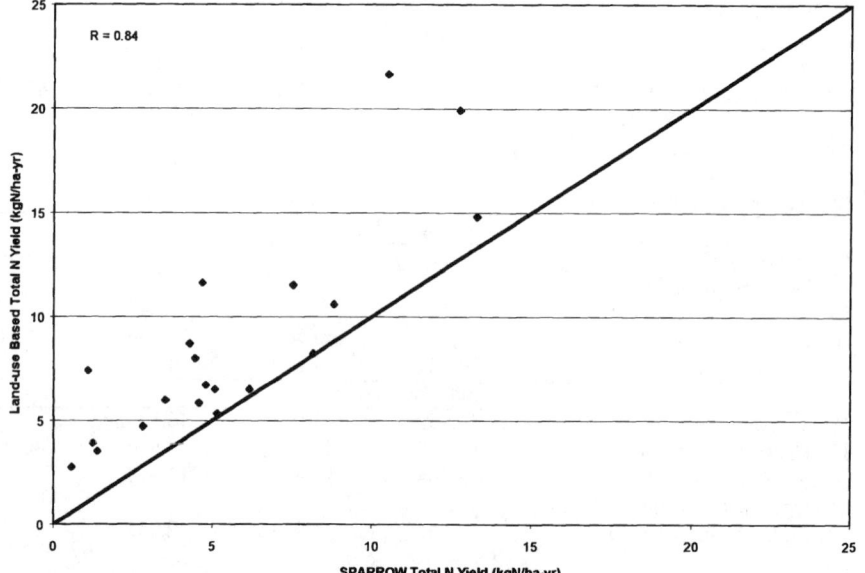

Figure 5. Comparison of SPARROW and land-use based total N yield estimates for 20 estuaries with fall line divisions defined by the land-use based group.

ROW estimates. While the area differences (Table 8) introduce a source of error as noted above for the land-use based and national comparisons (*e.g.*, the Casco Bay independent study included only one subsystem - Maquoit Bay), fourteen estuarine watersheds had total N yield reported by all three groups (Table 9). The land-use based estimates compared reasonably well with the independent estimates for some of the Northeast and Mid-Atlantic estuaries (Casco Bay, Long Island Sound, Narragansett Bay, Barnegat Bay, and Delaware Bay), varying by no more than 35% between estimates. But the land-use based estimates for most of the Southeast and Gulf estuaries were much higher than both the independent and the SPARROW estimates (Figure 6). Six of the 14 had a percent deviation greater than 70% higher for the land-use based estimates compared to the independent study data.

TABLE 9. Fourteen estuaries for which total nitrogen yields (kgN/ha-yr) were located in independent studies by the empirical estimate group and estimated by the land-based group and the national model group.

Estuary	Total N Yield from Watersheds (kgN/ha-yr)		
	Independent Studies	Land-based Group	National Model Group
1. Casco Bay	5.62	4.55	3.86
2. Great Bay	3.17	6.50	3.82
7. Narragansett Bay	18.62	21.64	10.51
9. Long Island Sound	7.02	10.59	8.81
11. Barnegat Bay	4.15	6.33	8.64
13. Delaware Bay	15.13	14.79	13.32
16. Chesapeake Bay	23.40	8.22	8.14
19. Charleston Harbor	1.29	7.41	1.07
24. Indian River	5.28	24.28	0.89
25. Charlotte Harbor	2.55	18.02	3.70
27. Tampa Bay	5.85	24.89	4.81
36. Galveston Bay	5.26	11.61	4.68
37. Matagorda Bay	0.71	3.91	1.23
38. Corpus Christi Bay	0.44	2.76	0.56
Minimum	0.44	2.76	0.56
25 percentile	2.55	6.33	1.23
Median	5.27	9.41	4.27
75 percentile	7.02	18.02	8.64
Maximum	23.40	24.89	13.32

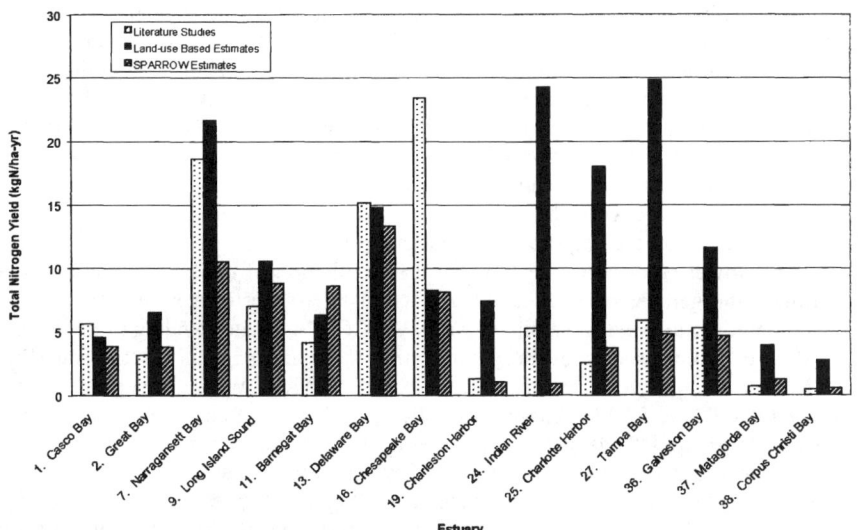

Figure 6. Total N export (kgN/ha-yr) delivered to 14 estuaries summarized in the literature compared to land-use based and SPARROW export estimates.

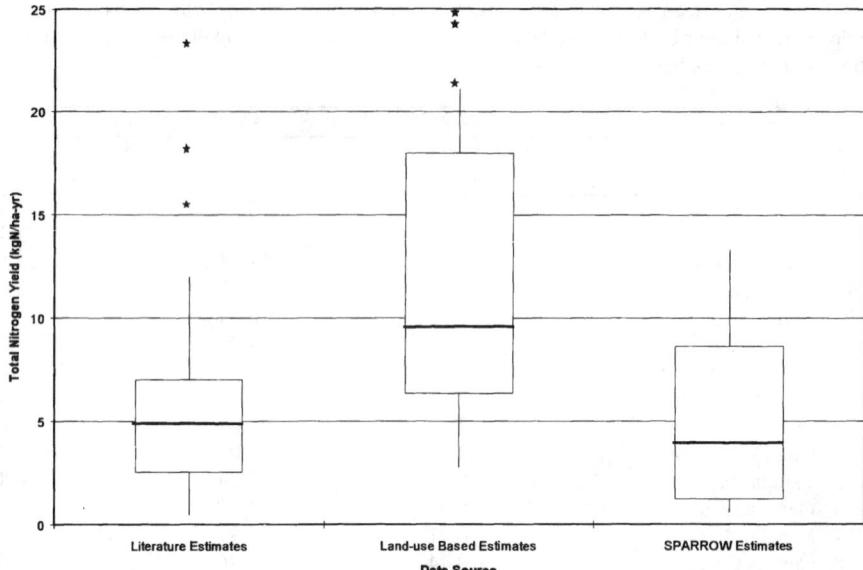

Figure 7. Comparison of independent, land-use based and SPARROW estimates of total N export rates from 14 estuaries (kgN/ha-yr). Boxes represent the interquartile range, the midline represents the median value, and the lines extend to the maximum and minimum values or +/- 1.5 x the interquartile range, whichever is less. Outlier values are marked with stars. The 14 estuaries represented are listed in Table 10.

SPARROW estimates compared more favorably with the independent estimates, with only one of the 14 (Indian River) having a percent deviation >70%. As noted earlier, there was a large difference between the Indian River watershed areas defined by the national model group compared to other estimates and the NOAA statistics that may explain this deviation (Table 8). Further, differences observed for the Texas estuaries may be a consequence of the relatively high total N export reported by the land-use based group [Chapter 4] compared to monitoring data.

The comparison between the independent studies and the SPARROW predictions is more favorable than the empirical estimate group conclusion that SPARROW estimates were 48% lower reported earlier. While, the distribution of the total N yield estimates confirmed the general relationship of the land-use based estimates being considerably higher than the independent estimates and SPARROW estimates tending to be lower, the SPARROW estimates were overall more similar to the independent assessments (Figure 7). Median total N yields for the land-use based, SPARROW, and independent studies were 9.41, 4.27 and 5.27 kgN/ha-yr, respectively (Table 9). As noted earlier, the empirical estimate group 48% deviation was based upon a difference in the means that was influenced by a relatively few outlying pairs of estimates with total N yields greater than 15 kgN/ha-yr. The interpretation in this summary (Figure 7) gives an equal weighting to each estuary and compares medians and interquartile ranges, which suggests that a majority of the pairs in the two datasets are much closer than a comparison of the means describes.

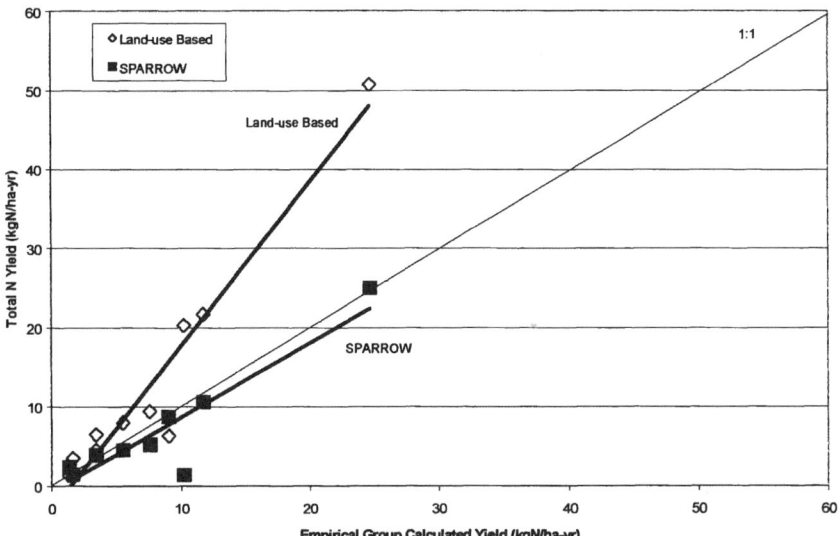

Figure 8. Comparison of total N yield as calculated from population density with land-use based and SPARROW total N yield estimates for nine Atlantic Coast estuaries.

Turner et al. [Chapter 5] also reported a strong coefficient of determination ($r^2 = 0.78$) for the relationship between population density and N yield from Atlantic Coast watersheds. Applying that relationship (y = 2.95x + 78, where x = persons/km^2 and y = kgN/km^2) to population estimates reported by the land-use based group for the smaller Atlantic Coast watersheds (<15,000 km^2; large rivers correlated differently) matched the SPARROW predictions very well (Figure 8). Buzzards Bay was the single outlier, which may be related to the much larger basin definition used by the national model group. The land-use based estimates of total N yield were much higher than the population-based predictions. The literature values and the population-based predictions appear to concur with the estimates from the SPARROW model.

Given the variation in source data, approach and estimate structure, it is not surprising that the estimates differed among the groups. In Chapter 7 several of the potential causes of this uncertainty were discussed in detail. Further, as the estimates in this volume are considered to be annual averages representing different base year conditions or averages of several years, Brock discussed some of the year-to-year variation that might occur in the field. In general, confidence bounds were estimated to be in the range of 70 to 85% of the value at the lower bound and 115 to 130% at the higher bound. Brock estimated lower and upper bounds for the estuary total export estimates compiled in the independent studies. Very few of the land-use based and SPARROW total N export estimates fell within the upper and lower bound ranges calculated by Brock. This comparison might be improved if yield were used instead of total export, as the SPARROW, land-use based, and independent study total N load estimates were not always based on equivalent watershed sizes.

Some of the differences in yield might be further explained if the estimates of input and export were available for both the land-use based approach and the SPARROW

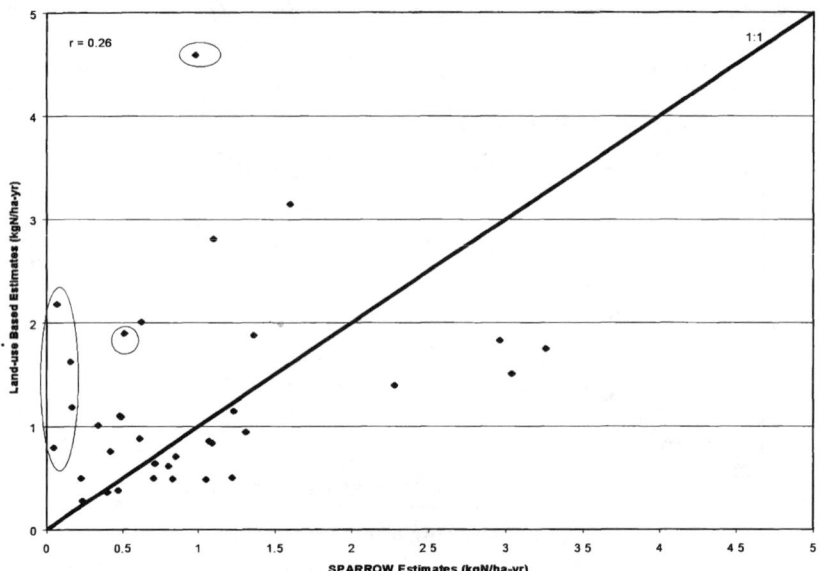

Figure 9. Comparison of SPARROW and land-use based atmospheric N yield estimates (kgN/ha-yr) to U.S. estuaries. The two estimate series were not well-correlated (r = 0.26). The circled outliers (>70% deviation) are (from top to bottom) Massachusetts Bay, Indian River, Tampa Bay, Buzzards Bay, Matagorda Bay, and Corpus Christi Bay.

model in equivalent categories, *e.g.*, point, urban, agricultural, etc. Unfortunately, they do not provide input and export of N and transport and attenuation coefficients in a complete and common manner to determine the sources of the observed differences in N yields. Only aggregated components (total N, atmospheric N, non-atmospheric N) could be practicably evaluated.

Comparison of Atmospheric Nitrogen Yield from the Watersheds

Watershed (indirect) atmospheric N yield estimates generated by the land-use based group and the SPARROW model were compared. Both groups estimated atmospheric total N yield in kgN/ha-yr and total N export in kgN/yr for 34 common estuarine drainage areas included in this study (Tables 4 and 5).

The correlation of atmospheric N yield between the land-use based and the SPARROW estimates was poor, having an r = 0.26 (Figure 9). The rank correlation test demonstrated a significant difference at the p >0.05 level. Recall the significant correlation between the national model and atmospheric deposition groups' (used by the land-use based group) NO_3^- deposition rates (Figure 3). In view of this, a better correlation might have been expected if both approaches processed the deposition similarly during transport through the watershed. However, there was no clear pattern to the differences in contrast to the total N yield comparison between the two datasets. About half of the SPARROW estimates were lower than the land-use based estimates but the land-use based median atmospheric yield was higher (0.97 kgN/ha-yr) than the SPARROW yield (0.76 kgN/ha-

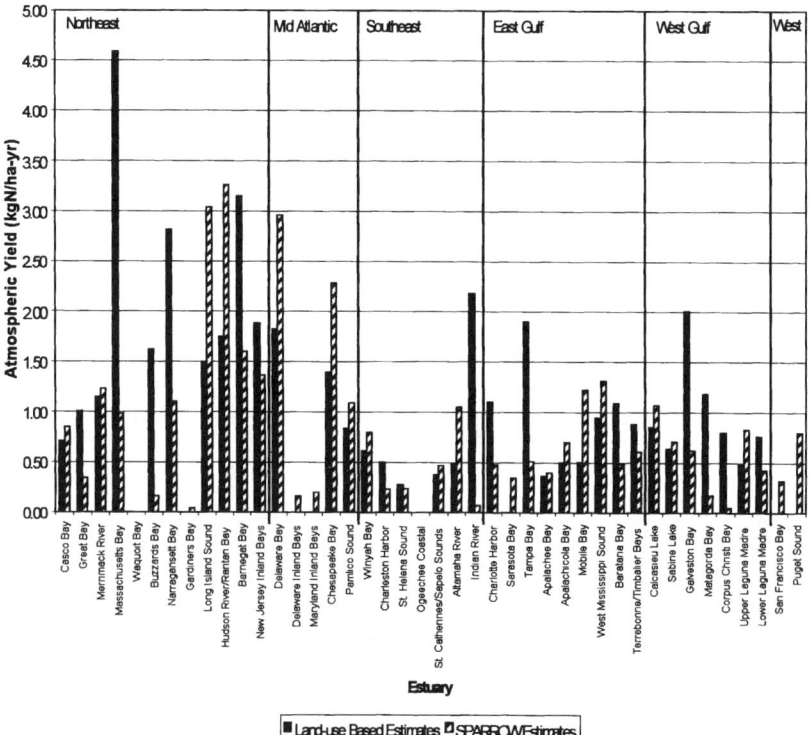

Figure 10. SPARROW and land-use based atmospheric yield (kgN/ha-yr) estimates for 42 estuarine watersheds.

yr) for the 34 common estuaries. Respective interquartile ranges were 0.61-1.75 kgN/ha-yr and 0.42-1.22 kgN/ha-yr and, while not greatly disparate, showed a tendency towards higher estimates by the land-use based approach. Also, outlying estimate pairs (>70% deviation) all exhibited higher yields reported by the land-use based group. They were Corpus Christi Bay, Matagorda Bay, Buzzards Bay, Tampa Bay, Indian River and Massachusetts Bay (Figure 9). One outlier that was especially prominent was Massachusetts Bay, which had a reported land-use based estimate nearly 5 times higher than the SPARROW estimate (Figure 10).

The differences in outlying estuaries may have been a consequence of the same factors that caused differences in the total N yield, *i.e.*, watershed size and attenuation. In addition, the land-use based work group reported some difficulties with Texas N flux estimates being much higher than monitored loads. Four of the six outlying estuaries were also disparate in watershed size reported by the two groups, as was discussed above for the total N yields. The SPARROW area for Buzzards Bay was more than three times larger than the area used by the land-use based group, Tampa Bay was 31% larger, Massachusetts Bay was 21% larger and Indian River was 38% smaller (Table 8). These differences in watershed definition may also have biased the atmospheric yields if land covers varied in the incongruent areas. In the cases of Corpus Christi Bay and Matagorda

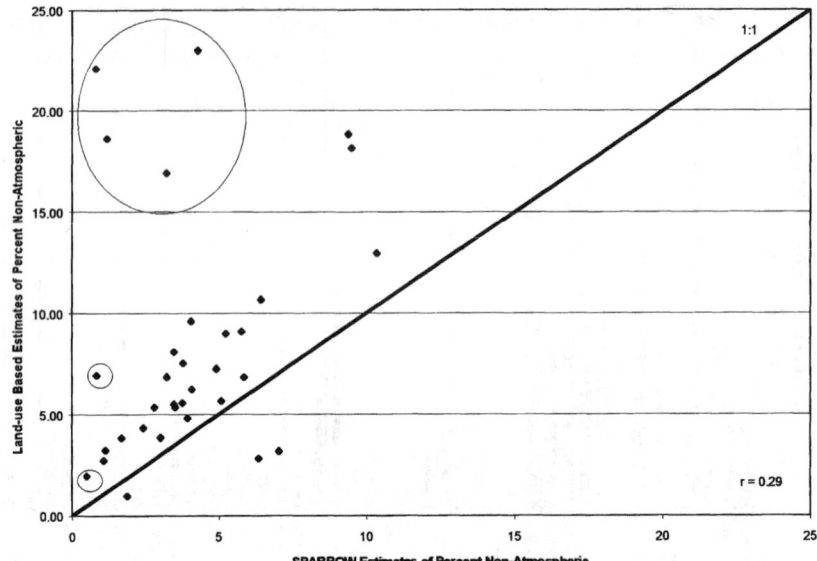

Figure 11. Comparison of land-use based and SPARROW estimates of non-atmospheric yield (kgN/ha-yr) to U.S. estuaries. Massachusetts Bay was excluded from the comparison for scaling purposes. The circled outliers (>70% deviation) are (from top to bottom) Tampa Bay, Indian River, Buzzards Bay, Charlotte Harbor, Charleston Harbor, and Corpus Christi Bay.

Bay, however, watershed sizes were nearly identical between the two groups, suggesting another factor, perhaps the Texas estimation problems noted by the land-use based group, may be causing some of the observed differences in atmospheric yield.

Removing the estuaries with disparate (>20%) size definitions and the Texas estuaries did not greatly improve the correlation between land-use based and SPARROW estimates (r = 0.33). However, the reduced datasets became significantly correlated using the rank correlation test at p <0.01 indicating the order of the relative estuarine atmospheric yields was very similar. Both the land-use based and national model approaches appear to agree very well on which estuaries have relatively higher atmospheric N yields.

Looking at atmospheric N yield for the 20 estuaries with land-use based group defined fall line divisions did not greatly improve the correlation, however, (r = 0.50 vs. r = 0.26) although the ranks were significantly correlated at the p <0.05 level of significance. As with total N yield, attenuation may be effecting some of the observed differences, although the relationship for atmospheric N yield is more erratic. It appears that the non-atmospheric component of N yield pushes the land-use based total N yields disproportionately higher. Comparisons of the non-atmospheric yield and the percent atmospheric yield support that conclusion (Figures 11 and 12). There is a fundamental difference in non-atmospheric estimates between the land-use based and SPARROW estimates that causes much of the total N yield difference. As many of the non-atmospheric outliers are the same as those observed for total N yield, and the land-use based estimates are consistently higher than the SPARROW predictions, disparate watershed delineations and attenuation rates are the likely causes. Analyses and explanations of atmospheric N differences would parallel those offered for total N yield earlier.

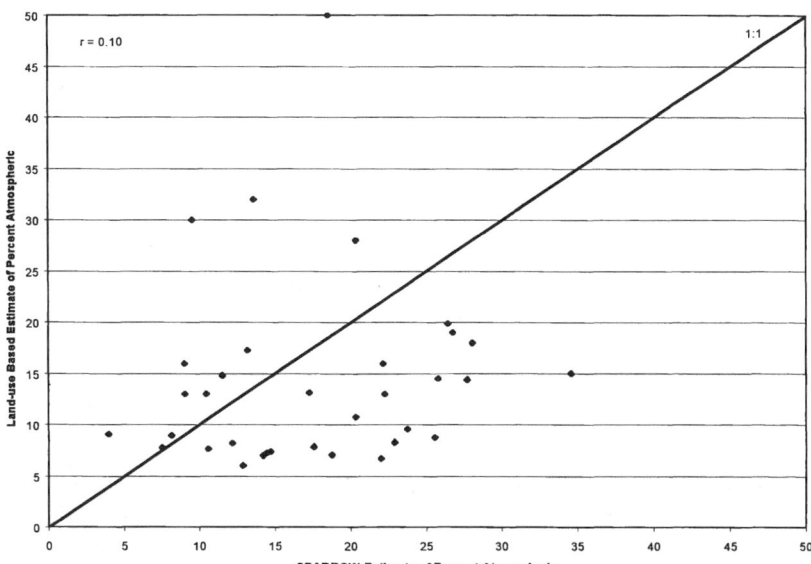

Figure 12. Comparison between SPARROW and land-use based estimates of percent atmospheric N component delivered to estuaries from watersheds.

Percent atmospheric contribution of the total N flux was generally lower for the land-use based estimates (median = 13%, interquartile range = 8 to 16%), indicating a higher non-atmospheric contribution to total N yield than estimated by the national model group (median = 17%, interquartile range = 12 to 23%; Figure 12). All but three estuaries (Upper Laguna Madre, Barnegat Bay and St. Catherines/Sapelo Sounds) had higher non-atmospheric yields for the land-use based estimates compared to the SPARROW estimates (Figure 11). This could be related to factors such as higher input estimates for point and non-atmospheric nonpoint loads or lower attenuation in the land-use based approach compared to the SPARROW model, or a combination of the two factors. Despite the closeness of median atmospheric N proportions between the land-use based and national model groups (13% and 17%, respectively), the poor correlation ($r = 0.10$) and insignificance of rank correlation ($p > 0.05$) reveal a lack of concurrence at the individual estuary level.

Direct Deposition

Although there were not independent estimates of direct atmospheric N deposition to compare, the contribution of direct deposition can be important in overall estuarine loading of atmospheric N. In particular, estuaries with small watershed to waterbody area ratios may derive a substantial percentage of anthropogenic N from atmospheric sources. To this point, atmospheric comparisons were of indirect contributions only. The land-use based and national model groups estimated indirect (watershed) atmospheric N median contributions at about 15% nationally. Maximum indirect contributions were 50 and 35%, respectively. To better define the role of direct atmospheric deposition in total N loadings to estuaries, the indirect atmospheric N contributions for the land-use based and national model groups' estimates were adjusted by adding in the direct loading estimated by the atmospheric deposition group. While the land-use based work group had already used this estimate and reported the results in Chapter 2, this application was used to consistently calculate effects on both the land-use based and SPARROW estimates. Using that adjustment, a percent atmospheric N contribution was recalculated based on the combined indirect and direct loading.

Atmospheric percent contributions rose by an average of 24 and 33% for the land-use based and SPARROW estimates, respectively, when both direct and indirect sources are included in the estimates (Table 10). The Terrebonne/Timbalier Bay estimate rose the most (68%) for the land-use based data and the Indian River estimate rose the most (89%) for the SPARROW predictions. For all estuaries, adding the direct atmospheric N deposition brought the land-use based median up to 16% from 13% and the SPARROW median up to 24% from 14%. Most of the difference between the two estimates as a percent can be attributed to the generally lower overall loads of N in the national model group's estimates. However, addition of the direct deposition does make an important difference in N source distribution for many estuaries. The Terrebonne/Timbalier Bay estimate for the land-use based group, for example, rose from 19% atmospheric N contribution to 60% with the addition of direct deposition. Gardiners Bay changed from 10% atmospheric N to 86% with the addition of direct N deposition to the SPARROW predictions (Table 10).

Regional Differences

All of the work groups identified similar regional relationships in N loading rates and yield. The geographic patterns of atmospheric N deposition, atmospheric N yield, total N yield, and direct atmospheric deposition to the estuarine water surface, were similar for all groups (Figure 13). Median N deposition and yield were higher in the Northeast and Mid-Atlantic regions for all four categories, except for the SPARROW estimates, which had the second largest median total N yield in the West Coast estimates (Figure 13(c)). (The land-use based group did not construct estimates for West Coast estuaries.) The Southeast and Gulf regions generally had lower rates in all categories, but West Coast atmospheric deposition rates for the two estuaries evaluated in that region were lowest in the atmospheric deposition group estimates (Figure 13).

By region, the national model group predictions generally had lower median atmospheric and total N yields than the land-use based estimates, although the range of observations within most regions overlapped considerably. There is concurrence between the land-use based and national model approaches that atmospheric yield was higher in the Northeast and Mid-Atlantic estuaries (Figure 13). Land-use based median yields were 1.68 and 1.39 kgN/ha-yr for the respective regions and SPARROW medians were 1.10 and 1.39 kgN/ha-yr, respectively (Table 11). Atmospheric yield was exceptionally high in the Massachusetts Bay watershed (4.6 kgN/ha-yr), according to the land-use based estimate. SPARROW atmospheric yield estimates were highest for six Northeast and Mid-Atlantic estuaries; the seventh highest was in the East Gulf region (West Mississippi Sound).

As a percent atmospheric contribution, comparing the land-use based and SPARROW estimates, the regional distribution was less consistent than was observed for the N yield data. Land-use based median percent atmospheric contributions had a small range for the

TABLE 10. Percent atmospheric contribution based on indirect and indirect plus direct atmospheric deposition for land-based and SPARROW estimates.

Estuary	Land-based estimates		SPARROW estimates	
	Indirect Deposition Only (%)	Indirect Plus Direct Deposition (%)	Indirect Deposition Only (%)	Indirect Plus Direct Deposition (%)
Northeast Region				
1. Casco Bay	16	30	22	33
2. Great Bay	16	17	9	11
3. Merrimack River	14	14	28	28
4. Massachusetts Bay	9	14	4	12
5. Waquoit Bay				
6. Buzzards Bay	8	23	12	51
7. Narragansett Bay	13	16	10	16
8. Gardiners Bay			11	86
9. Long Island Sound	15	19	35	39
10. Hudson River/Raritan Bay	9	10	26	27
11. Barnegat Bay	50	58	19	27
12. New Jersey Inland Bays	20	26	26	35
Mid-Atlantic Region				
13. Delaware Bay	13	16	22	25
14. Delaware Inland Bays			9	38
15. Maryland Inland Bays			8	24
16. Chesapeake Bay	18	23	28	33
17. Pamlico Sound	7	17	14	28
Southeast Region				
18. Winyah Bay	7	7	19	19
19. Charleston Harbor	7	7	22	23
20. St. Helena Sound	8	10	18	22
21. Ogeechee Coastal				
22. St. Catherines/Sapelo Snds	28	45	20	33
23. Altamaha River	8	8	23	23
24. Indian River	9	14	8	75
East Gulf Coast				
25. Charlotte Harbor	6	8	13	19
26. Sarasota Bay			11	26
27. Tampa Bay	8	11	11	22
28. Apalachee Bay	7	17	14	27
29. Apalachicola Bay	7	8	15	16
30. Mobile Bay	10	11	24	25
31. West Mississippi Sound	14	24	26	35
32. Barataria Bay	13	28	9	25
33. Terrebonne/Timbalier Bays	19	60	27	65

TABLE 10. Continued.

	West Gulf Coast			
34. Calcasieu Lake	13	16	17	20
35. Sabine Lake	11	11	20	21
36. Galveston Bay	17	18	13	16
37. Matagorda Bay	32	33	14	16
38. Corpus Christi Bay	30	30	10	16
39. Upper Laguna Madre	15	23	12	16
40. Lower Laguna Madre	8	12	8	18
	West Coast			
41. San Francisco Bay			5	6
42. Puget Sound			12	14

five regions studied, from 8 to 15% (Table 11). The Northeast and West Gulf had the highest medians at 15%. The median percent atmospheric contribution for the SPARROW data ranged from 9 to 20% for six regions and was highest in the Southeast (20%) followed by the Northeast (19%). Both groups identified a Northeast estuary as having the highest percent atmospheric contribution, 50% for Barnegat Bay by the land-use based group and 35% for Long Island Sound according to the SPARROW predictions.

In sum, the groups' estimates showed different degrees of concurrence depending on the category. Atmospheric NO_3^- deposition between the atmospheric deposition and national model groups correlated well, although the atmospheric deposition group's estimates were consistently higher. Atmospheric N yield from the watersheds tended to be higher in the land-use based estimates compared to the SPARROW estimates, but the proportions of atmospheric N flux from individual estuaries were not well-correlated. Land-use based total N yield estimates from all sources were, however, consistently and substantially higher than the SPARROW estimates. Much of that disparity was attributed to the non-atmospheric component of the estimates, and the dominant factors causing this difference appeared to be watershed size and attenuation rates. Land characteristics, input databases, and treatment of specific sources (*e.g.*, point, nonpoint) are likely sources for the differences but warrant further investigation.

Independent total N yield estimates compiled from the literature by the empirical estimate group were more similar to the SPARROW estimates, with the land-use based estimates about two times higher. An evaluation of uncertainty applied to the empirical estimate group's data summary did not resolve national model and land-use based group's disparity. However, using a population-based N loading approach for smaller Atlantic Coast estuaries reported by the empirical estimate group, the SPARROW total N yield estimates provided an excellent match. A more exhaustive look into uncertainty applied specifically to the national model and land-use based estimates might identify specific causes for the differences between land-use based and SPARROW estimates.

Finally, while not the dominant source of anthropogenic N in most estuaries, the contribution of atmospheric N to total estuarine loading is of managerial significance. The percent contribution from indirect atmospheric N sources has a median of around 15% but in some estuaries, it can range as high as 30 to 50% of the total N load and even higher (60% or more) if direct deposition to the estuary is included. Estuarine managers faced with difficult options to control N must consider the significance of atmospheric sources if water quality goals are to be met. Further, control of atmospheric N deposition will yield additional benefits in forest productivity, lake acidification, and tropospheric ozone control, lowering the overall cost-to-benefit ratio for management. Continued research into the sources and fate of atmospheric N and its role in comprehensive management plans to protect and improve U.S. estuaries is needed.

Conclusions

Biogeochemical and Ecological Implications

A summary of information from existing literature on the impacts and extent of atmospheric N deposition to coastal waters and their watersheds included the following:

- A national estuarine survey documented oxygen depletion in more than half of the 136 estuaries surveyed in the United States.

- Other impacts of nutrient enrichment include losses of seagrass beds, shifts in phytoplankton species composition, and harmful algal blooms.

- Atmospheric deposition is an important source of N entering the coastal zone and is, therefore, of research and managerial interest.

- Recent literature estimates report estuaries receive anywhere from about 1 to more than 40% of their total "new" N load from atmospheric sources.

- An estimated 20% of the dissolved inorganic N exported by rivers throughout the world is attributed to atmospheric sources.

Findings of This Report

- Four separate and independent approaches were used to examine the contribution of atmospheric deposition of N to estuarine waters and their watersheds. Findings from these approaches include:

- Total atmospheric N deposition rates to watersheds ranged from 3 kgN/ha-yr to 13.5 kgN/ha-yr. Examination of total wet and dry N deposition rates to 42 U.S. estuaries indicates that estimated rates were higher in the Northeast and Mid-Atlantic regions (from Massachusetts to the Chesapeake Bay) and in the central region of the Gulf of Mexico than in other regions of the country.

- A land-use based approach, incorporating retention and transport processes, provided estimates of combined N loads from indirect (from the watershed) and direct deposition to the water's surface. For the 34 estuaries examined, the total N load delivered to the estuaries attributed to atmospheric deposition from both direct and indirect sources ranged from 7 to 60% with a median of 16%. Indirect atmospheric sources from the watershed contributed from 6 to 50% of the total N flux with a median of 13%.

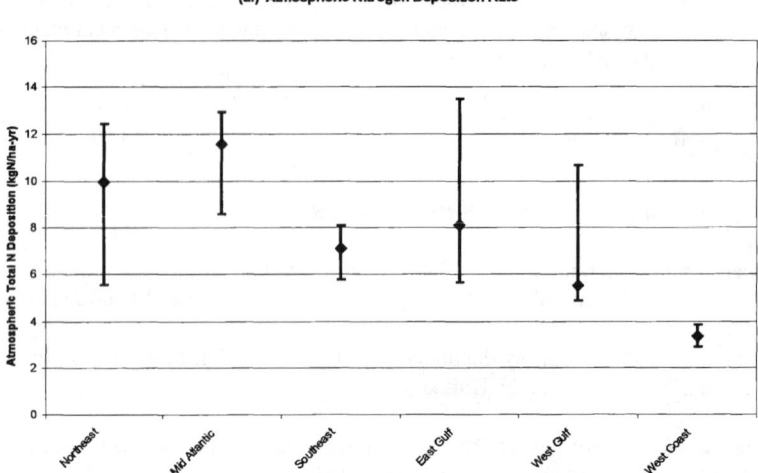

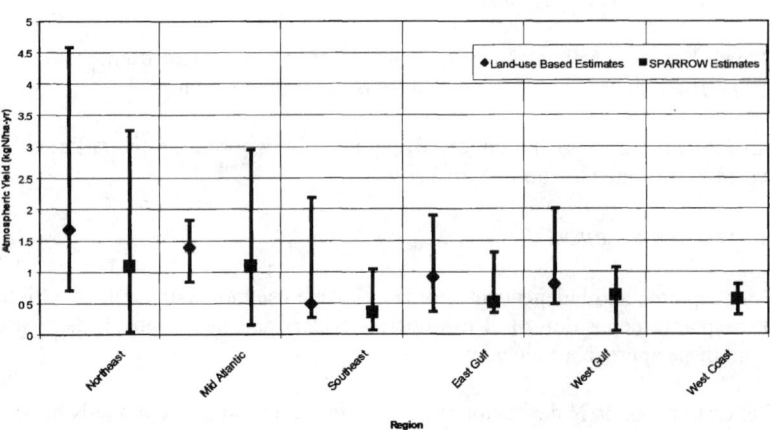

Figure 13. Median and range of N deposition and yield (kgN/ha-yr). (a) Atmospheric N deposition rates were estimated by the atmospheric deposition group; (b) atmospheric N yield from the watershed are land-use based and SPARROW estimates.

- A national model (USGS's SPARROW model) produced estimates of N loading from the watersheds of 40 estuaries. Estimated total N delivery rates to the estuaries were highest in the Northeast and Mid-Atlantic regions, but there was also a considerable range of estimated values within each region. Nitrogen yield for the atmospheric component to the estuaries estimated by the SPARROW model ranged from 0.04 to 3.26 kg/ha-yr, or from 4 to 35% of the total N yield with a median of 17%.

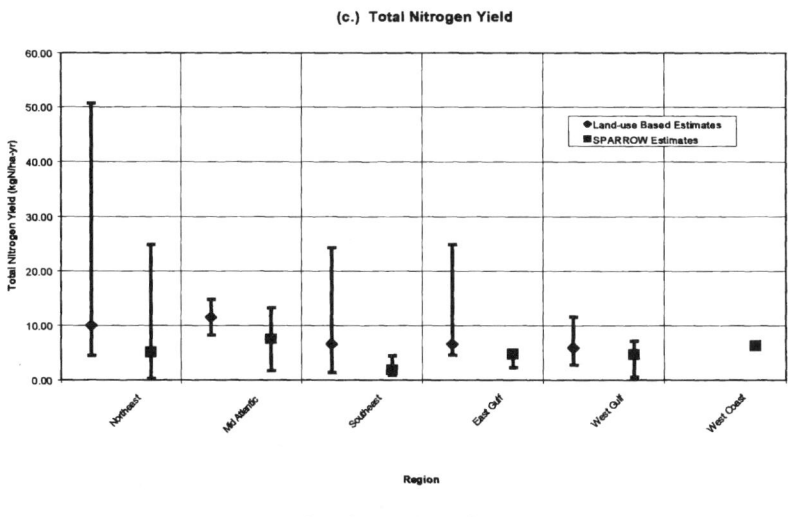

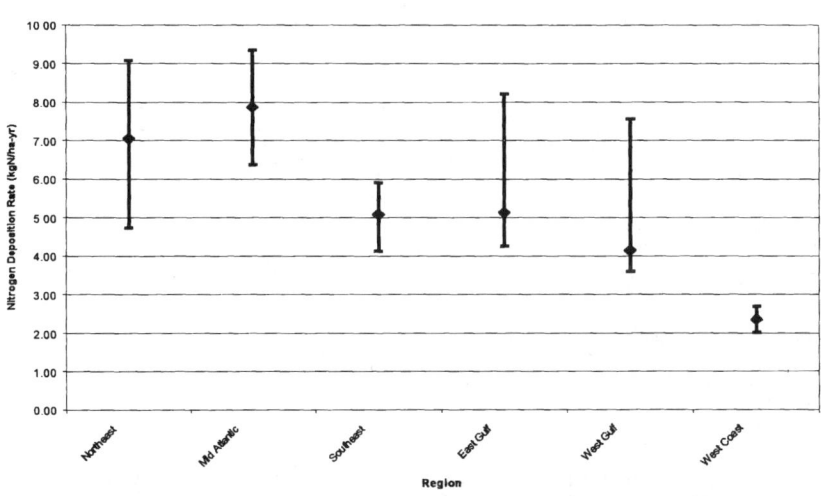

Figure 13. (Continued). Median and range of N deposition and yield (kgN/ha-yr). (c) total N export rates from the watershed are land-use based and SPARROW estimates; (d) direct atmospheric N deposition rates to the estuaries were estimated by the atmospheric deposition group.

- A compilation of independent studies from 27 U.S. estuaries identified 18 estuaries reasonably comparable to the 42 identified in this study. Total N yields for those 18 ranged from 0.44 to 27 kgN/ha-yr. The atmospheric N yields were not consistently identified in those studies and were not useful for comparisons with other approaches.

TABLE 11. Regional comparison of maximum, median and minimum values for direct atmospheric deposition, indirect atmospheric yield, total N watershed yield, direct deposition, and percent indirect atmospheric contribution of N estimated by the land-based and national model groups.

	Region					
	Northeast	Mid-Atlantic	Southeast	East Gulf	West Gulf	West Coast
Atmospheric Deposition Group Indirect Deposition (kgN/ha-yr)						
Maximum	12.43	12.91	8.08	13.50	10.65	3.84
Median	9.97	11.56	7.11	8.11	5.52	3.36
Minimum	5.56	8.55	5.80	5.66	4.89	2.87
Land-based Estimates of Indirect Atmospheric Yield (kgN/ha-yr)						
Maximum	4.59	1.83	2.18	1.90	2.01	
Median	1.68	1.39	0.49	0.91	0.80	
Minimum	0.71	0.84	0.28	0.37	0.49	
SPARROW Estimates of Indirect Atmospheric Yield (kgN/ha-yr)						
Maximum	3.26	2.96	1.05	1.31	1.07	0.80
Median	1.10	1.09	0.36	0.51	0.62	0.56
Minimum	0.04	0.16	0.07	0.35	0.05	0.32
Land-based Estimates of Total N Watershed Yield (kgN/ha-yr)						
Maximum	50.73	14.79	24.28	24.89	11.61	
Median	9.99	11.51	6.63	6.60	5.97	
Minimum	4.55	8.22	1.35	4.70	2.76	
SPARROW Estimates of Total N Watershed Yield (kgN/ha-yr)						
Maximum	24.89	13.32	4.57	5.41	7.17	6.77
Median	5.15	7.51	1.86	4.79	4.63	6.31
Minimum	0.38	1.74	0.89	2.89	0.56	5.85
Atmospheric Deposition Group Direct Deposition (kgN/ha-yr)						
Maximum	9.08	9.35	5.91	8.23	7.57	2.69
Median	7.05	7.87	5.08	5.13	4.14	2.35
Minimum	4.74	6.38	4.13	4.27	3.60	2.01
Percent Indirect Atmospheric Contribution from Land-based Estimates						
Maximum	50	18	28	19	32	
Median	15	13	8	9	15	
Minimum	8	7	7	6	8	
Percent Indirect Atmospheric Contribution from SPARROW Estimates						
Maximum	35	28	23	27	20	12
Median	19	14	20	14	13	9
Minimum	4	8	8	9	8	5

Comparison Between Efforts

- Reliance on the same (NADP) database to generate wet NO_3^- deposition estimates by the national model and atmospheric deposition groups resulted in a high level of correlation between the two estimates. However, the atmospheric deposition group estimates were consistently higher, suggesting a fundamental difference in data interpretation between the two groups. Differences appeared to be related to the base years used, interpretation of trends, and geographic interpolation of rates assigned to watersheds.

- Total N yields from all sources delivered to the estuaries from their respective watersheds as estimated by the national model and the land-use based groups indicated that the SPARROW estimates were consistently lower than the land-use based predictions. The difference was attributed more to the non-atmospheric portion of the estimates, possibly reflecting differences in watershed delineation and character, attenuation factors, or other processes within the approaches.

- Estimated total N yield for 14 Gulf and Atlantic watersheds reported in independent local studies of estuarine N loading were most similar to those generated by the SPARROW model. The median land-use based total N yield was 1.8 times greater than the median reported in the independent studies by the empirical estimate group.

- Population-based estimates of total N yields for smaller Atlantic Coast estuaries correlated very well with SPARROW predictions. Land-use based estimates were higher, reflecting the general tendency of those estimates to be higher.

- A comparison of total N yield from 20 estuaries in which the watershed was divided into above and below fall line segments by the land-use based group greatly improved the correlation with SPARROW predictions compared to the full set of 34, although they were higher. This suggests attenuation factors in the land-use based approach may be too low because attenuation was not applied below the fall line.

- The correlation of atmospheric N yield between land-use based and SPARROW estimates was poor, with the SPARROW yields generally lower than those estimated by the land-use based group. Although the number of land-use based estimates that was larger than the SPARROW estimates on an estuary-by-estuary basis equaled the number that was smaller, a rank correlation test was insignificant.

- The contribution of atmospheric N flux to the estuaries of the total N flux, exclusive of direct deposition to the water surface, ranged from about 6 to 50% in the land-use based estimates, and from 4 to 35% in the SPARROW estimates. When direct deposition of atmospheric N is included, the land-use based national median increases from 16 to 13% and the SPARROW national median increases from 14% to 24%.

- All evaluations identified regional differences in loading rates and total N loads. Median N deposition and yields were usually highest in the Northeast and Mid-Atlantic regions, and the Southeast and Gulf regions generally had lower rates in all categories.

- Although atmospheric sources of N do not dominate the anthropogenic nitrogen loading to most estuaries, they are of managerial significance.

References

Jordan, T.E. and D.E. Weller. 1996. Human contributions to terrestrial nitrogen flux: assessing the sources and fates of anthropogenic fixed nitrogen. Bioscience 46:655-664.

NOAA. 1996. NOAA's Estuarine Eutrophication Survey. Vol. 1. South Atlantic region. NOAA, Office of Ocean Resources Conservation and Assessment. Silver Spring, MD. 50 p.

NOAA. 1997a. NOAA's Estuarine Eutrophication Survey. Vol. 2. Mid-Atlantic region. NOAA, Office of Ocean Resources Conservation and Assessment. Silver Spring, MD. 50 p.
NOAA. 1997b. NOAA's Estuarine Eutrophication Survey. Vol. 3. North Atlantic region. NOAA, Office of Ocean Resources Conservation and Assessment. Silver Spring, MD. 45 p.
NOAA. 1997c. NOAA's Estuarine Eutrophication Survey. Vol. 4. Gulf of Mexico region. NOAA, Office of Ocean Resources Conservation and Assessment. Silver Spring, MD. 78 p.
NOAA. 1998. NOAA's Estuarine Eutrophication Survey. Vol. 5. Pacific Coast region. NOAA, Office of Ocean Resources Conservation and Assessment. Silver Spring, MD. 75 p.
Pacheco, P.A. 1999. Coastal Assessment and Data Synthesis (CA&DS) framework, 1999. National Coastal Assessments (NCA) Branch, Special Projects Office, National Ocean Service (NOS), National Oceanic and Atmospheric Administration (NOAA), Silver Spring, MD.
Snedecor, G.W. and W.G. Cochran. 1967. Statistical methods. Sixth ed. Iowa State Univ. Press, Ames, IA. 593 p.

Appendix 1

An Annotated Summary of Nitrogen Loading to US Estuaries

D. Stanley

Estuary: Casco Bay

Horsley & Whitten, Inc. [1996] developed a model to estimate present and future nitrogen loadings to Maquoit Bay from all principal sources. Maquoit Bay is one of the smaller bays within Casco Bay, with a surface area of approximately 13 km^2. The Maquoit Bay watershed area is 31.9 km^2. The loading model is based primarily on loading coefficients for each land use type. Some of the coefficients were derived indirectly by a series of computations, while others were based on more direct measurements. For example, nitrogen loading from lawns was calculated by assuming that the average lawn size in the watershed was 465 m^2, and that sixty percent of the applied nitrogen was available for transport (i.e., not taken up by the grass). On the other hand, nitrogen loading from agricultural fields was based on actual storm-water quality monitoring (measured nitrogen concentrations and runoff rates) at three test sites.

The authors calibrated their model to match the observed water quality conditions in one sub-watershed. The calibrated model was then applied to the other sub-watersheds for validation purposes. With one exception, the R^2 value between measured and modeled nitrogen concentrations was equal to or greater than 0.75.

The model estimated that, under existing conditions (circa 1996), the annual nitrogen loading to Maquoit Bay was 26,606 kg N. The largest source of nitrogen to the bay was direct precipitation (33%), followed by forested cover of the watershed (18%), agricultural fields (19%), livestock (12%), and septic systems (9%). The category "direct precipitation" refers to nitrogen loading directly onto the surface of the estuary via precipitation (8.7 kg N ha^{-1} y^{-1}).

TABLE 1. Annual N loading to Maquoit Bay, ME (kg N y^{-1}). From Horsley & Whitten, Inc. [1996]

Source	kg N y^{-1}	
Septic	2,285	
Lawns	1,340	
Agriculture/manure		5,055
Cows	3,136	
Forest	4,862	
Sludge disposal	144	
Road drainage	1,066	
Atmospheric deposition directly on bay**	<u>8,718</u>	
TOTAL	26,606	

**Wet deposition only

Reference

Horsley & Whitten, Inc. 1996. Identification and evaluation of nutrient and bacterial loadings to Maquoit Bay, Brunswick, and Freeport, Maine. Final report submitted to Casco Bay Estuary Project, Portland, Maine.

Estuary: Great Bay (Maine and New Hampshire)

Mosher [1995] assessed the relative importance of atmospheric nitrogen deposition (both wet and dry) to the Great Bay, NH watershed and estuary. Atmospheric sampling was conducted for six months during 1994. Gas phase dry deposition was estimated by a computational method to estimate the flux of nitric acid. The calculation depended on the atmospheric concentration of the gas, and also the dry deposition velocity, which was derived from experimental and modeling studies reported in the literature. Daily measurements of particulate nitrate and ammonia were made and the dry depositional flux of these species was calculated in a manner similar to that used for nitric acid. Wet deposition of nitrogen was calculated by multiplying individual event nitrate and ammonia concentrations by the event precipitation depth.

The fraction of atmospheric nitrogen deposited onto the watershed that actually reached the estuary was estimated by means of two transfer coefficients. The first coefficient dealt with the movement of deposited nitrogen from the terrestrial watershed into streams and rivers. The second coefficient described removal during transport in the streams and rivers prior to reaching the bay. The overall transfer coefficient was then calculated by multiplying the watershed/stream coefficient by the stream/estuary coefficient. Different coefficients were used for different land use categories (i.e., urban, forest, etc.).

The atmospheric nitrogen inputs (both direct to the bay and indirect via the watershed) were compared with other (non-atmospheric) terrestrial input data from NOAA and USEPA studies cited in Short [1992]. Methods used to generate those data were not given in the Mosher [1995] report.

TABLE 2. Annual nitrogen loading (kg N y^{-1}) to the Great Bay, NH estuarine system. From Mosher [1995].

Source	kg N y^{-1}
Non-point	
Atmospheric deposition directly onto bay	91,000
Indirect	
Urban	92,000
Forest	113,000
Other	35,000
Agriculture**	151,000
Urban**	206,000
Forest**	3,600
Point	
Wastewater treatment	208,600
Industries	11,800
TOTAL	912,000

** The author noted that these values may include some atmospherically derived nitrogen, but he cannot be sure because he was not aware of the methodology used in the studies cited in Short [1992].

References

Mosher, B. W. 1995. Assessment of atmospheric non-point source nitrogen input to the Great Bay watershed and estuary. Final report submitted to New Hampshire Coastal Program, Office of State Planning, Concord, NH.

Short, F. T. (ed.). 1992. The ecology of the Great Bay estuary, New Hampshire and Maine: an ecological profile and bibliography. NOAA Coastal Ocean Program Publication. 222 p.

Estuary: Waquoit Bay

Valiela et al. [1997] developed a nitrogen loading model from data of the Waquoit Bay Land Margin Ecosystems Research project (WBLMER), and from synthesis of published information. The model was then applied to the Waquoit Bay watershed, which drains 37.9 km^2. The model first estimates inputs by atmospheric deposition, fertilizer use, and waste waters to the surfaces of the major types of land use within the watershed (natural vegetation, turf, agricultural land, residential areas, and impervious surfaces). The model estimates the losses of nitrogen in the various compartments of the watershed ecosystem. For atmospheric and fertilizer nitrogen, the model allows losses in vegetation and soils, in the valdose zone, and in the aquifer. For wastewater nitrogen, the model allows losses in septic systems and effluent plumes, and it adds further losses that occur during diffuse transport through aquifers. The calculation of losses is done separately for each major type of land cover, because the processes and loss rates involved differ for different land uses. If groundwater flows into a freshwater body, then the model adds a loss of nitrogen while traversing a freshwater body and then subjects the remaining nitrogen to losses in the aquifer.

TABLE 3. Summary of total nitrogen loading to Waquoit Bay (kg N y^{-1}). Values represent only N from each source category that actually reaches the bay. Thus, the atmospheric deposition value does not include direct N deposition onto the bay. From Valiela et al. [1997].

Source	kg N y^{-1}	
Atmospherically derived input from watershed		6,923
Fertilizer		3,442
Wastewater	10,998	
Ponds up-grading		1,767
TOTAL	23,130	

Reference

Valiela, I., G. Collins, J. Kremer, K. Lajtha, M. Geist, B. Seely, J. Brawley, and C. H. Sham. 1997. Nitrogen loading from coastal watersheds to receiving estuaries: new method and application. Ecological Applications 7(2):358-380.

Estuary: Narragansett Bay

The authors of this study [Nixon et al., 1995] note that the boundaries of Narragansett Bay have been defined in different ways, and then state that "For this exercise we include Narragansett Bay proper (the East and West Passages from the mouth to Conimicut Point, including side bays and harbors), the Providence and Seekonk River estuary, and Mt. Hope Bay and the Taunton River estuary. We exclude the Sakonnet River, which has only a restricted hydraulic connection with the rest of the bay and receives very little fresh water inflow or sewage effluent. Defined in this way, the water area of the bay is 328 km^2."

The wet deposition of nitrate and ammonium from the atmosphere was measured at a station on Prudence Island during 1989 and part of 1990, and dry deposition was calculated from contemporaneous measurements of nitric acid vapor and nitrate aerosol [Fraher, 1991]. The deposition of dissolved organic nitrogen was estimated from nitrate deposition using data obtained during 1985 and 1986 at the Graduate School of Oceanography by Nowicki and Oviatt [1990].

Nitrogen concentrations were measured approximately every two weeks during 1982 and 1983 in four of the larger rivers flowing into the Bay. Data from the Taunton River were collected during 1988 and 1989. Daily discharge was obtained from U.S. Geological Survey gauging stations set at various distances upstream of the nitrogen sampling stations. Land area ratios were used to increase the measured discharges to account for the area of the watershed below each gauge. Direct groundwater discharge was not included because is was not thought to be significant in Narragansett Bay. The authors chose to use Beale's flow-weighted unbiased estimate to calculate annual fluxes from the biweekly concentration measurements and daily flow collected during one calendar year of measurement.

Concentrations of dissolved inorganic nitrogen (DIN) in urban storm water runoff from four land use categories in the upper Narragansett Bay watershed were measured through 12 storms during late 1979 and 1980. Hanson [1982] used these data to estimate annual DIN flux per unit of precipitation for each land use type. These nutrient loss coefficients were multiplied by the long-term mean precipitation around the bay, and by the total area of each land use type in the cities and towns that discharge their storm water directly into the bay to estimate a total input of DIN from this source. Finally, the authors calculated the total storm water nitrogen by applying the total N to DIN ratio reported by the National Urban Runoff Program.

Twenty-two sewage treatment facilities release their effluents into rivers and streams in the Narragansett Bay watershed, and their contribution to the bay is included in the rivers where measurements of nitrogen were obtained. For the ten plants that discharge directly into the bay, N loadings were estimated by applying Beale's unbiased estimation technique to measured (biweekly) effluent concentration data and reported daily effluent volumes.

TABLE 4. Total nitrogen loading to Narragansett Bay (10^3 kg N y^{-1}). From Nixon et al. [1995].

Source	10^3 kg N y^{-1}
Atmospheric deposition (direct to bay)	420
Rivers and streams	5,600
Urban runoff (direct to bay)	518
Wastewater treatment (direct to bay)	2,562
TOTAL	9,100

References

Fraher, J. 1991. Atmospheric wet and dry deposition of fixed nitrogen to Narragansett Bay. M.S. Thesis in Oceanography, University of Rhode Island, Narragansett, RI. 165 p.

Hanson, L. C. 1982. A preliminary assessment of nutrient loading into Narragansett Bay due to urban runoff. Masters in Marine Affairs, University of Rhode Island, Kingston, RI. 37 p.

Nixon, S. W., S. L. Granger, and B. L. Nowicki. 1995. An assessment of the annual mass balance of carbon, nitrogen, and phosphorus in Narragansett Bay. Biogeochemistry 31:15-61.

Nowicki, B. L. and C. A. Oviatt. 1990. Are estuaries traps for anthropogenic nutrients? Evidence from estuarine mesocosms. Marine Ecology Progress Series 66:131-146.

Estuary: Long Island Sound

The report by Stacey [1998] summarizes the origin, transport, and delivery of total nitrogen in the Long Island Sound drainage basin. The watershed is segmented according to a rather complicated scheme involving a mix of natural tributary boundaries and geopolitical boundaries. The report covers the entire watershed, including areas within Connecticut and New York as well as tributary rivers north of Connecticut. However, for purposes of the present study, the author updated the calculations so as to eliminate the Hudson River watershed portion of the basin [P. Stacey, personal communication]. This was done to facilitate comparison with other Long Island Sound loading estimates. Hence, the numbers in the summary table below are different from those in the original report. T methodology described below, however, is still appropriate.

Areas for twenty-three land cover types were determined from early 1990s LANDSAT imagery (for Connecticut), from a 1978 report (for New York), and from an article by

Jaworski et al. [1997] for tributaries north of Connecticut. Because export coefficients were not available for all twenty-three land cover categories, they were aggregated into three broad categories - - urban, agriculture, and forest. The area of the sound is 3,367 km^2. Four nitrogen source categories were considered:

1. Natural: based on what loads were when the basin was mostly forested, as in pre-Colonial days (estimated to be 2.9 kg N ha^{-1} y^{-1}). A natural load originates from all source categories except point sources;

2. Point sources: discharges from sewage treatment plants and industries (load calculated as discharge flows times concentrations);

3. Nonpoint sources: diffuse runoff from the land and through the ground and discharges from storm-water systems and combined sewer overflows. The loads were calculated as land areas times export coefficients. The coefficients used were as follows: (a) forest: 2.9 natural plus 1.4 atmospheric = 4.3 kg N ha^{-1} y^{-1}; (b) agriculture: 2.9 natural + 3.3 runoff + 1.4 atmospheric = 7.6 kg N ha^{-1} y^{-1}; (c) urban: 2.9 natural + 5.0 runoff + 5.5 atmospheric = 13.4 kg N ha^{-1} y^{-1}; and,

4. Atmospheric deposition, which was divided into two components: nitrogen falling on the land that is eventually delivered to Long Island Sound (attenuated by biological and other processes) and that which falls directly on the Sound's surface.

TABLE 5. Summary of total nitrogen loading to Long Island Sound (10^3 kg N y^{-1}). Modified from Stacey [1998].

Source	10^3 kg N y^{-1}	
From the watershed		
Atmospherically derived		6,360
Point	11,907	
Other Nonpoint		10,592
Atmospheric deposition directly onto Sound		3,475
TOTAL	32,334	

References

Jaworski, N. A., R. W. Howarth, and L. J. Hetling. 1997. Atmospheric deposition of nitrogen oxides onto the landscape contributes to coastal eutrophication in the northeast United States. Environmental Science and Technology 31:1995-2004.

Stacey, P.E. 1998. (Draft) Report on nitrogen loads to Long Island Sound. Connecticut Department of Environmental Protection, Bureau of Water Management, Hartford, CT.

Estuary: Peconic Bay

Most of this report by the Suffolk County, NY Department of Health Services [1998] is concerned with analyzing and discussing the sediment flux component of the nitrogen loading estimate. Unfortunately, no information is given in the report as to how the other source category loadings were derived, other than that contained in footnotes to Table 16 in the report. It appears that the point source loadings are based on flow and concentration data. The estuarine surface area is 534 km^2.

TABLE 6. Total nitrogen loading estimates (10^3 kg N y^{-1}) for Peconic Bay (adapted from Table 16 in report by Suffolk County, NY Department of Health Services [1998]).

Source	10^3 kg N y^{-1}	
Nonpoint Sources		
Sediment flux	2,298	
Groundwater	678	
Atmospheric deposition directly onto bay	1,171	
Storm-water runoff	22	
Point Sources		
Peconic River	23	
Sewage treatment plants		31
Meetinghouse Creek	18	
TOTAL	4,241	

Reference

Suffolk County, NY Department of Health Services. 1998. Peconic Estuary Program: Point and nonpoint source nitrogen loading overview (Draft report). Prepared by Suffolk County Department of Health Services, Division of Environmental Quality, Office of Ecology.

Estuary: Barnegat Bay

Moser et al. [1998] constructed a nitrogen budget for Barnegat Bay, NJ, using measurements of dissolved inorganic and particulate nitrogen and estimated water fluxes. River nitrogen input was calculated from concentration and flow measurements in the fresh water reaches of four tributaries of the drainage basin (Metedeconk River, Kettle Creek, Toms River, and Cedar Creek). Sampling was conducted twice monthly in the summer and monthly in the winter during 1990-1992. River flow was determined directly from United States Geological Survey (USGS) gaging station data for branches of the Toms and Metedeconk Rivers and Cedar Creek and estimated for ungaged river branches and drainages from the existing gaging data.

Data for atmospheric nitrogen inputs came from Scudlark and Church [1997] and from S. Seitzinger (unpublished data; derivation methods are not given). Ground water nitrogen concentrations were from Vowinkel and Battaglin [1997]. These data were used to make a model prediction of ground water nitrogen input, using methods from Nicholson and Watt [1996, 1997].

The Barnegat Bay watershed is 878 km^2 and the bay's surface area is 191 km^2.

TABLE 7. DIN and PN loading to Barnegat Bay, New Jersey (kg N y^{-1}). From Moser et al. [1998].

Source	kg N y^{-1}
Atmospheric deposition directly onto bay	267,000
Rivers	326,200
Ground water	37,800
TOTAL	631,000

References

Nicholson, R. S. and M. K. Watt. 1996. Ground-water flow in the unconfined aquifer of the northern Barnegat Bay watershed, New Jersey. The Barnegat Bay Ecosystem Workshop, Toms River, NJ, Rutgers Cooperative Extension, Rutgers University.

Nicholson, R. S. and M. K. Watt. 1997. Simulation of ground-water flow in the unconfined aquifer system of the Toms River, Metedeconk River, and Kettle Creek basins, New Jersey. Water Resources Investigation Report, U. S. Geological Survey.

Moser, F. C., S. P. Seitzinger, R. J. Murnane, and R. G. Lathrop. 1998. Local and regional nitrogen sources to a shallow coastal lagoon, Barnegat Bay, New Jersey. Unpublished manuscript submitted to Marine Ecology Progress Series.

Scudlark, J. R. and T. M. Church. 1993. Atmospheric input of nitrogen to Delaware Bay. Estuaries 16(4):747-759.

Vowinkel, E. F. and W. A. Battaglin. 1997. Relation of ground water quality to hydrogeology and landuse in the coastal plain of New Jersey. United States Geological Survey Water Supply Paper 2381-E.

Estuary: Delaware Bay

Nixon et al. [1996] synthesized published nutrient data for a number of estuaries, including Delaware Bay. Their goal was to use this information to assess the effects of biogeochemical exchanges in estuaries on the net transport of N and P from land onto the North Atlantic continental shelf and the net exchange between the shelf and the open sea.

Table 5 in the Nixon et al. [1996] paper summarizes the mass balance of total nitrogen for Delaware Bay. A variety of sources were used, and the authors state that the data are for the 1980s. The direct atmospheric deposition value was based on an estimate of direct wet dissolved inorganic nitrogen (DIN) deposition from Scudlark and Church [1994]. The authors also included in their value dry deposition, which was assumed to be equal to wet deposition. Other assumptions were that dissolved organic nitrogen (DON) deposited onto the estuary was equal to 50% of wet nitrate (NO$_3$) and that wet NO$_3$ was equal to 50% of wet DIN, based on Nowicki and Oviatt [1990]. The "land drainage and sewage" estimate was taken from Jaworski and Howarth [in press].

TABLE 8. Nitrogen inputs to Delaware Bay (10^3 kg N y^{-1}). From Nixon et al. [1996].

Source	10^3 kg N y^{-1}
Atmospheric deposition directly onto bay	2,785
Land drainage and sewage	50,123
TOTAL	52,908

References

Jaworski, N. and R. W. Howarth. (in press). Preliminary estimates of the pollutant loads and fluxes into the northeast shelf LME. In: Sherman, K. (ed.), The Northeast Shelf Large Marine Ecosystem. NOAA.

Nixon, S. W., J. W. Ammerman, L. P. Atkinson, V. M. Berounsky, G. Billen, W. C. Boicourt, W. R. Boynton, T. M. Church, D. M. Ditoro, R. Elmgren, J. H. Garber, A. E. Giblin, R. A. Jahnke, N. J. P. Owens, M. E. Q. Pilson, and S. P. Seitzinger. 1996. The fate of nitrogen and phosphorus at the land-sea margin of the North Atlantic Ocean. Biogeochemistry 35:141-180.

Nowicki, B. L. and C. A. Oviatt. 1990. Are estuaries traps for anthropogenic nutrients? Evidence from estuarine mesocosms. Marine Ecology Progress Series 66:131-146.

Scudlark, J. R. and T. M. Church. 1994. Atmospheric input of inorganic nitrogen to Delaware Bay. Estuaries 16(4):749-759.

Estuary: Delaware Inland Bays

The Delaware Inland Bays include three interconnected embayments: Indian River Bay, Rehoboth Bay, and Little Assawoman Bay. The Rehoboth Bay watershed drains to Rehoboth Bay, the Indian River, Indian River Bay, and Iron Branch watersheds drain to Indian River Bay, and the Little Assawoman watershed drains to Little Assawoman Bay. The Little Assawoman watershed is not entirely within the State of Delaware, a portion lies within the northern part of Maryland and was not assessed in this study of nitrogen loading by Horsley and Witten [1998].

Land uses within each watershed were defined and quantified using information provided by the Delaware Department of Natural Resources "from their GIS data layer" (a reference was not cited, but it was stated that the information was based on 1992 aerial photography). A spreadsheet nitrogen loading model was developed to tabulate the contribution of nitrogen loadings from each land use within the watershed, and to calculate the total mass that is discharged to the three bays.

Major sources in the analyses included point sources, such as wastewater treatment plants, and nonpoint sources, including septic systems, lawn and agricultural fertilizers, and road runoff. Nonpoint source loadings were computed by multiplying loading coefficients by the total acreage of each land use category. Coefficients were based on monitored data where available, and on literature values for similar land uses where no actual data were available. The loading rates used in the study reflect total nitrogen values.

Point source loadings from wastewater treatment plants were based on data from NPDES permits. Total nitrogen (TN) loading for most of the plants were calculated by multiplying the average concentrations of TN, from sampling periods between October, 1996 and August, 1997 by the flow rate of the particular plant. Discharge data from three plants were taken from an earlier study. All of the plants discharge directly into a bay or tributary or through spray irrigation of crops. Mitigating factors that may decrease the loadings from the spray irrigation (e.g., uptake by crops) were not considered.

Atmospheric loading was also computed, but only to the open surface waters of the Bays themselves (i.e., the loading coefficients for land uses were assumed to include atmospheric influences). The atmospheric coefficient used was 35.84 kg N ha^{-1} y^{-1}. Detailed loading spreadsheets are included in appendices to the report.

TABLE 9. Summary of nitrogen loading results for Delaware Inland Bays (10^3 kg N y^{-1}). From Horsley and Witten, Inc. [1998].

Source	10^3 kg N y^{-1}	
Watershed		
Rehoboth Bay	327	
Indian River Bay		1,021
Little Assawoman Bay**	375	
Atmospheric deposition directly onto bay		
Rehoboth Bay	133	
Indian River Bay		134
Little Assawoman Bay**	45	
TOTAL	2,035	

** does not include the Maryland portion of the watershed

Reference

Horsley and Witten, Inc. 1998. Assessment of nitrogen loading to the Delaware Inland Bays. Prepared for The Center for the Inland Bays, Nassau, DE

Estuary: Chesapeake Bay

Boynton et al. [1995] assembled and analyzed annual loadings for total nitrogen (TN) for Chesapeake Bay and three of its tributary estuaries (Potomac, Patuxent, and Choptank rivers). Three sources were included in the estimates: point, nonpoint (referred to herein as "diffuse", and "atmospheric". The time period is for 1985-1986.

Loading estimates for the Potomac, Patuxent and Choptank rivers included all of the areas of the estuaries and the associated watersheds, and extending to the mouths. However, the seaward boundary of the "Maryland Mainstem Bay" was set just upstream of the Potomac River mouth. Thus, only land areas above this line were (presumably) considered in arriving at loading estimates for the Maryland Mainstem Bay portion of the Chesapeake Bay.

Point source nitrogen loadings included all major (i.e., greater than 1 million gallons per day) point discharges. The load calculations were mostly based on flow gage and measured nitrogen concentration data, but some concentrations were estimated indirectly based on the level of treatment. Sources cited for the data are USEPA [1982] and Summers [1989]. Above fall-line (AFL) loadings were based on direct measurements (flows times concentrations) in the Patuxent, Potomac, Choptank, and Susquehanna rivers [data from Summers, 1989]. Below fall-line (BFL) values for N loading were based on an algorithm that adjusted AFL loads to land-use and size of BFL areas. Known point source loads were subtracted from the AFL loadings to give estimates of nonpoint-source loading. In the paper by Boynton et al. [1995], AFL and BFL loads are reported separately, but they are summed in Table 10.

TABLE 10. Summary of N loading for Chesapeake Bay ("Maryland Main Stem") and three tributary estuaries (Potomac, Patuxent, and Choptank rivers. These data are from Boynton et al. [1995]. In addition, data for all of Chesapeake Bay (bay surface area = 11,542 km^2) are included - - these are from Nixon et al. [1996].

Source	Estuarine surface area (km^2)	Drainage area (km^2)	N load (10^3 kg N y^{-1})
Point sources			
Chesapeake Bay			42,012
Maryland Mainstem Bay		19,150	
Potomac River		11,940	
Patuxent River		830	
Choptank River			140
Nonpoint sources			
Chesapeake Bay			91,297
Maryland Mainstem Bay	70,189	55,210	
Potomac River	29,940	20,163	
Patuxent River	2,393	640	
Choptank River		1,779	830
Atmospheric deposition direct onto bay**			
Chesapeake Bay	11,542		18,259
Maryland Mainstem Bay	3,942	6,240	
Potomac River	1,210	1,920	
Patuxent River	137	220	
Choptank River	361		570

** Wet deposition only

Atmospheric deposition includes only wet-fall to surface waters of the estuaries. Wet-fall collections were made at seven locations adjacent to the bay. Rainfall estimates were from southern, mid, and northern bay locations. Again, the data were taken from USEPA [1982] and Summers [1989].

Nixon et al. [1996] included a table summarizing nitrogen loading for the whole Chesapeake Bay. They cited Boynton et al. [1995] as their source. Apparently Nixon et al. [1996] extrapolated from the Boynton et al. [1995] numbers to arrive at their values for the whole bay.

References

Boynton, W. R., J. H. Garber, R. Summers, and W. M. Kemp. 1995. Inputs, transformations and transport of nitrogen and phosphorus in Chesapeake Bay and selected tributaries. Estuaries 18(1B):285-314.

Nixon, S. W., J. W. Ammerman, L. P. Atkinson, V. M. Berounsky, G. Billen, W. C. Boicourt, W. R. Boynton, T. M. Church, D. M. Ditoro, R. Elmgren, J. H. Garber, A. E. Giblin, R. A. Jahnke, N. J. P. Owens, M. E. Q. Pilson, and S. P. Seitzinger. 1996. The fate of nitrogen and phosphorus at the land-sea margin of the North Atlantic Ocean. Biogeochemistry 35:141-180.

Summers, R. M. 1989. Point and non-point source nitrogen and phosphorus loading to the northern Chesapeake Bay. Maryland Department of the Environment, Water Management Administration, Chesapeake Bay Special Projects Program. Baltimore, Maryland.

United States Environmental Protection Agency (USEPA). 1982. Chesapeake Bay Program, Technical Studies: A Synthesis. United States Environmental Protection Agency. Washington, D. C.

Estuary: Maryland Coastal Bays

Nitrogen loadings for seven bays making up the Maryland coastal bays system are included a report by Boynton et al. [1993] and in a subsequent journal publication [Boynton et al., 1996]. Three sources of nitrogen were considered: point, nonpoint (referred to by the authors as "diffuse"), and atmospheric.

The nonpoint source includes surface water inputs, groundwater inputs, and inputs from chicken-rendering plants, which the authors state are common in the drainage basin. Nonpoint source loadings were developed by Jacobs et al. [1993] and were based on multiplication of land areas times land-use-specific run-off coefficients. The nonpoint loadings are generally representative of average annual rainfall conditions rather than any specific annual period. Presumably the coefficients reflect conditions in the late 1980s or early 1990s. The fraction of atmospheric nitrogen deposition to watersheds which reaches streams is included in the nonpoint-source values.

TABLE 11. Summary of total nitrogen loading (10^3 kg N y^{-1}) to each of seven areas comprising the Maryland coastal bay system. From Boynton et al. [1993] and Boynton et al. [1996].

Source (10^3 kg N y^{-1})	Estuarine surface area (km^2)	Drainage area (km^2) / N load
Point sources		
Assawoman Bay		0
Isle of Wight Bay		0
St. Martin River		18.3
Turville Creek		0
Sinepuxent Bay		0.01
Newport Bay		36.9
Chincoteague Bay (Maryland portion)		0.03
Nonpoint sources		
Assawoman Bay	24.7	52.9
Isle of Wight Bay	17.5	13.0
St. Martin River	95.5	302.9
Turville Creek	34.3	78.2

TABLE 11. Continued.

Source	N load (10³ kg N y⁻¹)	Estuarine surface area (km²)	Drainage area (km²)
Sinepuxent Bay		26.7	22.6
Newport Bay		113	220.8
Chincoteague Bay (Maryland portion)		141	258.0
Atmospheric deposition direct to bay			
Assawoman Bay	22.5		39.8
Isle of Wight Bay	15.8		27.9
St. Martin River		8.4	12.4
Turville Creek	5.3		4.9
Sinepuxent Bay		24.6	35.8
Newport Bay	15.9		20.3
Chincoteague Bay (Maryland portion)	189		318.4

Loadings from point sources were also from Jacobs et al. [1993] and were based on information for the period 1990-1991 from the Maryland Department of Environment. Atmospheric inputs represent total nitrogen (TN) deposition in wet-fall directly to the surface of the bay waters. Input was calculated by multiplying precipitation nitrogen concentrations times rainfall depths. TN concentrations were from Smullen et al. [1982]. Rainfall data were from stations located in Snow Hill and Assateague Island National Seashore and were collected by National Oceanic and Atmospheric Administration, annual Climatological Summary (1980-1991). Average annual rainfall was 43.8 inches per year.

References

Boynton, W. R., L. Murray, W. M. Kemp, J. D. Hagy, C. Stokes, F. Jacobs, J. Browers, S. Souza, B. Krinsky, and J. Seibel. 1993. Maryland's Coastal Bays: an assessment of aquatic ecosystems, pollutant loadings, and management options. Report No. UMCEES CBL 93-053. Prepared for the Maryland Department of the Environment, Chesapeake Bay and Special Projects Branch, Baltimore, MD, by the Chesapeake Biological Laboratory, Solomons, MD.

Boynton, W. R., L. Murray, J. D. Hagy, C. Stokes, and W. M. Kemp. 1996. A comparative analysis of eutrophication in a temperate coastal lagoon. Estuaries 19(2B):408-421.

Jacobs, F., J. Bowers, S. Souza, B. Krinsky, and J. Seibel. 1993. Diagnostic assessments of terrestrial pollutant loadings, p. 2-1-3-16. In W. R. Boynton, et al. 1993. Maryland's Coastal Bays: an assessment of aquatic ecosystems, pollutant loadings, and management options. Part 2. Report No. UMCEES CBL 93-053. Prepared for the Maryland Department of the Environment, Chesapeake Bay and Special Projects Branch, Baltimore, MD, by the Chesapeake Biological Laboratory, Solomons, MD.

Smullen, J. T., J. L. Taft, and J. Macknis. 1982. Nutrient and sediment loads to the tidal Chesapeake Bay system, p. 147-258. In Chesapeake Bay Program. Technical Studies: A Synthesis. United States Environmental Protection Agency, Annapolis, Maryland.

Estuary: Pamlico River Estuary

Annual nitrogen loading to the Pamlico River estuary was estimated in 1994-1995 as part of a long-term study of eutrophication in the estuary [Stanley, 1997]. The Pamlico "River" is actually the estuary of the Tar River, which drains most of the 11,600 (km^2) basin area.

An automated water sampler (ISCOTM Model 3700) was in place at a station named Seine Beach on the lower Tar River near Grimesland, NC from June 1994 through September, 1995. The ISCO was programmed to pump one sample every 6 hours from mid-depth near the center of the river. Accumulated samples were retrieved at least every other day. The Seine Beach station integrates the drainage from approximately 85% of the Pamlico River estuary. The remainder is mostly low-lying coastal area. An effort was made to characterize the loading from this region by sampling in Durham Creek, the only USGS flow-gaged tributary stream in the area, and in the headwaters of Bath and South Creeks. Grab samples were taken every other week, and an ISCO was deployed for two one-month periods (summer and winter) in Durham Creek to give information on short-term variability. An estimate for total loading below the Tar sampling site was made, based on land-area-ratio extrapolation of the tributary creek values.

Daily loadings (kg N d^{-1}) were computed by multiplying the constituent concentrations times Tar River flows. The nearest USGS flow gaging station on the river is at Tarboro, NC, which is approximately 65 km upstream from the sampling site. It had been estimated that the sampling site flow lags that at Tarboro by about 2 days [D. Stanley, unpublished data]. However, a stage height recorder was installed at the Seine Beach sampling site, and the readings confirmed the 2-day lag. Thus, flow data were lagged to give more accurate estimates of Seine Beach flow.

For the one-year period 11 July 1995 through 10 July 1996, the computed loading at the Seine Beach location was 3.04 million kg N. The estimated N load below this station was 0.54 million kg N. Thus the total N load from river and stream discharge to the Pamlico River estuary was estimated to be 3.58 x 10^6 kg N y^{-1}. There has been no attempt made to quantify atmospheric N loading to this estuary.

TABLE 12. Summary of nitrogen loading results for Pamlico River (10^3 kg N y^{-1}). From Stanley [1997].

Source	kg N y-1
River and stream	3,580

Reference

Stanley, D. W. 1997. Water quality in the Pamlico River estuary: 1989-1996. Institute for Coastal and Marine Resources, East Carolina University, Technical Report No. 97-02. Greenville, NC.

Estuary: Neuse River Estuary

The Neuse estuary drains approximately 16,000 km^2, and the estuary covers approximately 400 km^2 [Boyer et al., 1994]. Annual nitrogen loadings to the Neuse River estuary were computed for the period 1985-1988, as part of a study of nitrogen cycling and primary productivity [Boyer et al., 1993, 1994]. Total nitrogen loads from above Cowpen Landing, a sampling station on the lower Neuse River, were estimated by multiplying N concentrations (biweekly grab samples) times daily flow rates, with linear interpolation of concentration data between sampling dates. The flows were from a United States Geological Survey (USGS) station above Cowpen Landing at Kingston, NC. The Kingston flows were scaled up by land-area ratioing (i.e., flow at Cowpen Landing was estimated to be equal to flow at Kingston times 1.45). A similar method was used to estimate loading from the Trent River, a tributary that joins the Neuse River at New Bern, NC. The Trent River nutrient concentration sampling was done at Pollocksville, NC, where there is a USGS flow gage. The multiplier for scaling Pollocksville loads up to the whole Trent River was 3.09.

About 22% of the total Neuse River estuary basin is downstream from the Cowpen Landing station on the Neuse or downstream from the Trent River. Nonpoint source loading from this area was estimated by land-area ratioing (i.e., these loads were assumed to be equal to 22% of the Cowpen Landing loads).

Nitrogen loading was computed separately for a pulp and paper mill downstream from the Cowpen Landing station. It is one of only two significant point sources below Cowpen Landing. The nitrogen loads from this source were estimated by multiplying effluent grab sample N concentrations times flows (the latter provided by the company operating the facility). Finally, loads were calculated for the New Bern, NC wastewater treatment plant. The method was the same as that used for the paper mill. The loading calculations were made for several nitrogen fractions, but are summarized here as total nitrogen. Also, loadings from all the geographic areas and the two point sources listed above are summed.

The fraction of atmospheric nitrogen deposited onto the Neuse watershed that reaches the Neuse estuary is included in the estimates described above, but there was no effort in this study to quantify atmospheric loading, either via the watershed or directly onto the Neuse estuary. However, D. Whitall and H. Paerl [personal communication] have estimated that between 18% and 35% of the total nitrogen reaching waterways draining into the Neuse is from the atmospheric. This conclusion is based on (both wet and dry) deposition data collected in North Carolina during the period 1996-1998. The total annual atmospheric deposition onto the watershed was estimated to be 13,255 10^3 kg N. The fraction of this nitrogen reaching the waterways was estimated by means of a routing model that is based on watershed retention factors that vary according to land use [Valigura et al., 1996].

TABLE 13. Summary of total nitrogen loading to the Neuse River estuary from all point and nonpoint sources in the watershed.

Year	10^3 kg N y^{-1}
1985	3,551
1986	2,704
1987	6,707
1988	2,964
AVERAGE	3,982

References

Boyer, J. N., R. R. Christian, and D. W. Stanley. 1993. Patterns of phytoplankton primary productivity in the Neuse River estuary, North Carolina, USA. Marine Ecology Progress Series 97:287-297.

Boyer, J. N., D. W. Stanley, and R. R. Christian. 1994. Dynamics of NH_4 and NO_3 uptake in the water column of the Neuse River estuary, North Carolina. Estuaries 17(2):361-371.

Valigura, R.A., W. T. Luke, R. S. Artz, and B. B. Hicks. 1996. Atmospheric nitrogen input to coastal areas: reducing the uncertainties. Decision Analysis Series No. 9. National Oceanic and Atmospheric Administration Coastal Ocean Program, Silver Spring, MD.

Estuary: North Inlet

Dame et al. [1991] assembled data to produce carbon, nitrogen, and phosphorus budgets for the Bly Creek system, a small tidal basin (only 0.6 km^2 water surface area) that is a tributary to the North Inlet estuary in South Carolina. The primary freshwater source for the basin is Bly Creek, which is a small blackwater stream draining about 395 ha of pine-oak forest. The study period was 1983-84. Four external nitrogen sources were considered: streamflow from the Bly Creek watershed, salt marshes surrounding the basin, ground water, and direct precipitation onto the basin (wet-fall only).

Nitrogen loading from the watershed was estimated by multiplying stream flows times ambient nitrogen concentrations. Nitrogen exchanges between the creek and the marsh were also estimated by multiplying flows times concentrations. The flow and concentration data were obtained by building a flume across a portion of the marsh near the upper reaches of the Bly Creek basin. Instantaneous mass fluxes were integrated over a complete tidal cycle to obtain the net nitrogen transport. Monthly and annual ground-water fluxes were calculated by multiplying material concentrations by groundwater discharge (as calculated by the Darcy equation). Data for this calculation were obtained from piezometers installed at varying depths along three transects across the Bly Creek basin. A weighing-type rain gauge was used for the precipitation N loading portion of the study. It was located 800 m west of the basin, and after each rainstorm, the researchers took water samples from the gauge and analyzed them for dissolved nitrogen concentrations. The collected rainwater and material concentrations were used to estimate monthly and annual N inputs.

TABLE 14. Summary of external N loading to Bly Creek estuary, 1983-1984 (kg N y^{-1}). From Dame et al. [1991].

Source	kg N y-1
Stream	904
Salt marsh	2,533
Ground water	215
Atmospheric deposition directly onto estuary	211
TOTAL	3,863

Reference

Dame, R. F., J. D. Spurrier, T. M. Williams, B. Kjerfve, R. G. Zingmark, T. G. Wolaver, T. H. Chrzanowski, H. N. McKellar and F. J. Vernberg. 1991. Annual material processing by a salt marsh-estuarine basin in South Carolina, USA. Marine Ecology Progress Series 72:153-166.

Estuary: Charleston Harbor

McKellar and Blood [1997] summarized recent information on nitrogen loading in the Cooper River/Charleston Harbor estuary. Three sources were considered in their study: the Cooper River above Lake Moultrie (the main fresh water source to the estuary), point sources downstream from Lake Moultrie, and nonpoint loading below Lake Moultrie.

Cooper River nitrogen loads were estimated by multiplying flows times concentrations. Point source loadings were based on NH_4 permit limits for the major dischargers in the Cooper River and Charleston Harbor. The authors stated that using this method may produce an over-estimate since most dischargers may not be discharging at their permitted level. However, they also noted that discharge monitoring reports for 1993 suggested that some of the major dischargers were discharging at levels approaching permit limits and that several major dischargers were not required to monitor N discharges and therefore not included in the estimate. Non-point source loading was estimated by means of a mathematical model (no further details given).

The report contains no information about atmospheric N loading, either directly to the estuary or indirectly via the watershed.

TABLE 15. Summary of total nitrogen loading to Charleston Harbor, SC (10^3 kg N y^{-1}). From McKellar and Blood [1997].

Source	10^3 kg N y^{-1}
Cooper River (above Lake Moultrie)	2,085
Point sources (below Lake Moultrie)	2,226
Nonpoint sources (below Lake Moultrie)	985
TOTAL	5,296

Reference

McKellar, H. and E. Blood. 1997. Nitrogen budget for Cooper River / Charleston Harbor estuary. Report to the Charleston Harbor Project, Office of Ocean and Coastal Resource Management, SC Department of Health and Environmental Control, Charleston, SC.

Estuary: Indian River Lagoon

This report by Woodward-Clyde Consultants [1994] is one of several prepared as support for the characterization of the Indian River Lagoon for the Indian River Lagoon National Estuary Program. The report quantified point and nonpoint source nitrogen loadings to the lagoon. For purposes of this study, three water bodies (the Indian River Lagoon, Banana River, and Mosquito Lagoon) were considered to be one linked system that was called the Indian River Lagoon complex, as opposed to the Indian River Lagoon proper, which refers to just the Indian River lagoon water body. The associated watershed is referred to as the Indian River Lagoon system.

Nonpoint-source total nitrogen loading (also referred to as "storm-water loading" in the report) was computed by a modeling technique that involved multiplying precipitation quantities times runoff coefficients taken from the published literature. Coefficients varied according to land use. The TN loads were adjusted (downward) if storm-water treatment was in place in a particular area. The study period was 1988-1990. For a few areas, the modeled loads were compared with calculations based on measured storm-water runoff and nitrogen concentrations in the runoff. The results were comparable, which the authors interpreted as a verification of their nonpoint-source model. Point source N loading was estimated by multiplying effluent flows times nitrogen concentrations in the effluent.

The authors of this report state that some of the nitrogen in precipitation is included in their estimates of nonpoint-source nitrogen, but the report did not include any quantification of the precipitation contribution to the total nonpoint-source N. Also, there are no data in the report on direct atmospheric deposition of N to the surface of the lagoon system.

TABLE 16. Summary of nitrogen loading to the Indian River Lagoon (kg N y^{-1}). From Woodward-Clyde Consultants [1994].

Source	10^3 kg N y^{-1}
Non-point source (storm water)	2,381
Point sources	220
TOTAL	2,601

Reference

Woodward-Clyde Consultants, Marshall McCully & Associates, Inc., and Natural Systems Analysts, Inc. 1994. Loading assessment of the Indian River Lagoon. Final Technical Report , Project No. 92F274C, prepared for Indian River Lagoon National Estuary Program, Melbourne, FL.

Estuary: Charlotte Harbor

Nitrogen loading to the Charlotte Harbor estuary was quantified in a study by Coastal Environmental, Inc. [1995]. The Charlotte Harbor watershed, as defined for this study, encompasses 8,547 km^2 and the open water estuary is 699 km^2 surface area. Nutrient source categories included in the study were:
1. point source (including domestic and industrial facilities),
2. nonpoint sources (including streamflow, baseflow, and direct runoff),
3. atmospheric (wet and dry deposition to the open water estuary only),
4. groundwater and springs, and
5. septic tanks, or on-site wastewater treatment systems.

Loadings were estimated for the period 1985-1991. Point-source loadings were estimated by multiplying effluent flows times effluent N concentrations. A combination of empirical (flows times concentration) and modeling methods were used to estimate nonpoint loads from different areas in the watershed, depending on how much streamflow and water quality data were available. The modeling involved runoff N concentration coefficients, land use data, and hydrologic variables.

Atmospheric nitrogen wet-deposition data were obtained from the National Atmospheric Deposition Program (NADP), and from the Tampa National Urban Runoff Study. Measured wet-fall concentrations were multiplied times 3.04 to account for the dry deposition component, yielding an estimate of total atmospheric deposition of TN.

Groundwater TN loads were estimated for the surficial aquifer, intermediate confined aquifer, and springs. Groundwater inflows were estimated using a flow net analysis and Darcy's equation. The flow net analysis is a graphical procedure used to identify groundwater flow paths based on water surface profiles. Nitrate-nitrogen concentrations in the groundwater were obtained from previously published reports of studies made in the 1980s. Section 4 of the report contains a very explicit, detailed description of the methodologies used in the study

TABLE 17. Summary of total nitrogen loading to the Charlotte Harbor estuary (10^3 kg N y^{-1}). From Coastal Environmental [1995].

Source	10^3 kg N y^{-1}
Point	264
Non-point	1,833
Atmospheric deposition directly onto estuary	547
Groundwater	1
Septic tanks	79
TOTAL	2,724

Reference

Coastal Environmental, Inc. 1995. Estimates of total nitrogen, total phosphorus, and total suspended solids loadings to Charlotte Harbor, Florida. Prepared for the Southwest Florida Water Management District, Tampa, Florida.

Estuary: Sarasota Bay

Heyl [1992] made an assessment of total nitrogen (TN) loading to Sarasota Bay from five sources:
1. storm-water runoff,
2. baseflow,
3. point-source discharges,
4. septic tanks, and
5. rainfall.

The study area extended from Anna Maria Island and Perico Island south to Casey Key. In addition to Sarasota Bay, the area included the smaller Roberts Bay, Little Sarasota Bay, Dryman Bay, and Blackburn Bay. Approximately 389 km^2 of land area and 135 km^2 of estuarine water surface comprise the study area.

Nitrogen loading via storm-water runoff was estimated by multiplying flows, which were estimated from rainfall by a modeling procedure, times storm event mean TN concentrations (EMCs) (i.e., total TN mass divided by total runoff volume). Different EMCs were used for each of 17 land-use categories. Point source loadings were computed by multiplying effluent flows times TN concentrations in the effluent. Loadings from septic tanks was based on literature data for TN concentrations in the effluent from the tanks and assumptions about nutrient retention and removal as the effluent travels through soil. Estimates of rainfall TN loading directly onto the water surface of the bay were based on measured values of precipitation amount and nitrogen concentration. Dryfall was not included in the estimate.

TABLE 18. Summary of annual TN loading to Sarasota Bay (10^3 kg N y^{-1}). From Heyl [1992].

Source	10^3 kg N y^{-1}
Storm-water runoff	267
Baseflow	51
Point sources	50
Septic tanks	56
Atmospheric deposition directly onto bay**	153
TOTAL	577

** Wet deposition only

Reference

Heyl, M. G. 1992. Point- and nonpoint-source loading assessment of Sarasota Bay, p.12.4-12.19 In Sarasota Bay National Estuary Program. Sarasota Bay: Framework for Action. Published by the Sarasota Bay National Estuary Program, Sarasota, FL.

Estuary: Tampa Bay

This objective of the report by Zarbock et al. [1996] was to update estimates of total nitrogen (TN) loadings to Tampa Bay, which has a watershed area of 5,895 km^2 and an open water surface area of 940 km^2. Loadings for the period 1985-91 had been previously

developed and were reported in Zarbock et al. [1994]. The estimates were for the period 1992-1994. Five major categories of sources of TN loading to the bay were considered:
1. nonpoint sources (storm-water runoff and base flow),
2. point sources (domestic and industrial),
3. groundwater and springs,
4. material losses (inadvertent losses of fertilizer products during handling and shipping), and
5. atmospheric deposition (wet deposition and dry deposition to the open water estuary).

Measured data were used for the loading estimates when possible. For those cases where measured data were not available, modeling techniques or other methods were used. Nonpoint source loads were calculated using measured streamflow and water quality data wherever possible. In some cases point sources discharged into the streams above the nonpoint-source sampling stations. In such cases, the point source loads (calculated by means described below) were subtracted from the total gaged nonpoint source loadings to avoid "double-counting" the point sources. Loads from point source were estimated using agency permit data. For each facility, TN loads were estimated by multiplying the reported monthly flow by the reported or estimated monthly effluent TN concentration.

Groundwater loads were estimated using groundwater quality data and current potentiometric surface maps. TN loadings from springs were estimated using measured water quality and discharge measurements. Material losses were estimated using data supplied by the phosphate industry and other methods of estimating loss based on the volume of material shipped from each facility. Atmospheric deposition loads were estimated using measured precipitation quantity and quality data, and locally-based ratios of dry deposition to wet deposition.

TABLE 19. Summary of nitrogen loading to Tampa Bay (10^3 kg N y^{-1}). From Zarbock et al. [1996].

Source	10^3 kg N y^{-1}
Nonpoint sources	1,566
Point sources	463
Groundwater	187
Material losses**	233
Atmospheric deposition directly onto bay	1,002
TOTAL	3,451

**Material losses = fertilizer products from loading docks at port facilities

References

Zarbock, H. W., A. Janicki, D. Wade, D. Heimbuch, and H. Wilson. 1994. Estimates of total nitrogen, total phosphorus, and total suspended solids to Tampa Bay, Florida. Final Report. Tampa Bay National Estuary Program Technical Publication #04-94, St. Petersburg, FL.

Zarbock, H. W., A. J. Janicki, and S. S. Janicki. 1996. Estimates of total nitrogen, total phosphorus, and total suspended solids to Tampa Bay, Florida. Technical appendix: 1992-94 total nitrogen loadings to Tampa Bay, Florida. Tampa Bay National Estuary Program Technical Publication #19-96, St. Petersburg, FL.

Estuary: Texas Coast (Nueces, Mission-Aransas, Guadalupe, Lavaca-Colorado, Trinity-San Jacinto)

Loading of nitrogen, phosphorus, and total organic carbon were compiled for five estuaries of the Texas coast as part of a study of the influence of freshwater inflows on Texas bays and estuaries [Longley, 1994]. The period covered by this work was 1977-1987. Additional work or refinements were subsequently completed for the Nueces, Guadalupe, and Galveston systems. These new data are unpublished.

Loadings of total nitrogen (TN) from the drainage basin were calculated from measured and simulated stream-flows and concentrations. USGS gaged flows were available for major rivers and streams. For coastal watersheds below the streamflow gage at head-of-tide, compilations included ungaged rainfall runoff, diversions, and wastewater return flow. Ungaged watershed streamflows were estimated by rainfall runoff models. Nitrogen concentration data were available from monthly and/or quarterly monitoring by the USGS and the Texas Natural Resource Conservation Commission.

Total nitrogen concentrations were usually the sum of $NO_3 + NO_2 + TKN$ from unfiltered samples. TN was estimated from DIN in cases for which some nitrogen species were unavailable. Values of nitrogen species which were reported as less than measurement detection limits were assigned a value, usually approximately half or less than the detection threshold. Missing month concentrations were estimated as average monthly values, except for final years of compilations for Guadalupe, Nueces, and Galveston Systems. The program FLUX, from the US Corps of Engineers was used to produce concentrations from flow-concentration relationships for the Nueces and Galveston systems. The missing values for the Guadalupe system were estimated by interpolation. For watersheds without concentration measurements, the concentrations from neighboring watersheds with similar land use and size were applied.

Wastewater return flows include volumes from many different sources with widely variant concentration profiles. Only those sources were considered which were not already included in gaged inflow volumes. Volumes of discharges were available through self-reporting records. Concentrations reported as part of waste discharge permit compliance often did not, however, cover all species of nitrogen. To represent wastewater concentrations applicable to inputs to Texas bays, trimmed averages were computed from data gathered during end-of-pipe samplings performed for special load-allocation studies, such as the Houston Ship Channel study [Kirkpatrick, 1986].

Atmospheric deposition reported is wet deposition to the bay surface only. Concentrations used were averages of period of record from local NADP stations. Volumes were calculated via a Thiessen network using data from local National Weather Service stations. Groundwater input was not considered to be a significant source of nitrogen to the estuary.

Table 20. Annual summaries of total nitrogen loading to Texas estuaries (10^3 kg N y^{-1}).

Estuary	Year	Gaged	Ungaged	Returns	Precipitation	Dry
Corpus Christi Bay System - Nueces Estuary						
	1977	1149.3	176.3	660.9	263.8	-
	1978	469.6	265.6	661.3	400.9	-
	1979	797.1	283.4	696.7	402.6	-
	1980	1036.0	484.4	645.3	328.9	-
	1981	2163.2	554.9	792.4	434.4	-
	1982	676.1	282.2	844.5	227.6	-
	1983	257.3	221.3	881.0	386.2	-
	1984	369.5	121.0	789.3	233.2	-
	1985	868.4	282.0	805.8	386.7	-
	1986	403.1	132.7	761.1	330.2	-
	1987	1934.8	342.5	725.9	320.3	-
	1988*	21.9	72.1	998.3	180.7	271.0
	1989*	13.6	71.4	1004.0	172.6	13.0
	1990*	508.0	567.7	1043.0	221.6	332.4
	1991*	223.4	275.3	1058.3	377.6	566.0
	1992*	1721.6	692.8	1025.6	400.6	600.0
	Average	788.3	301.6	837.1	316.7	345.4
Corpus Christi Bay System - Mission-Aransas Estuary						
	1977	269.1	661.3	42.4	350.6	
	1978	158.0	227.7	39.5	339.2	
	1979	423.1	1150.6	43.6	531.2	
	1980	337.4	915.6	26.4	329.6	
	1981	953.0	1959.6	35.1	574.0	
	1982	312.6	422.0	31.1	254.0	
	1983	554.3	2061.3	36.0	469.6	

Table 20. Continued

Estuary	Year	Gaged	Ungaged	Returns	Precipitation
	1984	134.5	435.8	38.9	276.3
	1985	253.5	907.8	42.4	401.7
	1986	92.9	202.0	36.7	340.5
	1987	266.2	525.8	38.0	365.7
	Average	341.3	860.9	37.3	384.8
The San Antonio Bay System - Guadalupe Estuary					
	1977	10693.4	514.9	1915.2	237.8
	1978	6032.4	296.5	2034.9	213.5
	1979	8813.6	873.8	1673.1	355.9
	1980	3877.8	1669.7	1544.1	166.7
	1981	9397.5	862.3	1175.9	319.2
	1982	4240.7	270.9	1194.0	182.4
	1983	4097.6	387.9	701.1	230.7
	1984	2488.8	156.1	890.4	204.8
	1985	5995.0	525.1	651.1	208.6
	1986	6350.8	297.4	563.8	246.4
	1987	15907.5	356.9	561.7	197.7
	1984*	2530.0	140.0	890.0	310.0
	1987*	14350.0	350.0	300.0	370.0
	Average	7290.4	564.7	1030.4	300.0

TABLE 20. Continued.

The Matagorda Bay system - Lavaca-Colorado Estuary

Year			
1977	2930.6	1713.8	475.8	389.9
1978	2325.1	1275.1	419.1	332.1
1979	5612.2	3845.3	412.8	519.6
1980	2009.8	1016.4	430.7	264.1
1981	5649.6	3524.4	411.2	480.8
1982	3603.9	1727.9	364.8	308.3
1983	4420.5	2827.4	264.5	400.8
1984	2089.8	1403.7	353.9	337.3
1985	4571.5	2805.9	303.2	325.3
1986	3349.8	1778.9	320.2	387.1
1987	6903.6	1461.1	304.2	285.5
Average	3951.5	2125.4	369.1	366.4

The Galveston Bay System - Trinity-San Jacinto Estuary

Year			
1977	15701.6	5103.5	5802.0	438.9
1978	9612.9	5013.5	6669.4	343.4
1979	34872.8	18651.2	7295.6	723.1
1980	14355.4	5913.1	6833.1	422.1
1981	22285.0	10181.8	6929.8	580.8
1982	24255.7	5532.9	6442.8	417.4
1983	25453.8	14162.5	6396.9	619.2
1984	18925.0	4927.3	6431.4	382.2
1985	27436.7	11069.3	6320.6	509.4
1986	30547.0	9669.9	6343.8	473.2
1987	22362.6	8695.2	6285.8	459.5
1988*	8360.0	3320.0	7250.0	570.0
1989*	25900.0	9590.0	7290.0	760.0
1990*	35090.0	5890.0	7570.0	700.0
Average	22511,3	8408.6	6704.4	528.5

* New, unpublished data

References

Kirkpatrick, J. 1986. Intensive Survey of the Houston Ship Channel System. IS 86-10. Texas Water Commission, Austin, Texas.

Longley, W. L. (ed.). 1994. Freshwater inflows to Texas bays and estuaries: ecological relationships and methods for determination of needs. Texas Water Development Board and Texas Parks and Wildlife Department, Austin, TX. 386 pp.

List of Contibutors

Richard Alexander
USGS, 413 National Center
12201 Sunrise Valley Drive
Reston, VA 20192

Walt Boynton
UMD/Chesapeake Biological Lab
1 Williams Street
Solomons, MD 20688

John W. Brakebill
U.S. Geological Survey
Water Resources Division
8987 Yellow Brick Road
Baltimore, MD 21237

David A. Brock
Texas Water Development Board
P.O. Box 13231
Austin, TX 78711-3231

Jaye E. Cable
Louisiana State University
Coastal Ecology Institute
Department of Oceanography & Coastal Sciences
Baton Rouge, LA 70803

Mark S. Castro
University of Maryland
Appalachian Laboratory
Gunter Hall, 301 Braddock Road
Frostburg, MD 21532

Tom Church
University of Delaware
College of Marine Studies
Newark, DE 19716-3501

Robin Dennis
US EPA
MD 80
Research Triangle Park, NC 27711

Charley Driscoll
Syracuse University
220 Hinds Hall
Syracuse, NY 13244

Jim Galloway
University of Virginia, Clark Hall
Environmental Sciences Department
Charlottesville, VA 22901

Holly Greening
Tampa Bay National Estuary Program
111 7th Avenue South
St. Petersburg, FL 33701

Thomas E. Jordan
Smithsonian Environmental Research Ctr.
647 Contees Wharf Road, P.O. Box 28
Edgewater, MD 21037

Jim Kremer
UCONN, Avery Point Campus
Department of Marine Sciences
1084 Shennecossett Road
Groton, CT 06340-6097

Tilden Meyers
NOAA/ATDD
P.O. Box 2456
Oak Ridge, TN 37830

Percy A. Pacheco
DOC/ NOAA/ NOS/ ORCA/ SEAD
1305 East West Highway
SSMC4, 9th Floor
Silver Spring, MD 20910

Hans W. Paerl
UNC-CH Institute of Marine Sciences
3431 Arendell Street
Morehead City, NC 28557

Dr. Jonathan R. Pennock
Dauphin Island Sea Lab
P.O. Box 369-370
Dauphin Island, Alabama 36528

David Peterson
USGS MS 496
345 Middlefield Road
Menlo Park, CA 94025

Stephen D. Preston
U.S. Geological Survey
Water Resources Division
8987 Yellow Brick Road
Baltimore, MD 21237

Nancy N. Rabalais
Louisiana Universities Marine Consortium
8124 Highway 56
Chauvin, LA 70344

Dr. William Reay
Virginia Institute of Marine Sciences
NOAA/ CBNERRVA
Greate Road
Gloucester Point, VA 23062

Gregory E. Schwarz
U.S. Geological Survey
Water Resources Division
413 National Center
Reston, Virginia 20192

Sybil Seitzinger
Institute of Marine & Coastal Science
Rutgers State University of New Jersey
71 Dudley Road
New Brunswick, NJ 08901-8521

Joe Sickles
EPA/ NERL/ ESD/ LCB
MD-56
Research Triangle Park, NC 27711

Richard A. Smith
U.S. Geological Survey
Water Resources Division
413 National Center
Reston, Virginia 20192

Dr. Raghavan Srinivasan
Blackland Research Center
808 E. Blackland Road
Tempe, TX 76502

Paul E. Stacey
CT Department of Environmental Protection
Bureau of Water Management
79 Elm Street
Hartford, CT 06106-5127

Donald W. Stanley
ICMR, Mamie Jenkins Building
East Carolina University
Greensville, NC 27858

Renee Styles
Institute of Marine & Science
Rutgers State University of New Jersey
71 Dudley Road
New Brunswick, NJ 08901-8521

Dave Tomasko
SWFWMD
115 Corporation Way
Venice, FL 34292

Gene Turner
Coastal Ecology Institute, Stadium Road
Louisiana State University
Baton Rouge, LA 70803

Richard Valigura
6771 Moonlight Circle
Sun Prairie, WI 53590

HARFORD COMMUNITY COLLEGE LIBRARY
401 THOMAS RUN ROAD
BEL AIR, MARYLAND 21015-1698